Northern Forested Wetlands

Ecology and Management

Northern Forested Wetlands

Ecology and Management

Edited by

Carl C. Trettin
Martin F. Jurgensen
David F. Grigal
Margaret R. Gale
John K. Jeglum

CRC LEWIS PUBLISHERS

Boca Raton   New York   London   Tokyo

Acquiring Editor:	Neil Levine
Project Editor:	Les Kaplan
Marketing Manager:	Greg Daurelle
Direct Marketing Manager:	Arline Massey
Typesetter:	Pamela Morrell
Cover design:	Dawn Boyd
Prepress:	Carlos Esser
Manufacturing:	Sheri Schwartz

Library of Congress Cataloging-in-Publication Data

Northern forested wetland : ecology and management / edited by Carl C.
Trettin ... [et al.].
    p.   cm.
    Includes bibliographical references and index.
    ISBN 1-56670-177-5
    1. Wetland forestry. 2. Forested wetlands—Management.
3. Wetlands—Management. 4. Wetland ecology. Forest ecology.
I. Trettin, Carl. C.
SD410.9.N67   1996
574.5'26325—dc20   96-26666
                                                        CIP

Preface

Wetlands are a significant feature of northern landscapes, providing important natural functions and essential values to society. For our purpose, northern wetlands are those that occur above the fortieth parallel. Within that region, wetlands are referred to by a variety of names including mire, bog, fen, swamp, and marsh. Historically, these lands have been used for timber and fiber products, hunting, fishing, trapping, gathering food and recreation, and those uses remain important today. Wetlands have also been converted to agricultural use and lost to urbanization. Loss of wetlands precipitated the widespread concern in the 1980s about the role of wetlands in the landscape, and the consequences of further conversions and management. Since then, considerable research has focused on understanding the role that different wetlands provide in the landscape and on learning how to manage wetlands while sustaining important functions.

While much of the attention to wetlands has concentrated on "southern" wetlands, particularly bottomland hardwoods and marsh systems, there has also been a considerable amount of work on northern wetlands. This work has often been done in the context of better understanding wetland processes as a basis for sustainable forest management practices. This volume results from the International Symposium on the Ecology and Management of Northern Forested Wetlands (24–31 August, 1994) that was organized to provide a forum for the presentation and discussion of the current state-of-the-understanding regarding the ecology and management of northern forested wetlands. We developed the program around the thematic areas of ecology, biogeochemistry, hydrology, resource characterization, and management, in both a plenary session and a field excursion format. The plenary session was held in Traverse City, Michigan and the 3-day field excursion was centered in Cochrane, Ontario. A total of 77 invited and contributed presentations, from eight countries (Finland, Sweden, Russia, Lithuania, China, Canada, United States, United Kingdom) provided the basis for informed and in-depth discussions of the ecology of northern wetlands and the opportunities for sustainable management. This volume presents selected papers that provide a current perspective on wetland studies. It was our intent to provide, in a single volume, a primary reference text for the study and management of northern forested wetlands, and to provide a forum for communicating information among researchers and managers for the many nations that contain northern wetlands.

This volume is divided into four sections that coincide with the thematic areas from the Symposium. The first section, Wetland Resources, contains two chapters that consider the distribution of northern forested wetlands in North America. Two other chapters address use and conservation of northern wetlands, and the final chapter in that section considers the use of satellite imagery for wetland inventories. The second section, Ecology and Vegetation, contains seven chapters that consider the interrelationships of vegetation community dynamics, soils and hydrology on both managed and undisturbed wetlands. The third section, Hydrology and Biogeochemistry, contains thirteen chapters covering hydrological and biogeochemical processes. The first two chapters provide a synthesis of the hydrology of natural and managed wetlands, while the third chapter focuses on headwater wetlands. The other chapters in this section address specifics of the hydrologic cycle, water quality, and element cycling with an emphasis on carbon. The final section, Wetland Management, focuses on management regimes of northern wetlands and the associated forest responses. The first chapter in this section provides a synthesis based on literature from undrained wetlands, and the other chapters consider site preparation, fertilization, water management, and best management practices. The information presented in each of these chapters is important and timely, and many of the questions and challenges that are set forth by the authors will engage wetland scientists in the years to come.

Carl C. Trettin
Charleston, SC

Acknowledgments

We would like to thank all the people who contributed their time and serious effort to developing this volume that is based on papers presented at the International Symposium on the Ecology and Management of Northern Forested Wetlands in Traverse City, MI, U.S.A and Cochrane, Ontario, Canada, 24–31 August, 1994. Many people were involved in the arduous tasks of coordinating the field excursion and symposium. We would first like to thank Sherry Sandretto from Michigan Technological University for her painstaking job of mailings, answering questions, producing programs, coordinating foreign travel, coordinating arrangements in Canada and the United States, registrations, and putting up with our extraordinary demands. Our thanks also go to Andrea Longhini, Janis Coleman-Plouff, and Kelly Bassett from Michigan Technological University for their help in mailings, accounting, and registration. We are also grateful for the assistance of Greg Kudray, Neil Wilkie, and Janet Silbernagel during the Symposium in Traverse City. We are deeply indebted to Tom Weldon, Fred Haavisto, and Paul Adams of Natural Resources Canada, Chris Davies and P. K. (Wally) Bidwell of the Ontario Ministry of Natural Resources, Rod Gemmel and Erik Turk of Abitibi-Price Ltd. and the Lake Abitibi Model Forest, and Sherri Cooper of the Cochrane Economic Development Corp. for their help and perseverance with the specific arrangements and programs for the field trip in Cochrane; and to Beth Nunnelley and Debra Duke Taylor of The Grand Traverse Resort for providing us with a comfortable setting for the plenary session in Traverse City.

We would like to thank not only the authors who contributed manuscripts for this volume but all the speakers for providing stimulating perspectives on important aspects of forested wetland ecology and management. We would also like to thank the participants of the field trip and symposium for adding to the discussions and debate during the long summer days and nights.

The reviewers of the chapters for this book were critical to making this an important addition to the science of northern forested wetlands. A special mention of gratitude is due to Dr. Elon S. Verry for the considerable time spent reviewing manuscripts and coordinating travel plans for foreign scientists. Lisa Jenkins, U.S.D.A. Forest Service Southern Research Station, was instrumental in handling the processing and revisions of the manuscripts and the correspondence between the editors and authors. Our gratitude is also extended to Neil Levine and Evelyn Krasnow, CRC Press, and the managing editorial staff, for their support and help in developing this project.

We would also like to thank our sponsors, the U.S. Environmental Protection Agency, Duluth, MN; U.S. Forest Service North Central Forest Experiment Station, Grand Rapids, MN; the National Council for the Paper Industry for Air and Stream Improvement; and Michigan Technological University who saw the need for the symposium and this publication, and who provided financial support for the program.

Our heartfelt gratitude goes out to all the above-mentioned persons and others who somehow contributed. We appreciate your time, effort, and unending interest in making this book an important contribution.

Carl C. Trettin
Martin F. Jurgensen
David F. Grigal
Margaret R. Gale
John K. Jeglum

Editors

Carl C. Trettin, Ph.D., is Project Leader for the Center for Forested Wetlands Research, USDA Forest Service, Charleston, South Carolina. The mission of the Center is research on the ecology and management of forested wetlands. Dr. Trettin received his B.S. in forest hydrology and M.S. in forest soils from Michigan Technological University, and his Ph.D. in forest wetland ecology from North Carolina State University.

He has held research appointments at the Environmental Sciences Division of Oak Ridge National Laboratory, and the School of Forestry and Wood Products and the Ford Forest Research Center at Michigan Technological University. He has led several interagency research cooperatives addressing wetland ecology, forest soils use and management, and landscape ecology. He has also participated as a subject expert on wetland resources in Finland and Russia. Dr. Trettin is active in academics at several universities, is on the adjunct faculty of Auburn University and Michigan Technological University, and is a member of numerous professional societies.

Dr. Trettin's research interests have focused on the effects of silvicultural practices on wetland functions, with specific emphases on carbon and nitrogen cycling, and water quality. In addition to those pursuits, his current work addresses forest wetland restoration and landscape functions of wetland communities. He has conducted research on water management systems, constructed wetlands, forest soil-site relationships, nutrient cycling, and peat resources. He has authored or co-authored over 60 technical papers in journals, proceedings, and other outlets.

Martin F. Jurgensen, Ph.D., is Professor of Forest Soils in the School of Forestry and Wood Products, Michigan Technological University. Prof. Jurgensen received his B.S. in forestry and M.S. in silviculture from the College of Forestry, State University of New York, and his Ph.D. in soil microbiology from North Carolina State University.

In addition to his current position, Prof. Jurgensen has held a faculty position at North Carolina State University, served as Acting Director of the Institute of Wood Research at Michigan Technological University, and been Management Agronomist for the US Air Force. Prof. Jurgensen teaches courses in forest soils, soil microbiology, and soil chemistry. He is a recipient of the New Zealand Senior Research Fellowship and the Michigan Technological University Faculty Research Award, and received a National Science Foundation Predoctoral Fellowship.

Prof. Jurgensen's research interests include microbiology of forested wetland soils, forest management impacts on soil organic matter and nutrient cycling, mycorrhizae relationships in forest stands, and ecological and forest management effects of coarse woody debris. He has authored or co-authored over 120 technical papers in journals, proceedings, and other outlets.

David. F. Grigal, Ph.D., is Professor of Forest Soils in the Department of Soil, Water, and Climate, University of Minnesota. Prof. Grigal obtained an As.S. in pre-forestry from Virginia Junior College, and a B.S. in forestry, an M.S in forest statistics, and a Ph.D. in soil science from the University of Minnesota.

In addition to his faculty position, Prof. Grigal has held research appointments at Oak Ridge National Laboratory and the University of Minnesota. He has taught courses in forest soils, forest hydrology, silviculture, peatland ecology, and soil management; he also maintains a large graduate research program. Prof. Grigal has contributed to many state, regional, and national committees and initiatives that focus on forest resource use, assessment, and management. He has also held positions within the American Society of Agronomy, and served as associate editor to the *Soil Science Society of America Journal* and the *Journal of Environmental Quality*.

Prof. Grigal's research interest is in the influence of natural and anthropogenic disturbances on natural ecosystems, especially forests. He has conducted research on vegetation classification using multivariate techniques on peatlands, nutrient cycling, soil pedology, and soil chemistry. His approach is directed toward a quantitative description of the phenomena under study. His work has concentrated on fire as a natural disturbance and on air pollution, including deposition of acids and mercury. He has authored or co-authored over 160 technical papers in journals, proceedings, and other outlets.

Margaret R. Gale, Ph.D., is Associate Professor of Production Ecology in the School of Forestry and Wood Products, Michigan Technological University. She obtained her B.S. in forestry and M.S. in forest ecology from Michigan Technological University, and the Ph.D. in production ecology from the University of Minnesota.

Prof. Gale teaches courses in wetland ecology and production ecology, and maintains a graduate research program. She was awarded a Faculty Fellowship with the Oak Ridge Institute in Education and Science for work on a nutrient cycling project with the Tennessee Valley Authority. She has also been active in many state and regional committees.

Prof. Gale's research interests lie in the areas of production ecology, belowground ecology, and modeling plant responses to varying environmental factors. Recent work has involved the study of wetland classification systems in Michigan, wetland plant community responses to perturbations, stress effects on carbon allocation in *Betula papyrifera* and *Pinus resinosa*, the role of coarse woody debris in old growth forests, and historical and present landscape structure and distribution of plant communities in the upper Great Lakes. She has authored or co-authored over 35 technical papers in journals, proceedings, and other outlets.

John K. Jeglum, Ph.D., is Professor of Forest Peatland Ecology, Department of Forest Ecology, Swedish University of Agricultural Science. Prof. Jeglum received his B.S. in botany and M.S. in plant ecology from the University of Wisconsin. He received his Ph.D. in plant ecology from the University of Saskatchewan.

He has held research appointments with the Canadian Forest Service, where he led a research program in peatland ecology, black spruce silviculture, forest drainage, and assessment of forestry impacts. He also held a faculty appointment at Eastern Illinois University. He has developed large multi-agency projects that have addressed sustainable forest productivity, forest ecosystem classification, and peatland forest management. He was awarded a sabbatical leave to Finland, and has organized international conferences on peatland sciences. He is a member of numerous international organizations.

Prof. Jeglum's research interests include forest classification, plant community dynamics, plant ecology, peatland management, and sustainable productivity. He has authored or co-authored over 15 technical papers in journals, proceedings, and other outlets.

Contributors

Erkki Ahti, D.For
Department of Forest Ecology
The Finnish Forest Research Institute
Vantaa, Finland

Lena ÅkerströmM.Sc.
Department of Forest Ecology
Swedish University of Agricultural
  Sciences
Umea, Sweden

Jukka Alm, M.Sc.
Department of Biology
University of Joensuu
Joensuu, Finland

Kenneth N. Brooks, Ph.D.
Department of Forest Resources
University of Minnesota
St. Paul, Minnesota

Thomas E. Dahl, M.S.
U.S. Fish and Wildlife Service
Wetland Status and Trends
St. Petersburg, Florida

Kevin J. Devito, Ph.D.
Department of Geography
Erindale College
University of Toronto
Mississauga, Ontario, Canada

Trae A. Forgette, M.S.
Environmental Researcher
Great Lakes Environmental Center
Traverse City, Michigan

Neil W. Foster, Ph.D.
Canadian Forest Service
National Resources Canada
Sault Ste. Marie, Ontario, Canada

W. E. Frayer, D.For.
School of Forestry and Wood
  Products
Michigan Technological University
Houghton, Michigan

Margaret R. Gale, Ph.D.
School of Forestry and Wood
  Products
Michigan Technological University
Houghton, Michigan

Michelle Garneau, Ph.D.
Geological Survey of Canada
Natural Resources Canada
Ste.-Foy, Quebec, Canada

David F. Grigal, Ph.D.
Department of Soil, Water, and
  Climate
University of Minnesota
St. Paul, Minnesota

V. F. Haavisto, M.Sc.
Canadian Forest Service
Natural Resources Canada
Sault Ste. Marie, Ontario, Canada

Björn Hånell, Ph.D.
Department of Silviculture
Swedish University of Agricultural
  Sciences
Umea, Sweden

Paul Hazlett, M.S.
Canadian Forest Service
Natural Resources Canada
Sault Ste. Marie, Ontario, Canada

Alan R. Hill, Ph.D.
Department of Geography
York University
North York, Ontario, Canada

Graham R. Hillman, Ph.D.
Department of Natural Resources
Canadian Forest Service
Northern Forestry Centre
Edmonton, Alberta, Canada

Hannu Hökkä, D.For.
Finnish Forest Research Institute
Rovaniemi Research Station
Rovaniemi, Finland

John K. Jeglum, L.For.
Canadian Forest Service
Natural Resources Canada
Sault Ste. Marie, Ontario, Canada

Samuli Joensuu, D.For.
Forestry Development Centre Tapio
Helsinki, Finland

Martin F. Jurgensen, Ph.D.
School of Forestry and Wood
  Products
Michigan Technological University
Houghton, Michigan

Seppo Kaunisto, D.For.
Parkano Research Station
Finnish Forest Research Institute
Parkano, Finland

Inez M. Kettles, M.Sc.
Terrain Sciences Division
Geological Survey of Canada
Ottawa, Ontario, Canada

Vladimir A. Klimanov, D.Sci.
Institute of Geography
Moscow, Russia

Gregory M. Kudray, Ph.D.
School of Forestry and Wood
  Products
Michigan Technological University
Houghton, Michigan

Erkki Lähde, Ph.D.
Vantaa Research Centre
Finnish Forest Research Institute
Vantaa, Finland

Olavi Laiho, D.For.
Parkano Research Station
Finnish Forest Research Institute
Parkano, Finland

Raija Laiho, Ph.D.
Department of Forest Ecology
University of Helsinki
Helsinki, Finland

Jukka Laine, Ph.D.
Department of Forest Ecology
University of Helsinki
Helsinki, Finland

Dawn M. Majewski, M.F.
School of Natural Resources and
  Environment
University of Michigan
Ann Arbor, Michigan

Laurie J. Manor, M.S.
School of Natural Resources and
  Environment
University of Michigan
Ann Arbor, Michigan

Pertti J. Martikainen, Ph.D.
Laboratory of Environmental
  Microbiology
National Public Health Institute
Kuopio, Finland

James W. McLaughlin, Ph.D.
Cooperative Forestry Research Unit
University of Maine
Orono, Maine

Kari Minkkinen, M.Sc.
Department of Forest Ecology
University of Helsinki
Helsinki, Finland

Ian K. Morrison, Ph.D.
Canadian Forest Service
Natural Resources Canada
Sault Ste. Marie, Ontario, Canada

Yrjö Norokorpi, Ph.D.
Rovaniemi Research Station
Finnish Forest Research Institute
Rovaniemi, Finland

Hannu Nykänen, M.Sc.
Laboratory of Environmental
  Microbiology
National Public Health Institute
Kuopio, Finland

Juhani Päivänen, Ph.D.
Department of Forest Ecology
University of Helsinki
Helsinki, Finland

Pia Papadopol, M.Sc.
Canadian Forest Service
Natural Resources Canada
Sault Ste. Marie, Ontario, Canada

Bijan Payandeh, Ph.D.
Canadian Forest Service
Natural Resources Canada
Sault Ste. Marie, Ontario, Canada

Timo Penttilä, M.Sc.
Rovaniemi Research Station
Finnish Forest Research Institute
Rovaniemi, Finland

Michael J. Phillips, Ph.D.
Minnesota Department of Natural
  Resources
Division of Forestry
St. Paul, Minnesota

André P. Plamondon, Ph.D.
Centre de Forestière
Université Laval
Ste.-Foy, Québec, Canada

François Quinty, M.A.
Département de Phytologie
Université Laval
Ste.-Foy, Québec, Canada

Cynthia W. Radcliffe, M.S.
School of Natural Resources and
  Environment
University of Michigan
Ann Arbor, Michigan

Stephen D. Robinson, M.A.
Department of Geography
McGill University
Montreal, Quebec, Canada

Line Rochefort, Ph.D.
Département de Phytologie
Université Laval
Ste.-Foy, Québec, Canada

Vincent Roy, B.Sc.A.
Centre de Recherche en Biologie
  Forestière
Université Laval
Ste.-Foy, Québec, Canada

Clayton D. A. Rubec, M.Sc.
Water and Habitat Conservation Branch
Canadian Wildlife Service
Environment Canada
Ottawa, Canada

Juozas Ruseckas, D.Habil.
Department of Forest Soils and
  Hydrology
Lithuanian Forest Research Institute
Kaunas-Girionys, Lithuania

Timo Saksa, Ph.D.
Suonenjoki Research Station
Finnish Forest Research Institute
Suonenjoki, Finland

Ilkka Savolainen, Ph.D.
VTT Energy
Espoo, Finland

John A. Shuey, Ph.D.
Director of Conservation Biology
The Nature Conservancy
Indianapolis, Indiana

Matti Siipola
Rovaniemi Research Station
Finnish Forest Research Institute
Rovaniemi, Finland

Deborah J. Silcock, Ph.D.
Macaulay Land Use Research Institute
Craigiebuckler, Aberdeen, U.K.

Jouko Silvola, Ph.D.
Department of Biology
University of Joensuu
Joensuu, Finland

Jukka Sinisalo, M.Sc.
VTT Energy
Espoo, Finland

Andrej A. Sirin, Ph.D.
Institute of Forest Research
Russian Academy of Sciences
Uspenskoye, Moscow Region, Russia

Yuanxin Teng, Ph.D.
Canadian Forest Service
Natural Resources Canada
Sault Ste. Marie, Ontario, Canada

Terence M. Tompkins, M.S.
Comprehensive Studies Program
University of Michigan
Ann Arbor, Michigan

Carl C. Trettin, Ph.D.
Center for Forested Wetlands
  Research
U.S.D.A. Forest Service
Charleston, South Carolina

Harri Vasander, Ph.D.
Department of Forest Ecology
University of Helsinki
Helsinki, Finland

Elon S. Verry, Ph.D.
U.S.D.A. Forest Service
Grand Rapids, Minnesota

Stanislav E. Vompersky, Ph.D.
Institute of Forest Research
Russian Academy of Sciences
Uspenskoye, Moscow Region,
  Russia

William W. Whipps, M.S.
School of Natural Resources and
  Environment
University of Michigan
Ann Arbor, Michigan

Michael J. Wiley, Ph.D.
School of Natural Resources and
  Environment
University of Michigan
Ann Arbor, Michigan

Berwyn L. Williams, Ph.D.
Macaulay Land Use Research
  Institute
Craigiebuckler, Aberdeen, U.K.

Stephen C. Zoltai, B.Sc.F.
Canadian Forest Service
Natural Resources Canada
Edmonton, Alberta, Canada

Table of Contents

Section 3 — Hydrology and Biogeochemistry

Northern Forested Wetlands

Ecology and Management

Wetland Resources

CHAPTER 1

Forested Northern Wetlands of North America

Thomas E. Dahl and Stephen C. Zoltai

CONTENTS

INTRODUCTION

Wetlands in the United States are defined as areas that are transitional between terrestrial and aquatic systems, where the water table is usually at or near the surface or the land is covered by shallow water (Cowardin et al., 1979). The definition of wetlands in Canada expresses a similar concept: wetlands are lands that have the water table at, near, or above the land surface or that are saturated for a long enough period to promote wetland or aquatic processes, as indicated by hydric soils, hydrophytic vegetation, and various kinds of biological activity that are adapted to the wet environment (Tarnocai, 1980).

1-56670-177-5/97/$0.00+$.50

A great variety of wetlands are formed under different hydrological, water quality, climatic, or edaphic conditions, each with its distinctive vegetation and biota. In this chapter, we are considering forested wetlands only, where a closed-canopy tree cover (>5 m) forms the dominant vegetation, or if immature, the trees have the potential of becoming closed-canopy forests. The hydraulic conditions are generally not severe, and a seasonally aerated rooting zone permits the growth of trees that are adapted to growing under wet conditions. The moderate severity of waterlogging also makes these wetlands ready targets for conversion to non-wetland uses through human intervention.

HISTORICAL EXTENT OF NORTHERN WETLAND FORESTS

Prior to European settlement, forested wetlands may have composed as much as 65 to 70 percent of the total wetland area in the northern United States, including Alaska, and southern Canada. Two distinct types of wet forests were found to dominate the land area: boreal conifer swamps (Zoltai et al., 1988) and mixed hardwood swamps.

The boreal conifers, primarily black spruce (*Picea mariana*) and tamarack (*Larix laricina*), as well as northern white cedar (*Thuja occidentalis*) and balsam fir (*Abies balsamea*) in the east, occupied the glaciated lowland portions of the North American continent. This area extended from the interior of Alaska southeast in an arc across the Canadian provinces. It included the continental (subhumid) and humid high and mid-boreal wetland regions, and the low and northern Atlantic boreal wetland regions of Canada (National Wetlands Working Group, 1986). Elements of the boreal coniferous wetlands also extended into the eastern temperate zones of the St. Lawrence Lowlands of southern Ontario, Quebec, and New Brunswick, as well as the Laurentian Mixed Forest Province as it bisects the upper states of the Great Lakes and New England. Soils in these wet boreal forests are typically histosols (Bailey, 1980), called organic soils in Canada (Canada Soil Survey Committee, 1978). Peat deposits are common, and surface flooding or ponding of water is rare, as the organic substrates tend to remain saturated for much of the growing season.

In Alaska the coniferous forested wetlands are dominated by black spruce. Tamarack is found in association with black spruce in wet lowland sites. A smaller concentration of forested wetlands is found along Alaska's Pacific coast. These areas are dominated by western hemlock (*Tsuga heterophylla*) and Alaskan cedar (*Chamaecyparis nootkatenis*).

The boreal forested wetlands in the subhumid portion of Canada (approximately west of the Manitoba/Ontario border) are composed of black spruce occurring in dense stands, with a carpet of feathermosses. White birch (*Betula papyrifera*) may become established following fires, but they seldom persist into the next generation. In the more humid climate of eastern Canada, black spruce dominates the wetland forests, sometimes in association with balsam fir. These black spruce swamps are the backbone of the pulp and paper industry of northern

Table 1   Wet Forest Species Associations in the Northern United States

Forest associations	Predominant state occurrences
Northern spruce — fir forest (*Picea — Abies*)	Maine, New Hampshire, New York, Vermont
Northern hardwoods (*Acer — Betula — Fagus — Tsuga*)	Maine, New Hampshire, New York, Michigan, Wisconsin
Northern hardwoods — spruce forest (*Acer — Betula — Fagus — Picea — Tsuga*)	Maine, New Hampshire, New York, Vermont
Beech — maple forest (*Fagus — Acer*)	New York, Ohio, Michigan
Elm — ash forest (*Ulmus — Fraxinus*)	Ohio, Wisconsin
Conifer bog (*Larix — Picea — Thuja*)	Michigan, Wisconsin, Minnesota
Northern hardwoods — fir forest (*Acer — Betula — Abies — Tsuga*)	Michigan, Wisconsin
Maple — basswood forest (*Acer — Tilia*)	Wisconsin, Minnesota
Black spruce — fir forest (*Picea — Abies*)	Minnesota, Wisconsin

From Küchler, A. W., Potential Natural Vegetation of the Conterminous United States, Special Publication No. 36, American Geographical Society, New York, 1964. With permission.

Ontario and Quebec. Eastern white cedar forms dense forests on the margins of many peatlands in the southern parts of Canada. Tamarack, being a shade-intolerant species, seldom occurs in large numbers in the forested wetlands of this region.

In the northern conterminous United States, wet boreal conifers are dominated by black spruce, tamarack, northern white cedar, and balsam fir. The boreal forest is not known for towering stands of trees. Typical black spruce may be as much as 20 m tall and 23 cm in diameter (breast height) on a good site (Van Hees, 1990).

In contrast with the boreal conifer swamps, the other major northern forested wetland type is the mixed hardwood swamp forest that is characterized by a mix of both hardwood and conifer species (Table 1). In the past this region occupied portions of southern Canada and extended as far south as the state of Ohio. Figure 1 illustrates the region where northern wet forests may have existed prior to European settlement. Historical data indicate that the southern extent of the northern wet forest was close to 40° north latitude. This area approximates the southernmost point of the Wisconsin glaciation and can be distinguished from other forest types by containing species unique to the northern forest community (Gordon, 1969). It encompasses an area in Canada mostly within the eastern temperate region of the St. Lawrence Lowlands, extending from southwestern Ontario to the eastern portions of Quebec. In the United States the demarcation is not quite so clear. Here the deciduous swamp forests intermix with wet conifers in a transition within the Laurentian Mixed Forest Province. Historically, northern hardwood swamps also occurred within the Eastern Deciduous Forest Province as it exists throughout central Minnesota, southern Wisconsin, southern Michigan, northeastern Indiana, and northern Ohio.

In general terms the mixed hardwood swamp forests of the eastern temperate wetland region are typified by maple-ash (*Acer* spp.–*Fraxinus* spp.) associations. These swamps require reasonably warm temperatures during the growing season, moderately high rainfall, and relatively mild winters (Glooschenko and Grondin, 1988). Unlike the boreal conifer swamp forests, which generally occur on peat soils,

Figure 1   Region of northern wet forests in the United States and Canada prior to European settlement. (Adapted from Küchler, A. W., *Potential Natural Vegetation of
the Conterminous United States,* Special Publication No. 36, American Geographical Society, New York, 1964 (map scale 1:3,168,000). With permission;
from National Wetlands Working Group, Canada wetland regions (map), The
National Atlas of Canada, 5th ed., Energy, Mines and Resources Canada,
Geographical Services Division, Surveys and Mapping Branch, Ottawa, Ontario,
1986. With permission; and from Viereck, L. A. and Little, E. L., *Atlas of United
States Trees,* Volume 2 — *Alaska Trees and Common Shrubs,* Miscellaneous
Publication No. 1293, U.S. Department of Agriculture, Forest Service, Washing-
ton, D.C., 1975. With permission.)

the mixed hardwood swamp forests of the region are generally flooded by surface water seasonally (or for longer periods) and occur primarily on mineral soils or clays. In both the northern United States and southeastern Canada these wetlands are dominated by hardwood species including maple (*Acer* spp.), beech (*Fagus grandifolia*), elm (*Ulmus americana*), ash (*Fraxinus* spp.), and birch (*Betula* spp.).

Although northern forested wetlands were extensive in the past, the great majority of them supported only scattered, stunted trees of black spruce and tamarack. Here the water table was too high to allow vigorous tree growth. In addition, the high precipitation in the eastern and oceanic regions maintained a high water table, restricting the growth of tree species to low shrub sizes. A large proportion, perhaps as much as 35 to 40 percent of the total original wetland area, has remained in nontreed wetland communities since the last glacial period. These other types of wetland area include graminoid fens; small open lakes, ponds, and potholes; emergent marshes either in isolated kettles or found in conjunction with rivers and streams; and shrub wetlands including the ericacious shrub bogs and seepage fens or springs.

HISTORICAL TRENDS AND CURRENT STATUS OF THE WET BOREAL CONIFERS

Alaska and Western Canada

Unlike the northern conterminous United States and the southern and eastern portions of Canada, the impacts of human activities on Alaska's wetlands have been relatively minor. It is estimated that Alaska has lost about 81,000 ha of all wetland types over the past 200 years (Dahl, 1990). This accounts for a net change of less than 0.1 percent. Of the 70.4 million ha of wetlands that occur in Alaska, only 7 percent (roughly 5.2 million ha) are forested (Hall et al., 1994). The majority of this area lies south of the Alaskan tundra within the interior part of the state (Figure 2).

The original extent of Canada's forested wetland area is difficult to estimate, because forest inventories did not differentiate between forest stands growing in wetlands or on mineral soils. The best estimates show a considerable amount of commercially productive wetland forests. In Ontario, 18 percent, or nearly 80,000 km^2, of the total 434,000 km^2 of peaty wetlands support commercial black spruce forests (Ketcheson and Jeglum, 1972). Recent estimates (Zoltai and Martikainen, in press) show similar proportions of forested wetlands in other parts of the boreal zones (Table 2), but neither the northern regions nor the maritime regions have extensive forested wetlands. In all, approximately 18.5 million ha of forested conifer swamps occur in Canada.

Situations similar to that in Alaska exist in the boreal wet forests of western Canada where human pressures such as drainage to improve forest productivity and development for other purposes have had only minor impacts (Beazley, 1993). In eastern Canada, mainly in the Clay Belt of Ontario and Quebec, south of Hudson Bay, large areas were cleared for agriculture beginning in the 1920s.

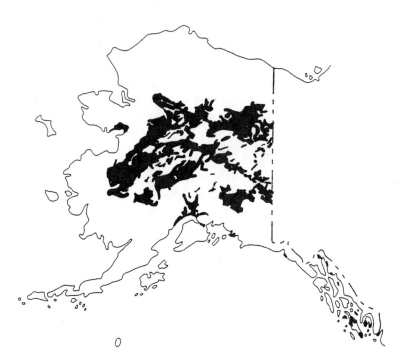

Figure 2   Area where Alaska's wetland resources are concentrated. (Adapted from Hall,
J. V., Wetland Resources of Alaska (map), U.S. Fish and Wildlife Service, Wash-
ington, D.C., 1991 (map scale 1:2,500,000). With permission; and from Viereck,
L. A. and Little, E. L., *Atlas of United States Trees,* Volume 2 — *Alaska Trees
and Common Shrubs,* Miscellaneous Publication No. 1293, U.S. Department of
Agriculture, Forest Service, Washington, D.C., 1975. With permission.)

Although some forested peatlands were cleared, limited success was achieved on
these cold, wet soils, and many abandoned fields have reverted to forests.

At present large areas of conifer swamps are being logged for pulpwood,
especially in northern Ontario and Quebec. As the demand for quality pulpwood
increases, the logging pressures on this wetland forest type also increase. In some
instances this creates a problem, as the water table often rises after an area is
clear cut, allowing the invasion of reed grass (*Calamagrostis* spp.) (Heikurainen,
1967). This prevents the reestablishment of spruce seedlings, resulting in the
deterioration of the site. Special silvicultural practices, including supplemental
plantings, are necessary to assure the regeneration of the cutover swamps.

Many northern conifer swamps are prone to damage by fires that spread into
the swamps from the adjacent uplands. The age of the stand is not as important
in determining this phenomenon as is the fire history of the area. Other swamps
are protected from wildfires by the adjoining wet peatlands, allowing the individual trees to reach ages up to 200 years. However, because the trees are
shallowly rooted, such old stands are prone to windthrow, which creates small

Table 2   Estimated Extent of Forested Wetlands in Canada

Province/territory	Wetland area (thousand ha)[a]	Forested wetland area (thousand ha)
Alberta	12,673	1,457
British Columbia	1,289	200
Manitoba	20,664	3,513
New Brunswick	120	10
Newfoundland–Labrador	6,429	5
Northwest Territories	25,111	25
Nova Scotia	158	2
Ontario[b]	43,400[b]	8,000[b]
Prince Edward Island	8	—
Quebec	11,713	3,800
Saskatchewan	9,309	1,400
Yukon Territory	1,298	65
Canada	132,172	18,477

[a] From Tarnocai, C., Peat Resources of Canada, NRCC No. 24140, National Research Council of Canada, Division of Energy, Ottawa, Ontario, 1984, 17 pp. With permission.
[b] From Ketcheson, D. E. and Jeglum, J. K., Estimates of Black Spruce Peatlands in Ontario, Information Report, O-X-172, Canadian Forestry Service, Sault St. Marie, Ontario, 1972, 29 pp. With permission.

openings in the forest. New seedlings often become established in the openings, resulting in uneven aged stands.

Southern Canada and the Conterminous United States

Within the United States and in the southern portions of Canada, the adverse impacts to forested wetlands began in the early 1800s. During the period from 1830 to 1885, logging and settlement activities began to drastically alter the northern forested regions both at the forest stand scale (species composition and structure) and at the landscape level (Vora, 1994).

Prior to 1900, white pine (*Pinus strobus*) was the prize tree species for many commercial loggers in this region. Other desirable stands of Norway (red) pine (*Pinus resinosa*) and white spruce (*Picea glauca*) were also targets of early logging activity. These species (commercial grade stands) were primarily confined to the uplands, so that the direct impacts on forested wetlands were not significant in the boreal northwoods until sometime after 1900. By the early 1900s white cedar was also in demand (Ryan, 1976) for use as fence posts, railroad ties, and telephone, telegraph, and electric light poles. Since cedar occurs mainly in swamps (Eyre, 1980), these activities greatly impacted the forested wetlands of Minnesota, Wisconsin, and northern Michigan. Ryan (1980) indicates that black spruce was seldom cut commercially until after 1900. At that time black spruce started to supplement demand as the white pine stands were becoming depleted.

Tamarack was also used for fence posts and for some log home construction because it resisted rotting.

In the United States, cutting virgin timber on a large scale continued through the early 1940s. Today, old growth forests are scarce in the Great Lakes states, and much of the remaining forest lands are secondary growth or disturbed areas. An estimated 369,000 ha of primary forest remain, of which 50 percent is coniferous wetland forest (Frelich, 1994). This indicates that the remaining northern forested wetlands make up a significant proportion of the remaining old growth forest in the northern United States. The ecological shift from a preponderance of old growth upland forests dominated by white pine to old growth wetland forests dominated by black spruce, tamarack, and fir may have unrecognized consequences for plant and animal biodiversity. Few long-term studies have been undertaken to examine changes in cover type at the landscape level in forested communities. Although a sizable amount of old growth remains in wetland forest types (northern white cedar and other swamp conifers), a very low percentage is reserved from timber harvesting (Vora, 1994). Table 3 indicates the amount of forest land subject to timber harvest in the United States and Canada.

CLEARING OF HARDWOODS IN THE NORTHERN SWAMP FORESTS

The southernmost temperate area of forests represents the area that has experienced the most change over time. Within the United States extensive areas of

Table 3   Forest Land Resources of the United States and Canada

Country	Hectares
United States	
Reserved forest land	7,907,287
Noncommercial forest land	92,414,575
Commercial forest land	195,546,550
Deferred forest land	1,872,874
(Subtotal)	297,741,286
Canada	
Timber productive forest lands	236.7 million
Reserved forest lands	8.7 million
Timber unproductive forest lands	167.5 million
Unclassified forest lands	3.3 million
(Subtotal)	416.2 million
United States and Canada estimated total	713.9 million
Reserved from timber harvesting	16.6 million (2.3%)

From Canadian Council of Forest Ministers, Compendium of Canadian Forestry Statistics, 1993, Natural Resources Canada, Ottawa, Ontario, 1993, 152 pp. With permission; from Conway, S., *Logging Practices: Principles of Timber Harvesting Systems,* Miller Freeman Publications, Inc., San Francisco, CA, 1982, 432 pp. With permission; from Council on Environmental Quality, Environmental Trends, Executive Office of the President, Washington, D.C., 1989, 152 pp. With permission; and Stenzel, G., Walbridge, T. A., Jr., and Pearch, J. K., *Logging Practices and Pulpwood Production,* John Wiley & Sons, New York, 1985, 385 pp. With permission.

Table 4   Estimates of Presettlement Forested Wetland Area
for the Great Lakes States and New England

Northern state	Presettlement wetland area[a] (hectares)	Presettlement forested wetland area[a] (hectares)
Minnesota	6,101,200	2,279,500
Wisconsin	3,967,600	2,192,100
Michigan	4,534,400	2,947,400
Indiana	2,267,200	294,700
Ohio	2,024,300	657,900
Pennsylvania	456,300	29,700
New York	1,037,200	74,200
Vermont	138,100	89,700
New Hampshire	89,100	55,000
Maine	2,615,400	1,615,000
Estimated total	23,230,800	10,235,200

[a] The total land area for all states is not considered as part of these estimates. Northern forested wetland area was approximated by the area shown in Figure 1.

wet forests have been cleared and developed. Southern Michigan, northern Ohio, and eastern Indiana are now largely deforested but originally contributed substantially to the estimated 11 million ha of forested wetlands that bordered the Great Lakes and New England north of the fortieth parallel (Table 4).

Much of the historical conversion that occurred targeted the wet hardwoods of the temperate region of the Great Lakes states and Ontario. These activities used the hardwood forest products for fuel and construction purposes as well as cleared the land and capitalized on the fertile mineral soils of these forested areas for agricultural production. By the early 1900s much of the northern swamp forests of the Great Lakes states were gone, and the land had been converted to agriculture. In some parts of the Great Lakes region of the United States, conversion of forested wetlands has greatly diminished extensive stands of forest resources (Table 5). One of the most notable examples of this is the Black Swamp in Ohio.

The northern swamp forests of Ohio, Indiana, and southern Michigan contained a mix of birch, ash, elm, several species of oaks (*Quercus* spp.), cottonwoods (*Populus* spp.), poplars (*Populus* spp.), maples, basswoods (*Tilia* spp.), and hickories (*Carya* spp.). The Black Swamp, located in northwestern Ohio, contained many of these trees and covered an immense area. The estimated size of the Black Swamp was 193 km long and 64 km wide (Gordon, 1969; Ohio Department of Natural Resources, 1988). From 1860 to 1885 intense timber cutting and land clearing virtually eliminated the Black Swamp (Dahl and Allord, in press). By the twentieth century this area was Ohio's richest agricultural region and the least wooded (Whitney, 1994).

Similar to the United States, the area of temperate wet forests of southern Ontario and Quebec have been greatly reduced. In Canada agricultural conversion, industrial development, energy development (e.g., hydroelectric projects), recre-

Table 5   Estimated Change in Total Forested Wetland Area and Current Amount of Northern Forested Wetland for Key States in the United States

State	Original forested wetland area by state (hectares)	Current forested wetland area by state (hectares)	Percent change	Estimated area — northern forested wetland (hectares)
Minnesota	2,279,473	1,258,380	–45	1,258,380
Wisconsin	2,192,105	677,085	–69	677,085
Michigan	2,947,736	515,222	–82	515,222
Indiana	294,736	5,190	–98	Minimal
Ohio	657,894	39,554	–94	16,200
Pennsylvania	296,578	98,461	–67	24,700
New York	674,736	86,396	–87	45,750
Vermont	89,716	16,518	–82	15,380
New Hampshire	54,979	14,129	–74	12,150
Maine	1,615,020	578,785	–64	578,785
Alaska		5,393,643	0	5,393,643

Table 6 Projected Demand for Lumber and Forest
Products in the United States

Forest product	2000 estimate	2030 estimate
Lumber (billion bd. meters)	18.3	20.5
Plywood Particleboard	9.2	13.1
Hardboard Insulation board	8.3	11.4
Paper and paper products (million metric tons)	111.9	176.4

From Conway, S., *Logging Practices: Principles of Timber Harvesting Systems,* Miller Freeman Publications, Inc., San Francisco, 1982, 432 pp. With permission.

ational development and forestry have been cited as major activities resulting in losses of forested wetlands (Glooschenko and Grondin, 1988; Rubec et al., 1988, Snell, 1987). In areas where deciduous forested wetlands persist, many have been logged during previous decades. Those that have remained untouched are generally regarded as having low to no commercial timber value.

PRESENT AND FUTURE IMPACTS

Silviculture is the major threat to northern forested wetlands (United States Department of the Interior, 1994). Currently, timber harvest levels are near historically high levels (Council on Environmental Quality, 1989). The projected demand for timber and wood products in the United States is presented in Table 6.

It is possible that southern and western softwood forested stands will be insufficient to meet the increased needs of timber harvesters, habitat conservation areas and expanding urban land uses of the future. Commercial foresters are already implementing intensive management practices, including aerial application of herbicides and pesticides on southern pine stands, in an effort to obtain a 17-year rotation (Okeefenokee National Wildlife Refuge, personal communication). If this trend continues, increased tree harvesting may be expected in the future, especially for pulpwood and fiber-based construction panels and related products (Powell et al., 1993).

In Canada conversion of noncommercial, stagnant treed areas, now growing on wet peatlands, to productive conifer swamps through partial drainage is a possibility. Such drainage improvement has been practiced in Fennoscandia and the former U.S.S.R. on a large scale, but only on an experimental scale in Canada (Hillman, 1987). Substantial increases in growth have been obtained through lowering the water table, but currently the economics of such treatments cast doubt on their large-scale application.

In the past 20 years the forest and wood products industry has made better and more efficient use of all forest products. By-products or nonmarketable timber that was once left in the field can now be collected and chipped, pulped, mulched, or made into pressed wood products. Data indicate that the amount of residue left in the field has steadily decreased since 1986 (Powell et al., 1993). Improving wood utilization technologies combined with increasing prices for wood products have led to practices that not only utilize heretofore wasted products, but may create a marketable use for timber products previously considered to have low or no commercial value. As a result northern forested wetlands may be increasingly threatened, not as primary targets for lumber but as a source of fiber for other wood and paper products.

Timber harvesting practices are not solely responsible for the loss or conversion of forested wetlands. Other activities that threaten northern forested wetland resources include hydrodevelopment projects that have drowned wetland forests in both the United States and Canada. In Canada these impacts have been due primarily to hydroelectric development. In the United States, flood control projects have taken a toll on forested wetlands. Dutch elm and other forest diseases have severely impacted species such as American elm, a common wetland tree in the temperate regions of the United States and Canada. Wildlife management projects have also had an impact on forested wetlands as species management priorities for waterfowl or furbearers often take precedence. Efforts aimed directly at increasing waterfowl habitat in the northern breeding grounds do little to protect or restore forested wetlands, and some habitat "improvement" projects may permanently impound wetland trees and convert forested wetland areas to open water ponds or marshes.

Beaver can create wetlands by impounding surface water or increase the amount of open canopy through flooding and cutting of trees. Beaver have been known to create ponds ranging in size from 0.5 to 30 ha and typically can cause significant alteration to the ecosystem (Hammerson, 1994). Overall, the impacts of beaver on forested wetland area are unquantified.

Agricultural conversion has also greatly diminished forested wetlands in both the United States and Canada. Historically agricultural conversion of forest land to cropland was a major cause for the loss of wetlands, especially in the eastern temperate zones of Ontario and the Great Lakes states in the United States. Hopefully, agricultural conversions have slowed, and the impacts to forested wetlands will be minimized in the future. Urban development and expansion have also destroyed forested wetlands in both countries. Within Ontario, expansion of urban centers such as Ottawa, Niagara, Toronto, Thunder Bay, and Windsor have resulted in losses from 30 to 90+ percent of the original wetland area (Rubec et al., 1988).

Evidence from the U.S. Wetland Status and Trends study suggests that while the rate of wetland loss is declining overall, the rate of loss of forested wetlands has accelerated. The results of a study of wetland losses between the 1950s and the 1970s revealed that 54 percent of freshwater wetland losses were losses of forested wetland (Frayer et al., 1983). A similar study covering the mid-1970s to mid-1980s timeframe revealed that 95 percent of freshwater wetland losses were forested wetland losses (Dahl and Johnson, 1991).

SUMMARY

While northern forested wetlands are extensive throughout the northern latitudes, historical losses have been significant, especially in the mixed hardwood swamp forests of the Great Lakes states and Ontario. The ecological consequences of these changes cannot be fully recognized at this time as there are myriad things that can have impacts on the extent, distribution, and function of forested wetlands.

Fortunately, trees are a renewable resource if managed scientifically. However, restoration of mature forested wetlands is a long-term proposition. This is especially true in the northern latitudes where tree growth is slow (Van Hees, 1990). As a result, development interests that are beginning to investigate utilization of northern woodlands may pose a significant threat to these wetland resources if they cannot be managed on sustainable basis.

REFERENCES

Bailey, R. G., Description of the Ecoregions of the United States, Miscellaneous publication No. 1391, U.S. Department of Agriculture, Forest Service, Washington, D.C., 1980, 77 pp.

Beazley, M., *Wetlands in Danger,* Reed International Books Limited, London, England, 1993, 187 pp.

Canada Soil Survey Committee, The Canadian System of Soil Classification, Publication No. 1646, Canada Department of Agriculture, Research Branch, Ottawa, Ontario, 1978.

Canadian Council of Forest Ministers, Compendium of Canadian Forestry Statistics, 1993, Natural Resources Canada, Ottawa, Ontario, 1993, 152 pp.

Conway, S., *Logging Practices: Principles of Timber Harvesting Systems,* Miller Freeman Publications, Inc., San Francisco, 1982, 432 pp.

Council on Environmental Quality, Environmental Trends, Executive Office of the President, Washington, D.C., 1989, 152 pp.

Cowardin, L. M., Carter, V., Golet, F. C., and LaRoe, E. T., Classification of Wetlands and Deepwater Habitats of the United States, FWS/OBS-79-31, U.S. Department of the Interior, Fish and Wildlife Service, Washington, D.C., 1979, 103 pp.

Dahl, T. E., Wetlands — Losses in the United States 1780's to 1980's, U.S. Department of the Interior, Fish and Wildlife Service, Washington, D.C., 1990, 13 pp.

Dahl, T. E. and Allord, G. J., History and Trends of Wetlands in the United States, National Water Summary Report, United States Geological Survey, Reston, VA, in press.

Dahl, T. E. and Johnson, C. E., Status and Trends of Wetlands in the Conterminous United States, Mid 1970's to Mid-1980's, U.S. Department of the Interior, Fish and Wildlife Service, Washington, D.C., 1991, 28 pp.

Eyre, F. H., Ed., Forest Cover Types of the United States and Canada, Society of American Foresters, Washington, D.C., 1980, 148 pp.

Frayer, W. E., Monahan, T. J., Bowden, D. C., and Graybill, F. A., Status and Trends of Wetlands and Deepwater Habitats in the Conterminous United States, 1950's to 1970's, Colorado State University, Fort Collins, CO, 1983, 31 pp.

Frelich, L. E., Old forest in the lake states now and then, *Nat. Areas J.*, in press.

Glooschenko, V. and Grondin, P., Wetlands of eastern temperate Canada, in *Wetlands of Canada*, Rubec, C. D. A., Ed., Polyscience Publications Inc., Montreal, Quebec, 1988, pp. 201–248.

Gordon, R. B., The natural vegetation of Ohio in pioneer days, *Bull. Ohio Biol. Surv.*, III (2), 1969, 113 pp.

Hall, J. V., Wetland Resources of Alaska (Map), U.S. Fish and Wildlife Service, Washington, D.C., 1991, (map scale 1:2,500,000).

Hall, J. V., Frayer, W. E., and Wilen, B. O., Status of Alaska Wetlands, U.S. Fish and Wildlife Service, Anchorage, AK, 1994, 33 pp.

Hammerson, G. A., Beaver (*Castor canadensis*): ecosystem alterations, management, and monitoring, *Nat. Areas J.*, 14 (1), 44–57, 1994.

Heikurainen, L., Effect of cutting on the ground water level of drained peatlands, in *International Symposium on Forest Hydrology, August 29 to September 10, 1965, University Park, PA*, Sopper, W. E. and Lull, H. W., Eds., Pergamon Press, Oxford, England, pp. 345–354, 1967.

Hillman, G. R., Improving Wetlands For Forestry in Canada, Information Report, NOR-X-228, Canadian Forestry Service, Northern Forestry Centre, Edmonton, Alberta, 1987, 29 pp.

Ketcheson, D. E. and Jeglum, J. K., Estimates of Black Spruce Peatland Areas in Ontario, Information Report, O-X-172, Canadian Forestry Service, Sault Ste. Marie, Ontario, 1972, 29 pp.

Küchler, A. W., Potential Natural Vegetation of the Conterminous United States, Special Publication No. 36, American Geographical Society, New York, 1964, (map scale 1:3,168,000).

National Wetlands Working Group, Canada wetland regions (map), The National Atlas of Canada, 5th ed., Energy, Mines and Resources Canada, Geographical Services Division, Surveys and Mapping Branch, Ottawa, Ontario, 1986.

Ohio Department of Natural Resources, Ohio Wetlands Priority Conservation Plan: an Addendum to the 1986 Ohio Statewide Comprehensive Outdoor Recreation Plan, Office of Outdoor Recreation Services, Columbus, OH, 1988, 67 pp.

Powell, D. S., Faulkner, J. L., Darr, D. R., Zhu, Z., and MacCleary, D. W., Forest Resources of the United States, 1992, General Technical Report RM-234, U.S. Department of Agriculture, Forest Service, Fort Collins, CO, 1993, 132 pp.

Rubec, C. D. A., Lynch-Stewart, P., Wickware, G. M., and Kessel-Taylor, I., Wetland utilization in Canada, in *Wetlands of Canada*, Rubec, C. D. A., Ed., Environment Canada and Polyscience Publications Inc., Montreal, Quebec, pp. 381–412, 1988.

Ryan, J. C., Early Loggers in Minnesota, Vol. II, Minnesota Timber Producers Assoc., Duluth, MN, 1976, 63 pp.

Ryan, J. C., Early Loggers in Minnesota, Vol. III, Minnesota Timber Producers Assoc., Duluth, MN, 1980, 63 pp.

Snell, E. A., Wetland Distribution and Conversion in Southern Ontario, Working Paper No. 48, Inland Waters/Lands Directorate, Environment Canada, Ottawa, Ontario, 1987.

Stenzel, G., Walbridge, T. A., Jr., and Pearce, J. K., *Logging Practices and Pulpwood Production*, John Wiley & Sons, New York, 1985, 358 pp.

Tarnocai, C., Canadian wetland registry, in Workshop in Canadian Wetlands, Ecological Land Classification Series No. 12, Rubec, C. D. A. and Pollett, F. C., Eds., Lands Directorate, Environment Canada, Ottawa, Ontario, 1980.

Tarnocai, C., Peat Resources of Canada, NRCC No. 24140, National Research Council of Canada, Division of Energy, Ottawa, Ontario, 1984, 17 pp.

United States Department of the Interior, The Impact of Federal Programs on Wetlands, Vol. II, A Report to Congress by the Secretary of the Interior, U.S. Department of the Interior, Washington, D.C., 1994, 333 pp.

Van Hees, W. W. S., Boreal forested wetlands — what and where in Alaska, *Forest Ecol. Manage.*, Jackson, B. D., Ed., 33/34 (1–4), pp. 425–438, 1990.

Viereck, L. A. and Little, E. L., Atlas of United States Trees, Volume 2 — Alaska Trees and Common Shrubs, Miscellaneous Publication No. 1293, U.S. Department of Agriculture, Forest Service, Washington, D.C., 1975.

Vora, R. S., Integrating old-growth forest into managed landscapes: a northern Great Lakes perspective, *Nat. Areas J.*, 14 (2), pp. 113–123, 1994.

Whitney, G. G., *From Coastal Wilderness to Fruited Plain. A History of Environmental Change in Temperate North America, 1500 to the Present*, Cambridge University Press, Cambridge, United Kingdom, 1994, 451 pp.

Zoltai, S. C. and Martikainen, P. J., The role of forested peatlands in the global carbon cycle. NATO Advanced Research Workshop, Sept. 12–16, 1994, Banff, Alberta, in press.

Zoltai, S. C., Taylor, S., Jeglum, J. K., Mills, G. F., and Johnson, J. D., Wetlands of Boreal Canada, in *Wetlands of Canada*, Rubec, C. D. A., Ed., Environment Canada and Polyscience Publications Inc., Montreal, Quebec, pp. 98–154, 1988.

CHAPTER 2

Status and Trends of Forested Wetlands in the Northern United States

W. E. Frayer

CONTENTS

INTRODUCTION

Total acreage of forested wetlands in the northern United States in the 1980s was 24.7 million acres. Of this figure, 11.4 million acres are in the northern conterminous states, and 13.3 million acres are in Alaska. Estimated net losses of forested wetlands in the northern conterminous states over a 9-year period from 1974 to 1983 averaged 22,700 acres per year. For the remainder of the lower 48 states, the average annual net loss of palustrine forested wetlands during the same period was 355,500 acres.

The U.S. Fish and Wildlife Service, through its National Wetlands Inventory Project, has been conducting inventories of the nation's wetlands and associated deepwater habitats since 1974. The purpose is to develop and disseminate comprehensive data concerning the characteristics and extent of wetlands.

Results of a National Wetlands Inventory study of wetland gains and losses between the 1950s and 1970s were published by Frayer et al. (1983) and Tiner (1984). Between the 1950s and 1970s, there was a net loss of about 6 million

1-56670-177-5/97/$0.00+$.50

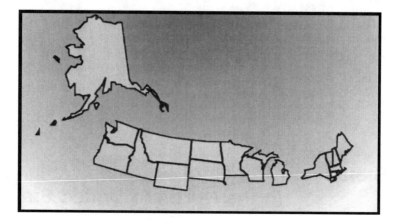

Figure 1   Northern United States.

acres of forested wetlands. This 20-year loss equates to an average annual net loss of approximately 300,000 acres of forested wetlands.

A later study documented trends in wetlands and deepwater habitats from the 1970s to 1980s (Frayer, 1991; Dahl and Johnson, 1991). During that 9-year period, there was an average annual net loss of about 378,000 acres of palustrine forested wetlands. This increase in the net loss rate is particularly noteworthy because the average annual net losses for all wetlands for the 1950s to 1970s and 1970s to 1980s were 485,000 acres and 290,000 acres, respectively. This means that the loss of palustrine forested wetlands in the more recent period was larger than the loss figure for all wetlands combined. This is because some wetland categories increased in size, partially masking the accelerated losses in forested wetlands. Of the wetlands at the time of settlement in the area now comprising the 48 contiguous states, only 47 percent remained in the 1980s (Dahl, 1990).

This paper documents the status of the forested wetlands of the northern United States. For the purposes of this paper, the states included are Alaska, Washington, Oregon, Montana, Idaho, Wyoming, North Dakota, South Dakota, Minnesota, Wisconsin, Michigan, New York, Massachusetts, Connecticut, Rhode Island, Vermont, New Hampshire, and Maine (Figure 1). For all of the states except Alaska, the status and trends of forested wetlands were estimated from the same data used to prepare the latest national report (Frayer, 1991). The first wetlands status report for Alaska has only recently been prepared (Hall et al., 1994), so there are no trend estimates. While this paper provides information on abundance and change, it does not provide information on the quality of forested wetlands.

CLASSIFICATION SYSTEM

The definitions, classifications, and categories of wetlands and deepwater habitats used are those described by Cowardin et al. (1979). Although this chapter deals primarily with forested wetlands, it is informative to give brief descriptions of other categories, because some are used in later discussion of trends.

In general terms, wetland is land where saturation with water is the dominant factor determining the nature of soil development and the types of plant and animal communities living in the soil and on its surface. Technically, wetlands are lands transitional between terrestrial and aquatic systems where the water table is usually at or near the surface or the land is covered by shallow water. Wetlands must have one or more of the following three attributes: (1) at least periodically, the land predominantly supports hydrophytes; (2) the substrate is predominantly undrained hydric soil; and (3) the substrate is nonsoil and is saturated with water or covered by shallow water at some time during the growing season of each year. Common terms used to describe various wetlands include *marshes, swamps, bogs, muskegs, tundra, fens, small ponds, sloughs, potholes, river overflows, mud flats,* and *wet meadows.*

Deepwater habitats consist of certain permanently flooded lands. In saltwater areas, the separation between wetland and deepwater habitat coincides with the elevation of the extreme low water of spring tide. In other areas, the separation is at a depth of 2 m (6.6 ft) below low water. This is the maximum depth in which emergent plants normally grow. Common names used for deepwater habitats include *bays, lakes, sounds, fjords, lagoons, inlets, rivers,* and *reservoirs.*

Within the classification structure that follows, wetlands and deepwater habitats are grouped according to systems. A system consists of environments of similar hydrological, geomorphological, chemical, and biological influences. Each system is further divided by the driving ecological force, such as ebb and flow of tide, and by substrate material and flooding regimes, or by vegetative life form. Groupings of categories were made to accommodate the special interests of the study and the detail to which aerial photography could be interpreted.

The *marine system* extends from the outer edge of the continental shelf to the high water of spring tides or to the boundary of other systems as defined later. The deepwater habitats in the marine system are not included in this study. Marine intertidal wetlands are those in which the substrate is exposed and flooded by tides, including the associated splash zone.

The *estuarine system* consists of deepwater tidal habitats (bays) and adjacent tidal wetlands, which are usually semi-enclosed by land but have open, partially obstructed, or sporadic access to the open ocean and in which ocean water is at least occasionally diluted by freshwater runoff from the land. Offshore areas with typically estuarine plants and animals, such as mangroves and oysters, are also included.

The *lacustrine system* includes wetlands and deepwater habitats situated in topographic depressions or dammed river channels. Each area must exceed 20 acres or be deeper than 6.6 ft or have an active wave-formed or bedrock shoreline feature. Lacustrine areas were grouped together as deepwater habitats in this study.

The *riverine system* includes wetlands and deepwater habitats contained within a channel. Riverine subsystems were grouped together as deepwater habitats in this study.

The *palustrine system* includes all nontidal wetlands not included within any of the other four systems and does not include any deepwater habitats. For this study, palustrine wetlands include the following groups: *nonvegetated,* which contains no emergent vegetation but includes vegetation in the form of aquatic beds and *vegetated,* which includes *emergent, forested,* and *scrub/shrub.* Emergent contains primarily those erect, rooted herbaceous plants typically found in wet environments. *Forested* is characterized by the presence of trees, and *scrub/shrub* includes areas dominated by shrubs and small or stunted trees.

All remaining surface area (area not classed as wetland or deepwater habitat) was placed in three categories. These are *agriculture, urban,* and *other.* These correspond to classes described by Anderson et al. (1976) at their Classification Level I. *Other* includes Anderson's Level I classes of forest land, rangeland, and barren land.

This is only a brief discussion of the classification used in this study. It is difficult to differentiate the categories further without introducing highly technical terms. More detailed discussions, exact definitions, and fuller descriptions are presented by Cowardin et al. (1979) and Anderson et al. (1976).

SURVEY AND ANALYSIS PROCEDURE

The objectives of the study were to develop statistical estimates for forested wetlands and associated categories of wetlands and deepwater habitats for the 1970s and the 1980s, and changes for the period. For Alaska, only the more recent estimate was possible.

A stratified random sampling design was used, with the basic strata in the northern conterminous states being formed by state boundaries and the physical subdivisions described by E. H. Hammond (1970). Additional strata specific to the study were special coastal strata encompassing most of the marine and estuarine categories used in the study and other strata encompassing the Great Lakes. In Alaska, a stratified random sampling design was used, with 25 physical subdivisions (strata) being formed by modification of the major land resource areas described by Rieger et al. (1979) and four specially developed coastal strata encompassing areas in the marine and estuarine categories used in the study.

Sample units were allocated to strata in proportion to expected amounts of wetlands and deepwater habitats as estimated by U.S. Fish and Wildlife personnel. The total number of sample units used in this study was 3239 (673 in the northern conterminous states and 2566 in Alaska).

Each sample unit was a 4 mi^2 area, 2 mi on each side. The units were plotted on U.S. Geological Survey topographic maps. Also, 1:80,000 scale black and white aerial photography had been obtained for the 1970s. The 1980s aerial photography was then obtained, which was primarily 1:58,000 scale color infrared transparencies. The mean years of photography were 1974 and 1983 in the northern conterminous states, a 9-year period. The mean year of the Alaska photography was 1980.

The 1980s photography was interpreted and annotated in accordance with the classification system described earlier and with procedures developed by the U.S. Fish and Wildlife Service's National Wetlands Inventory Project. The results were compared with the 1970s photography for the plots in the northern conterminous states, and any changes in classification were annotated and measured. Both the recent classification and the classification for the 1970s were recorded for each change.

RESULTS AND DISCUSSION

Results for forested wetlands and some of the other categories discussed in the classification system are given in Table 1 for the northern conterminous states. Similar data for Alaska are given in Table 2. Historically, Alaska's wetlands have not been impacted by human activities as severely as the wetlands in the lower 48 states. The U.S. Fish and Wildlife Service estimates that during the 200-year period between 1780 and 1980, approximately 0.1 percent of the original wetland acreage in Alaska was lost. During the same period, more than half of the original acreage total was lost in the lower 48 states (Dahl, 1990).

The status and trends results presented in the remainder of this section are based on information found in Tables 1 and 2 and other data collected during the study.

There was a net loss of 204,400 acres of palustrine forested wetlands in the northern conterminous states since the 1970s. The 1980s estimates are 11.4 million acres of palustrine forested wetlands, with an average annual net loss of 22,700 acres. The forested wetlands in the northern conterminous states comprise 22.0 percent of the forested wetlands in the lower 48 states. However, only 6.0 percent of the loss of forested wetlands since the 1970s occurred in the northern states. Nationally, over 2 million acres of palustrine forested wetlands were converted to non-wetlands in the period, with 1.0 million acres lost to agriculture and another 1.0 million acres converted to the "*other*" category (Frayer, 1991). Northern states with significant losses between the 1970s and 1980s include Michigan (57,900 acres), Massachusetts (67,600 acres), and Maine (8,900 acres).

Forested wetlands comprise 13.3 million acres in Alaska. Although the total acreage of forested wetlands is somewhat similar for Alaska and other combined northern states, the proportionate amounts they represent are markedly different. Wetlands comprise a total of 5.2 percent of the total surface area of the lower 48 states (Frayer, 1991). Forested wetlands represent 2.6 percent of the total area, or half of the wetlands. In the northern conterminous states, wetlands represent 5.4 percent of the total surface area, and forested wetlands represent 1.7 percent of the area, or one-third of the wetlands. In Alaska the trends continue with increased area covered by wetlands (43.3 percent). Forested wetlands represent 3.3 percent of the total surface area, which is only 7.6 percent of the wetlands.

Based on the data from this study, we may infer several things about northern forested wetlands.

Table 1 Area, in Thousands of Acres, of Wetlands and Deepwater Habitats in the Northern Conterminous United States

	Current classification										
	Marine and estuarine wetlands	Palustrine wetlands					Deepwater habitats	Agriculture	Urban	Other	Total surface area
		Nonvegetated	Emergent	Forest	Scrub/shrub	All wetlands					
Marine and estuarine wetlands	369.1	0	0	0	0	—	0.2	0	0.1	0	369.4
Original palustrine wetlands											
Nonvegetated	0	1065.0	2.0	0	0.1	—	0	1.0	0.2	5.1	1073.4
Emergent	0	8.8	13258.4	1.6	34.2	—	3.8	290.9	6.0	44.5	13648.2
Forest	0	8.8	63.5	11229.3	213.3	—	0.2	45.0	2.4	5.6	11568.1
Scrub/shrub	0	6.2	109.3	132.8	7867.6	—	0	124.7	5.6	1.1	8247.3
Classification											
All wetlands	—	—	—	—	—	34370.0	4.2	461.6	14.3	56.3	34906.4
Deepwater habitats	0.2	0	11.9	0	1.9	14.0	52065.3	0	0	3.1	52082.4
Agriculture	0	31.7	367.8	0	10.2	409.7	2.2	157570.6	841.5	694.3	159518.3
Urban	0	0	0	0	0	0	0	1.4	11473.7	1.8	11476.9
Other	0.2	26.4	1.3	0	0	27.7	2.7	1033.2	870.5	390050.6	391984.7
Total surface area	369.3	1146.9	13814.2	11363.7	8127.3	34821.4	52074.4	159066.8	13200.0	390806.1	649968.7
Change	-0.1	73.5	166.0	-204.4	-120.0	-85.0	-8.0	-451.5	1723.1	-1178.6	0

Table 2   Area, in Thousands of Acres, of Alaska Wetlands and Deepwater Habitats

Marine and estuarine wetlands	**2180.5**
Palustrine wetlands	
Nonvegetated	2670.2
Emergent	42000.8
Forest	13322.3
Scrub/shrub	114510.1
All wetlands	174683.9
Deepwater habitats	29870.4

1. The remoteness of many northern forested wetlands, especially in Alaska, has undoubtedly been one causal factor in the lesser rates of human-induced conversion compared to conversions in the southern states. It is interesting to note that there was an estimated 221.1 million acres of wetlands in the lower 48 states in 1780 (Dahl, 1990). There are now 103.3 million acres of wetlands in the lower 48 states and 174.7 million acres in Alaska. Remoteness is certainly a factor in this major reversal. Another factor, probably of more importance, is discussed under item 4.

2. Forested wetlands represent a smaller proportion of surface area in northern states. The manner by which land was formed determines to a large extent whether forests and wetlands exist. The glaciated nature of much of the north country resulted in diverse topography which, coupled with the climate, limits the extent of forested wetlands.

3. Forested wetlands represent a smaller proportion of wetlands in northern states. In the far north, tundra and other forms of scrub/shrub wetlands are predominant.

4. Forested wetlands in the north have undergone only minor human-induced conversions compared to the huge decreases in the southern United States. Some obvious human-induced changes to agriculture, urban, and other non-wetland categories have occurred, but not nearly at the rates experienced farther south. Much of the loss of wetlands in the southern United States has been from forested wetlands, and the vast majority of these losses have occurred where the vegetation has had economic importance in the form of forest products, such as high-valued bottomland hardwoods, but even more so when the land has been drained and planted with agricultural crops such as soybeans.

5. A largely heretofore unknown change from forested status to other palustrine vegetated categories (emergent and scrub/shrub) has occurred in the northern conterminous states at about the same rate as in the states farther south. In the 1970s there were 55.2 million acres of forested wetlands in the lower 48 states. By the 1980s, 1.8 million acres or 3.3 percent of these wetlands had converted to other palustrine emergent categories (emergent and scrub/shrub) (Frayer, 1991). In the northern conterminous states, there were 11.6 million acres of forested wetlands in the 1970s. By the 1980s, more than 275,000 acres or 2.4 percent of these wetlands had converted to palustrine emergent or scrub/shrub. What are the reasons behind these changes? Perhaps a significant portion are actually human-induced as a result, for example, of changes in subsurface flows due to road construction as well as clearcutting on forested wetlands.

CONCLUSION

The results reported are based on a designed study of the wetlands and deepwater habitats of the lower 48 states and Alaska. Losses of northern forested wetlands have occurred, but are minor compared to changes in the south. Continual monitoring of surface area use and changes in use is needed to provide the basis for wise decisions. This study is the result of one such method of monitoring initiated by the U.S. Fish and Wildlife Service.

REFERENCES

Anderson, J. R., Hardy, E. E., Roach, J. T., and Witmer, R. E., A Land Use and Cover Classification System for Use with Remote Sensor Data, U.S. Geological Survey Professional Paper 964, 1976, 22 pp.

Cowardin, L. M., Carter, V., Golet, F. C., and LaRoe, E. T., Classification of Wetlands and Deepwater Habitats of the United States, U.S. Fish and Wildlife Service, 1979, 103 pp.

Dahl, T. E., Wetland Losses in the United States, 1780's to 1980's, U.S. Fish and Wildlife Service, 1990.

Dahl, T. E. and Johnson, C. E., Status and Trends of Wetlands in the Conterminous United States, Mid-1970's to Mid-1980's, U.S. Fish and Wildlife Service, 1991, 28 pp.

Frayer, W. E., Status and Trends of Wetlands and Deepwater Habitats in the Conterminous United States, 1970's to 1980's, Michigan Technological University, 1991, 32 pp.

Frayer, W. E., Monahan, T. J., Bowden, D. C., and Graybill, F. A., Status and Trends of Wetlands and Deepwater Habitats in the Conterminous United States, 1950's to 1970's, Colorado State University, 1983, 32 pp.

Hall, J. V., Frayer, W. E., and Wilen, B. O., Status of Alaska Wetlands, U.S. Fish and Wildlife Service, 1994, 32 pp.

Hammond, E. H., Physical subdivisions of the United States, in *National Atlas of the United States*, U.S. Geological Survey, 1970, 417 pp.

Rieger, S., Schoephorster, D. B., and Furbush, C. E., Exploratory Soil Survey of Alaska, U.S. Department of Agriculture, Soil Conservation Service, 1979, 213 pp.

Tiner, R. W., Jr., Wetlands of the United States: Current Status and Trends, U.S. Fish and Wildlife Service, 1984, 59 pp.

CHAPTER 3

Forested Mires as a Renewable Resource — Toward a Sustainable Forestry Practice

Juhani Päivänen

CONTENTS

INTRODUCTION

Forested wetlands form an important renewable resource of wood. Some of these sites support tree growth already in their natural state, but forest drainage has made wood production the most extensive mode of utilizing forested wetlands sustainably. Currently, about 15 million ha of wetlands have been drained for forestry in the boreal and temperate zones. The purpose of the paper is to give a synthesis on this management practice aiming at sustainable wood production.

Wetlands are areas that are inundated or saturated by surfacewater or groundwater, at such a frequency and duration that under natural conditions they support organisms adapted to poorly aerated and/or saturated soil (Lugo, 1990). The Canadian definition even includes shallow open water in wetlands (Zoltai and Pollett, 1983). *Forested wetland* is defined to be any wetland with a significant component of woody vegetation, regardless of the size of the plants (Lugo, 1990).

In the boreal zone, wetland ecosystems are often characterized by the accumulation of organic matter, which is produced and deposited at a greater rate than it is decomposed, leading to the formation of *peat* (Gore, 1983). Thus, the vegetation has a decisive role in modifying edaphic conditions. These sites — called *peatlands* or *mires* — are usually supported by a humid climate and high water table levels, leading to a low level of microbial activity in the soil. The term *mire* can be considered a slightly wider concept than *peatland*, because it encompasses all peat-forming habitats, irrespective of the peat thickness. However, the terms *peatland* and *mire* are often used as synonyms. In an ecological context the term *mire* has been preferred, whereas in connection with forest amelioration or management, the more traditional term *peatland* has been used (Paavilainen and Päivänen, 1995).

A mire site is *minerotrophic* when the nutrient concentration in the surface peat and mire water is significantly higher than that of precipitation. The term *ombrotrophic* describes a site receiving nutrients only in the form of precipitation or windborne dust (e.g., Eurola et al., 1984). *Bogs* are ombrotrophic sites where the peat is predominantly *Sphagnum* moss peat. *Fens* are at least slightly minerotrophic sites where the peat is formed mainly from *Carex, Phragmites*, and other residues.

A mire site may be forested, sparsely forested, or treeless (Laine et al., 1986). The sparsely forested sites have also been referred to as *treed fens* and *treed bogs* (Jeglum, 1991a).

In this chapter, the term *wetland forest* is used to designate tree stands either on mires (in the presence of peat-forming plants) or peatlands (on a peat deposit) or wet mineral soil sites (excess water in the substrate). This is more or less comparable to the term *swamp*, which is in common use in Canada (Jeglum, 1991a).

FORESTED WETLAND RESOURCE

There is no agreed minimum thickness of peat for defining mires or peatlands. In Finland, for example, mire or peatland is defined by the presence of peat-forming plants; in other countries a minimum value for peat thickness is usually between 20 to 45 cm. From the point of view of wetland forestry, however, it is not always necessary to separate the sites from each other according to the peat thickness; excessive water in the substrate is the common, most important growth-limiting factor (Paavilainen and Päivänen, 1995).

Because the definitions of wetland and peatland (or mire) vary from country to country, the estimates of the global coverage of these ecosystems differ considerably. Quite recently, however, it has been agreed that the final figure for the total mire area would be somewhere between 500 million (Heathwaite, 1993; Pfadenhauer et al., 1993) and 550 million ha (Paavilainen and Päivänen, 1995). The last-mentioned estimate includes, in the boreal zone, both peatlands and wet mineral soils.

The statistics do not separate forest-covered mires or wetlands from the treeless ones; thus, the forested wetland area remains unclear.

In the northern hemisphere, the countries having the highest percentage cover of mires are as follows (Paavilainen and Päivänen, 1995): Finland — 31 percent, Estonia — 21 percent, Sweden — 20 percent, Canada — 19 percent, Latvia — 17 percent, Republic of Ireland — 17 percent, and Russia — 11 percent.

EXTENT OF DRAINAGE FOR FORESTRY

General Features

Some wetlands support tree growth naturally (Figure 1), but forest drainage enables economical wood production from a much more extensive wetland area. In wetland sites, excessive water in the substrate checks root growth and microbial activity, and may even lead to unfavorable biochemical phenomena. Thus, the most important objective of drainage is to adjust the water content of the soil to a level that ensures sufficient aeration. On the other hand, ombrotrophic raised bogs are too nutrient-poor to support tree growth, even after amelioration measures (Figure 2). Forest drainage in waterlogged areas is now part of the normal forestry practice in Fennoscandia, the Baltic countries, Russia, the British Isles, and some parts of the United States. In Canada there are large-scale experiments, but forest drainage is still scanty in practice. Currently, about 15 million ha of peatlands and wetlands have been drained for forestry in the boreal and temperate zones. More than 90 percent of this is situated in Fennoscandia (Finland, Sweden, and Norway) and the former U.S.S.R. (Päivänen and Paavilainen, 1990; Paavilainen and Päivänen, 1995).

Regional Review

The current practice and extent of drainage in different countries varies significantly, depending on several issues, e.g. the potential wetland area, the land ownership relations, the demand for raw wood, national economic considerations, etc. In the following, a short review of forest drainage in countries where this kind of forest amelioration is considered important will be given (see also Paavilainen and Päivänen, 1995).

Figure 1   Herb-rich spruce swamp in natural state growing Norway spruce (*Picea abies*) with an uneven-aged stand structure, southern Finland. (Photo: J. Päivänen.)

Figure 2   Ombrotrophic raised bog in the southern boreal zone, southern Finland. The site is not suitable for forest drainage. (Photo: J. Päivänen.)

In *Finland* the forest drainage activity developed into a nationwide campaign to increase forest growth in the 1960s. The area annually drained increased steadily up to 1969, when 295,000 ha were drained. The total forest drainage area according to forestry statistics is about 6.0 million ha (Aarne, 1993). Forest improvement legislation, together with low interest government loans and direct subsidies, have had a decisive impact on the level and scope of drainage in the private forest sector in the country. The legislation has also come to include both employment and social policy aspects. Forest improvement, particularly the draining of peatlands and waterlogged mineral soil sites, has tremendously increased the country's growing stock increment and has thereby contributed to the sustainable forestry practiced. The era of ditching for forestry is coming to its end, and emphasis is now shifting to activities aimed at maintaining the productivity of the drained sites — first thinning and improvement ditching (cleaning of old ditches + complementary ditching) (Päivänen, 1994).

In *Sweden*, about 15 percent (1.5 million ha) of the country's peat-covered wetlands have been drained for forestry. At present, about 17 percent of the current annual increment in Swedish forests comes from peatland forests. Since 1986, due to environmental issues, traditional forest drainage can be carried out only with a special permit. During the last 2 decades a new type of drainage, called *remedial ditching*, has come into practice in Sweden. The function of this activity is more for site preparation for regeneration than ditching for forest drainage (Hånell, 1990, 1991).

In *Norway* the total area drained for forestry is about 420,000 ha, which is about 19 percent of the total wetland area below the timberline. Drainage is now directed toward wet mineral soil sites and forested peatlands, but complementary ditching and ditch cleaning on old drainage areas are given priority (Braekke, 1990).

In the Baltic countries there has been a long tradition in forest drainage. The total area drained for forestry has been reported to be 463,000 ha for *Estonia* (Valk, 1988; Zobel, 1992), 500,000 ha for *Latvia* (Zalitis, 1990), and 600,000 ha for *Lithuania* (Ruseckas, 1991).

In the *former U.S.S.R.* the total area drained for forestry was about 5.5 million ha, which accounted for 2.2 percent of the total area of wetlands (Sabo, 1988; Vompersky, 1991). The rate of forest drainage reached its annual maximum of more than 260,000 ha in the 1980s (Vompersky, 1991). When taking into consideration the new independent countries, the forest drainage area in *Russia* can be estimated at about 3.8 million ha. However, published data are not available for every country.

In the *United Kingdom* drainage of peatlands and wet mineral soils in connection with afforestation has been vast. The two main purposes of drainage are to improve tree growth, particularly in the first 10 years, and to increase the depth of rooting in order to reduce the risk of windthrow (Pyatt, 1990a). The total area drained for forestry may be around 600,000 ha, including 505,000 ha in Great Britain (Cannell et al., 1993) and about 80,000 ha in Northern Ireland (Taylor, 1983). The long-term prospects for forests on peatland are reported to be good.

Yields from the second rotation seem to be higher than the first generation, even using less fertilizer (Pyatt, 1990b).

In the *Republic of Ireland* wetland forestry is totally dependent on drainage and afforestation due to the humid climate and open ground conditions. The area of man-made forests on peatlands may be around 210,000 ha, corresponding to less than 18 percent of the total mire area of the republic. Industrial cut-away peatlands also offer great prospects for forestry. It is assumed that a significant proportion of the more than 80,000 ha of peatlands currently harvested for fuel peat will later become afforested (Farrell, 1990).

In the *United States* a very small area of mires has been drained using traditional forest drainage techniques. In the Lake States, the so-called prescription drainage system has been practiced on poorly drained mineral soil sites. This low-intensity ditch system is composed of open ditches with which the natural drainage patterns are opened and the locally high water table controlled (Trettin, 1986). Prescription drainage has been an important forest management practice for sustaining commercially viable hardwood forests in northern Michigan (Trettin et al., 1991). In the southeastern states, wetland sites have been converted into the most productive forest land in connection with afforestation by applying a site-specific water management system, which includes drainage, bedding, and fertilization treatments (Duncan and Terry, 1983).

In *Canada* drainage and amelioration of wetlands for forestry purposes have rarely been practiced, although the potential wetland area is high. Until now, extensive and untapped forest resources have been readily available, the harvesting of which has been cheaper than creating additional forest area or increasing the growth of stagnant tree stands. However, this situation may be changing gradually. The areal extent of forest drainage activities has recently been summarized thoroughly on a provincial and territorial basis (see Haavisto and Jeglum, 1991). According to this, in only three provinces is the drained area more than 1000 ha.

In *Alberta* it has been estimated that an area of 4 million ha could be drained to improve tree growth or to convert the area into productive forest land (Hillman, 1987). However, only some systematic peatland drainage trials have been established to ascertain the most suitable ditch depths and spacings on a variety of sites (Hillman, 1991).

In *Ontario* there are 22.6 million ha of peatlands (25 percent of the total land area). It has been estimated that 8 million ha of these are supporting productive black spruce (*Picea mariana*) (Ketcheson and Jeglum, 1972). The earliest drainage trials were established in 1929 (see Payandeh, 1973), but the first properly planned operational drainage experiment was not carried out until 1984 (Koivisto, 1985; Jeglum, 1991b).

In *Quebec* the first drainage experiment was established in 1968 (Stanek, 1968). Operational scale forest drainage activity started in the 1980s. By the end of 1988, a total of 13,500 ha had been drained (Trottier, 1991). The province has given monetary support for private forest landowners to undertake drainage on peatlands.

In the other provinces the forest drainage has been very limited (see, for example, Päivänen and Wells (1977) and Wells (1991) for Newfoundland, and Neily (1991) for Nova Scotia). Altogether, less than 25,000 ha have been drained for forestry purposes in Canada (Jeglum, 1990; Haavisto and Jeglum, 1991). The earlier estimate — 100,000 ha — given by Kivinen and Pakarinen (1981) was evidently far too high.

FOREST DRAINAGE TECHNIQUES

Drainage Requirements

Recently the drainage requirements for forest growth in connection with peatlands and wet mineral soils have been reviewed and discussed by Paavilainen and Päivänen (1995). As a very generalized synthesis, it can be said that the objective should be to lower the groundwater table to the depth of >35 cm below soil surface in ombrotrophic, undecomposed peats and to >55 cm in minerotrophic, well-decomposed peats. These are achieved by using a suitable combination of ditch depth and spacing. In wetland forestry, soil drainage is conducted almost exclusively with open ditches. Drainage ditches should be 80 to 90 cm deep. Shallower water furrows (<60 cm) are used on open mires mainly for soil preparation for planting and only secondarily for local drainage.

Forest drainage is usually achieved by using the cross-drainage principle; the lateral ditches (contour ditches) form a more or less systematic network of parallel ditches, crossing the contours preferably at a small angle. If lateral ditches are placed on both sides of the main ditch, they form a "herringbone" system (for the principles, see Päivänen and Wells, 1977; Haavisto and Päivänen, 1987; *Guide…*, 1989; Rosen, 1989).

Working Methods

Up to World War II, ditches were dug manually. After the war, the ditch-digging techniques and machinery have developed somewhat differently, depending on the country. During the 1960s, especially in Finland, the bulk of the ditches were done with forest drainage plows pulled by heavy crawler tractors by means of a winch. During the last 2 to 3 decades, tractor diggers (Figure 3) or heavy excavators have mainly been used. The forest ditches are usually dug longitudinally with a bucket scoop conforming to the standard forest ditch dimensions (Päivänen and Wells, 1977; Päivänen, 1984; Rosen, 1989).

The ditches in the earlier drainage experiments in Canada were dug with some modified, local equipment not necessarily well-fitted for forest ditch digging (Haavisto and Atkinson, 1977; Hillman, 1987). In the former U.S.S.R., ditches used in forest drainage have been deep and wide, mainly dug with excavators along the ditch side (e.g., Valk, 1988). In the maritime climates, on treeless mires the initial drainage for large-scale planting schemes is done with water furrow

Figure 3   A tractor digger digging a ditch into deep peat soil with a bucket scoop conforming to the standard forest ditch dimensions. (Photo: J. Päivänen.)

plows. The cut-off drains and collecting drains are then dug with excavators (Thompson, 1979; Pyatt, 1990a).

Contracting

In Finland most drainage has taken place on private land. An essential feature of Finnish forest improvement has been the principle of joint projects. This has been of great importance, considering the country's fragmented forest ownership, especially in the private sector. Both forest improvement and water legislation have heavily promoted joint undertakings.

Both ditch digging and ditch cleaning are done by private contractors. The prices, based on work studies and cost analysis, used to be agreed upon annually between the association of contractors and employers. However, the new European antitrust legislation calls for total freedom of competition.

Water Protection Questions

Erosion of ditches is the most frequent cause of organic matter and nutrient loading to watercourses after ditching or ditch cleaning. The water from a drainage area should be led into a natural watercourse by a single main ditch, using a small gradient. The load of suspended solid particles can be limited by proper planning. The key is to slow down the water flow in the ditches. If necessary, special sedimentation ponds must be planned and constructed (Joensuu, 1992, 1997).

Maintenance of Drainage

In respect to its water regime, a mire that has been drained for forestry is an unstable ecosystem, and its state of drainage requires continuous follow-up (Heikurainen, 1980b). Several factors have an effect on the groundwater level (drainage norm) reached after drainage. Immediately after drainage, ditch depth and spacing and peat soil properties are the most important factors. Drainage increases the growth of the stand, which results in an increase in the standing volume. Later interception and evapotranspiration of the tree stand have an important role in controlling the drainage norm and compensating for the negative effects of ditch deterioration. This is the case on sites with fast initial stand development.

However, if the tree stand at the date of drainage is sparse, deterioration of the ditches leads to decreased tree growth and a need for *ditch cleaning* or *complementary ditching* before the final felling (Päivänen and Ahti, 1988; Ahti and Päivänen, 1997). The need and form for drainage improvement are evaluated based on the tree growth and the conditions of the ditches (see Ahti et al., 1988; Paavilainen and Päivänen 1995). It must be stressed that poor tree growth may also be due to the low nutrient status of the site or other kinds of damage. These cannot be improved by increasing the effect of drainage.

Three main types of ditching machines have been tested and used in Finland for ditch cleaning and complementary ditching: tractor diggers, special ditch cleaning diggers equipped with a hydraulic power transmission, and ordinary excavators on their own base. Problems connected with the maintenance of forest ditches have also been discussed in other countries like Norway (Braekke, 1990), Estonia (Valk, 1988), Russia (Vompersky, 1991) and Great Britain (Pyatt, 1990a).

FOREST MANAGEMENT ON VIRGIN MIRES

In the following the forest management on virgin mires is dealt with in two cases, one from northern Europe, Finland, and the other from North America, mainly Ontario, Canada.

In *Finland* some of the forested mire site types support commercial-size tree stands. Minerotrophic site types are usually dominated by Norway spruce (*Picea abies*), although the proportion of pubescent birch (*Betula pubescens*) in volume may be considerable. Ombrotrophic site types are dominated by Scots pine (*Pinus sylvestris*). Typically for the virgin mire tree stands, the number of stems is very high in small diameter classes and decreases abruptly with increasing diameter (Gustavsen and Päivänen, 1986). This means also that tree stands on virgin mires have a highly uneven age structure. The increment is eliminated by natural mortality, mainly of older trees.

Very little tree harvesting is carried out on virgin, forested mires in Finland. The cuttings, if any, should be "light"; otherwise, there is a risk that the site will become wetter due to the groundwater rise (Päivänen, 1982). Virgin, forested mires may be one of the few cases where management to promote uneven-aged stand structure might be recommended (so called single tree selection for continuous cover, see Lähde et al., 1992). However, the harvesting of single trees is not usually economical.

In *Canada* black spruce (*Picea mariana*) is the predominant tree species on mires and occurs alone or mixed with tamarack (*Larix laricina*) or eastern white cedar (*Thuja occidentalis*). In western parts of Canada, lodgepole pine (*Pinus contorta*) is also of economic importance on mires (Haavisto and Jeglum, 1991).

Harvesting and regeneration of black spruce are significant, especially in Ontario and Quebec. Black spruce is a major source of fiber for the pulp and paper industry and a large proportion of the volume of black spruce comes from forested mires (Haavisto and Jeglum, 1991). Guidelines for the management of black spruce stands on mires are under development (Haavisto et al.,1988). Efforts to minimize damage to the soil are important because forest drainage is not practiced in operational scale (Groot, 1987). The preservation of advance growth, mainly black spruce layerings, has been considered to be an essential feature of regeneration in wetland habitats. On sites without sufficient advance growth, good results have been achieved with both seed-tree groups and clearcut strips (Haavisto, 1979; Haavisto et al., 1988; Jeglum and Kennington, 1993).

Typical for the peatland forestry in Ontario is that intermediate cuttings and tending of young stands are hardly practiced at all. However, Canadian forestry is gradually changing from exploitation toward a planned utilization of forest resources on a renewable basis (Jeglum, 1991b; Jeglum and Kennington, 1993).

FOREST MANAGEMENT ON DRAINED MIRES

Mineral soil sites and drained mires differ from each other in many respects (physical, chemical, and biological soil properties; microclimate; hydrological features, etc., see Paavilainen and Päivänen, 1995). These differences must be taken into consideration when managing tree stands on these two totally different site categories.

In maritime climates such as the British Isles, drainage and afforestation is a common practice on treeless mires (Taylor, 1983; Farrell, 1990; Malcolm, 1991; Cannell et al., 1993). The long-term prospects for forests on drained mires are reported to be good. Yields from the second rotation seem to be even higher than the first generation (Pyatt, 1990b). In Fennoscandia afforestation of mires is now restricted to cut-away peatlands and abandoned farmland on peat soil (Kaunisto, 1997).

Most of the drainage in Fennoscandia, Russia, and the Baltic nations has been practiced on the extensive areas of naturally tree-covered mires. The profitability of draining mires is dependent on the fertility of the site, the volume of the tree stand capable of response at the time of draining, the geographical location of the site, and the stumpage price. As a generalization, it may be said that drainage becomes more profitable with increasingly fertile sites, the more timber there is on the site, and the further south the site is located (Päivänen, 1984).

Figure 4 Minerotrophic mire site 50 years after drainage supporting a mixed stand of *Betula pubescens* and *Picea abies*, southern Finland. (Photo: J. Päivänen.)

The forests growing on drained mires in Finland are mainly seedling stands or young thinning stands, and the proportion of more mature tree stands is small (Keltikangas et al., 1986). Thus there is an urgent need for treatment of young stands and first commercial thinnings; in most cases the regeneration actually does not occur until after some decades. Minerotrophic mire site types will develop mixed stands after drainage (Figure 4), the more nutrient-poor sites supporting mainly Scots pine (Figure 5). About 15 percent of the drained area is considered unfit for maintenance, either in terms of site type or climatic region (Keltikangas et al., 1986).

Only mire forests, where conifers are the dominant tree species, are fertilized in Finland. The silvicultural condition of the stand should be satisfactory, the

Figure 5   Scots pine (*Pinus sylvestris*) stand on slightly minerotrophic deep peat site 40 years after drainage, southern Finland. (Photo: J. Päivänen.)

trees should have recovered after drainage, and the stocking density should be high enough. The temperature sum should be at least 800°C (5°C threshold). Fertilization with potassium and phosphorus usually improves tree growth for 10 to 15 years. Because of the deficiency of boron in most mires in northern Europe, this micronutrient has been added to some fertilizers. For literature concerning forest fertilization in mire forests, see Paavilainen and Päivänen (1995).

An attempt has also been made to estimate the profitability of forest drainage by calculating the inputs (all cost factors) and outputs (the increase in allowable cut evaluated with stumpage prices). Accordingly, the internal profitability percentage in the Finnish forest drainage activity would lie somewhat above 5 percent (Heikurainen, 1980a). The Canadian calculations show that drainage of an existing stand is economical if it can reduce the rotation age by 30 years or more (Payandeh, 1988).

FOREST DRAINAGE AND ENVIRONMENT

According to a popular belief, virgin mires are considered to act as hydrological buffers. Several studies have, however, shown this so-called sponge theory to be wrong (Eggelsmann, 1972; Boelter and Verry, 1977; Verry, 1988). The very thinness of the acrotelm, which is the thin, nonwaterlogged layer at the surface, means that peat-covered catchments are poor suppliers of base flow (Burt et al., 1990; Eggelsmann et al., 1993). Thus the conservation of mires for the function of storing water in the catchment is a reasonable argument only for the aapa mires with heavy formation of strings, flarks, and fen pools. In any case, these open mires do not have high priority for forestry.

Obviously one of the immediate effects of forest drainage is that peak flows are increased. The actual changes depend to an essential degree on the character of the rainfall or snowmelt, the effectiveness of drainage, and the vegetation, particularly the tree stand, of the area concerned (Starr and Päivänen, 1981). The hydrological changes due to drainage, especially in runoff, decrease with time (Braekke, 1970; Ahti, 1988). Peak flows and total runoff may resume their predrainage level within 10 to 20 years from drainage. Low runoff may remain higher from drained than from undrained mires for several decades (Seuna, 1980). This can even be considered beneficial for several reasons.

In the planning and execution of forest drainage, much effort is devoted to reducing the load of suspended solids, organic material, and nutrients in receiving waters (see Paavilainen and Päivänen, 1995).

After forest drainage, plant species richness and diversity usually decrease in the field layer but are more likely to increase in the ground layer. On the spoil banks at the edges of ditches, the species richness may also increase. These changes may happen rather rapidly. Later on mire species are gradually substituted by the forest species that already dominate in the surrounding mineral soil sites. This means that the regional diversity is diminished (Vasander, 1990; Laine et al., 1995b; Vasander et al., 1997). The effect of forest drainage on the carbon balance of a mire ecosystem and on the greenhouse impact is being studied intensively (e.g. Laine et al., 1997).

CONCLUSIONS

Mires have a great potential for increased wood production. The stage of realization of this potential differs between countries depending on raw wood demand, silvicultural management practice and tradition, infrastructure of the countryside, etc. Forest amelioration activities on forested mires and waterlogged mineral soils are part of the normal forestry practices in Fennoscandinavia (Finland, Sweden, and Norway), the Baltic countries, Russia, the British Isles, and the southeastern United States. In other countries such as Canada and China, there are experiments going on, but forest drainage on an operational scale is still scanty.

Forest drainage has been shown to be profitable, if only directed toward appropriate sites. This man-made disturbance in the mire ecosystem must be maintained if we wish to use mire sites for sustainable wood production. Environmental consequences must be taken into consideration both in planning and realization, and measures must be taken to minimize harmful effects on the site, on stream water, and even globally on the atmosphere. In addition to increasing tree growth on mires, we must also reserve untouched mires and old-growth forests for future science, teaching, and recreation.

REFERENCES

Aarne, M., Ed., *Metsatilastollinen vuosikirja 1992* (*Yearbook of Forest Statistics 1992*), Finnish Forest Research Institute, Helsinki, 1993, 317 pp.

Ahti, E., Hydrologia ja turpeen vesitalous, *Metsantutkimusl. tied.*, 308, 29, 1988.

Ahti, E. and Päivänen, J., Response of stand growth and water table level to maintenance of ditch networks within forest drainage areas, in *Northern Forested Wetlands: Ecology and Management*, Trettin, C. C. et al., Eds., Lewis Publishers, Boca Raton, FL, 1997.

Ahti, E., Päivänen, J., and Vuollekoski, M., Kunnostusojitus, *Metsantutkimusl. tied.*, 308, 46, 1988.

Boelter, D. H. and Verry, E. S., Peatland and water in the northern Lake States, *Forest Service, U.S. Department of Agriculture, General Technical Report NC-31*, 1, 1977.

Braekke, F. H., Myrgrøfting for skogsproduksion. Inflytelse på vannhusholdning og flomfare, *Tidskr. for Skogbruk.*, 78, 227, 1970.

Braekke, F. H., Peatland and paludified forest on mineral soil in Norway — potentials for forest production, in *Biomass Production and Element Fluxes in Forested Peatland Ecosystems.* Hånell, B., Ed., Swedish University of Agricultural Science, Department of Forest Site Research, 1990, 29 pp.

Burt, T. P., Heathwaite, A. L., and Labadz, J. C., Runoff production in peat-covered catchments, in *Process Studies in Hillslope Hydrology,* Anderson, M. G. and Burt, T. P., Eds., John Wiley & Sons, England, 1990, 463 pp.

Cannell, M. G. R., Dewar, R. C., and Pyatt, D. G., Conifer plantations on drained peatlands in Britain: a net gain or loss of carbon, *Forestry,* 66(4), 353, 1993.

Duncan, D. V. and Terry, T. A., Water management, in *The Managed Slashpine Ecosystem,* Stone, E. L., Ed., School of Forest Resources and Conservation, University of Florida, Gainesville, 1983, 91 pp.

Eggelsmann, R., The water balance of lowlands areas in north-west coastal region of the FRG, Int. Symp. Hydrol. Marsh-Ridden Areas, Minsk, Byelorussian SSR, 1972, 15 pp.

Eggelsmann, R., Heathwaite, A. L., Grosse-Brauckmann, G., Kuster, E., Naucke, W., Schuch, M., and Schweickle, V., Physical processes and properties of mires, in *Mires: Process, Exploitation and Conservation*, Heathwaite, A.L., Ed., John Wiley & Sons, Chichester, 1993, chap. 4.

Eurola, S., Hicks, S., and Kaakinen, E., Key to Finnish mire types, in *European Mires,* Moore, P. D., Ed., Academic Press, London, 1984, chap. 2.

Farrell, E. P., Peatland forestry in the Republic of Ireland, in *Biomass Production and Element Fluxes in Forested Peatland Ecosystems.* Hånell, B., Ed., Swedish University of Agricultural Science, Department of Forestry Site Research, 1990, p. 13.

Gore, A. J. P., Introduction, in *Ecosystems of the World,* Vol. 4A, *Mires: Swamp, Bog, Fen and Moor,* Gore, A. J. P., Ed., Elsevier, Amsterdam, 1983, chap. 1.

Groot, A., Silvicultural consequences of forest harvesting on peatlands: site damage and slash conditions, Canadian Forestry Service, Great Lakes Forest Research Centre, Information Report O-X-384, 1987, p. 1.

Guide sur le drainage sylvicole, Gouvernement du Quebec, 1989, 53 pp.

Gustavsen, H. G. and Päivänen, J., Luonnontilaisten soiden puustot kasvullisella metsa-maalla 1950-luvun alussa (Summary: Tree stands on virgin forested mires in the early 1950's in Finland), *Folia For.,* 673, 1, 1986.

Haavisto, V. F., Some considerations for regenerating black spruce on peatlands in the Northern Clay Forest Section, Ontario, Canadian Forestry Service, Report O-X-295, 1979, p. 1.

Haavisto, V. F. and Atkinson, G. T., Capabilities of a modified "Timberjack" and a "Terrain-Master" for ditching on peatlands, Canadian Forestry Service, Great Lakes Forest Research Centre, Report O-X-264, 1977, p. 1.

Haavisto, V. F. and Jeglum, J. K., Peatland potentially available for forestry in Canada, in *Peat and Peatlands — Diversification and Innovation,* Vol. 1, Jeglum, J. K. and Overend, R. P., Eds., Canadian Society of Peat and Peatlands, 1991, pp. 30–37.

Haavisto, V. F., Jeglum, J. K., and Groot, A., Management practices — black spruce ecosystem, in *The Ecology and Management of Wetlands,* Vol. 2, Hook, D. D., et al., Eds., Croom Helm Ltd, Timber Press, Portland, OR, 1988, pp. 195–202.

Haavisto, V. F. and Päivänen, J., Planning operational drainage in boreal forest peatlands, American Society of Agricultural Engineering, Paper 87-2517, 1987, pp. 1–13.

Hånell, B., Present situation and future possibilities of peatland forestry in Sweden, in *Biomass Production and Element Fluxes in Forested Peatland Ecosystems,* Hånell, B., Ed., Swedish University of Agricultural Science, Department of Forestry Site Research, 1990, pp. 45–48.

Hånell, B., Peatland forestry in Sweden, in *Peat and Peatlands — Diversification and Innovation,* Vol. 1, Jeglum, J. K. and Overend, R. P., Eds., Canadian Society of Peat and Peatlands, 1991, pp. 19–25.

Heathwaite, A. L., Ed., *Mires: Process, Exploitation and Conservation,* John Wiley & Sons, Chichester, 1993, 506 pp.

Heikurainen, L., Input and output in Finnish forest drainage activity, Proc. 6th Int. Peat Congr., 1980a, pp. 398–402.

Heikurainen, L., Kuivatuksen tila ja puusto 20 vuotta vanhoilla ojitusalueilla (Summary: Drainage condition and tree stand on peatlands drained 20 years ago), *Acta For. Fenn.,* 167, 1, 1980b.

Hillman, G. R., Improving wetlands for forestry in Canada, Canadian Forestry Service, Northern Forestry Centre, Information Report NOR-X-228, 1, 1987.

Hillman, G. R., The Canada-Alberta wetlands drainage and improvement program for forestry: an update, in *Peat and Peatlands — Diversification and Innovation,* Vol. 1, Jeglum, J. K. and Overend, R. P., Eds., Canadian Society of Peat and Peatlands, 1991, pp. 54–61.

Jeglum, J. K., Peatland forestry in Canada: an overview, in *Biomass Production and Element Fluxes in Forested Peatland Ecosystems,* Hånell, B., Ed., Swedish University of Agricultural Science, Department of Forestry Site Research, 1990, pp. 19–28.

Jeglum, J. K., Definition of trophic classes in wooded peatlands by means of vegetation types and plant indicators, *Ann. Bot. Fenn.,* 28, 175, 1991a.

Jeglum, J. K., The Wally Creek area forest drainage project in Ontario's Clay Belt: progress report, in *Peat and Peatlands — Diversification and Innovation,* Vol. 1, Jeglum, J. K. and Overend, R. P., Eds., Canadian Society of Peat and Peatlands, 1991b, pp. 47–53.

Jeglum, J. K. and Kennington, D. J., Strip Clearcutting in Black Spruce: A Guide for the Practicing Forester, Forestry Canada, Ontario Region, Great Lakes Forestry Centre, 1993, 102 pp.

Joensuu, S., Peatland forest ditch maintenance and runoff water quality: the effect of sedimentation ponds, Proc 9th Int. Peat Congr., Uppsala, Sweden, 2, 423, 1992.

Joensuu, S., Factors affecting sediment accumulation in sedimentation ponds, in *Northern Forested Wetlands: Ecology and Management,* Trettin, C. C. et al., Eds., Lewis Publishers, Boca Raton, FL, 1997.

Kaunisto, S., Peatland forestry in Finland: problems and possibilities from the nutritional point of view, in *Northern Forested Wetlands: Ecology and Management,* Trettin, C. C. et al., Eds., Lewis Publishers, Boca Raton, FL, 1997.

Keltikangas, M., Laine, J., Puttonen, P., and Seppala, K., Vuosina 1930-1978 metsäojitetut suot: ojitusalueiden inventoinnin tuloksia (Abstract: Peatlands drained for forestry during 1930-1978: results from field surveys of drained areas), *Acta For. Fenn.,* 193, 1, 1986.

Ketcheson, D. E. and Jeglum, J. K., Estimates of black spruce and peatland areas in Ontario, Canadian Forestry Service, Information Report O-X-172, 1972.

Kivinen, E. and Pakarinen, P., Geographical distribution of peat resources and major peatland complex types in the world, *Ann. Acad. Sci. Fenn. Series A III Geologia — Geographica,* 132, 1, 1981.

Koivisto, I. J., Kokemuksia Wally Creek — metsäojitusprojektista Pohjois-Ontariossa, Kanadassa (Summary: The Wally Creek area forest drainage experiment in northeastern Ontario, Canada), *Suo,* 36, 52, 1985.

Lähde, E., Laiho, O., Norokorpi, Y., and Saksa, T., Alternative silvicultural treatments as applied to advanced stands — research plan, in *Silvicultural Alternatives,* Hagner, M., Ed., Swedish University of Agricultural Science, Department of Silviculture, Report 35, 1992, p. 66.

Laine, J., Minkkinen, K., Sinisalo, J. Savolainen, I., and Martikainen, P., Greenhouse impact of a mire after drainage for forestry, in *Northern Forested Wetlands: Ecology and Management,* Trettin, C. C. et al., Eds., Lewis Publishers, Boca Raton, FL, 1997.

Laine, J., Päivänen, J., Schneider, H., and Vasander, H., Site types at Lakkasuo mire complex. Field guide, *Publ. Dept. Peatland For., Univ. Helsinki,* 8, 1, 1986.

Laine, J., Vasander, H., and Laiho, R., Long-term effects of water level drawdown on the vegetation of drained pine mires in southern Finland, *J. Applied Ecol.,* 32, 785, 1995.

Lugo, A. E., Introduction, in *Ecosystems of the World,* Vol. 15, *Forested Wetlands,* Lugo, A. E., Brinson, M., and Brown, S., Eds., Elsevier, Amsterdam, 1990, chap. 1.

Malcolm, D. C., Afforestation in Britain — a commentary, *Scottish For.,* 45(4), 259, 1991.

Neily, P., Forest drainage in Nova Scotia, in *Peat and Peatlands — Diversification and Innovation,* Vol. 1, Jeglum, J. K. and Overend, R. P., Eds., Canadian Society of Peat and Peatlands, 1991, pp. 45–46.

Paavilainen, E. and Päivänen, J., *Peatland Forestry. Ecology and Principles,* Springer-Verlag, Berlin-Heidelberg-New York, 1995.

Päivänen, J., Hakkuun ja lannoituksen vaikutus vanhan metsäojitusalueen vesitalouteen (Summary: The effect of cutting and fertilization on the hydrology of an old forest drainage area), *Folia For.,* 516, 1, 1982.

Päivänen, J., Increasing the land base and yield through drainage, University of Alberta, Forest Industry Lecture Series, 13, 1, 1984.

Päivänen, J., Metsänparannuslaki ojituksen ja ojitusaluemetsien kunnostuksen suuntaajana (Summary: Forest improvement act in steering drainage and treatment of stands on drained sites), *Työtehoseuran metsatiedote,* 530, 1, 1994.

Päivänen, J. and Ahti, E., Ditch cleaning and additional ditching in peatland forestry — effect on ground water level, *Publ. Acad. Finl.,* 4, 184, 1988.

Päivänen, J. and Paavilainen, E., Managing and protecting forested wetlands, IUFRO World Congress, Montreal, Canada, Div. 1, Vol. 1, 1990, p. 432.

Päivänen, J. and Wells, E. D., Guidelines for the development of peatland drainage systems for forestry in Newfoundland, Newfoundland Forestry Research Centre, St. John's, Newfoundland, Information Report N-X-156, 1977, 1.

Payandeh, B., Analyses of a forest drainage experiment in northern Ontario. I. Growth analysis, *Can. J. For. Res.,* 3, 387, 1973.

Payandeh, B., Economic evaluation of forest drainage and fertilization in northern Ontario peatlands with an investment decision model, *New For.,* 2, 145, 1988.

Pfadenhauer, J., Schneekloth, H., Schneider, R., and Schneider, S., Mire distribution, in *Mires: Process, Exploitation and Conservation,* Heathwaite, A. L., Ed., John Wiley & Sons, Chichester, 1993, chap. 2.

Pyatt, D. G., Forest drainage, Forestry Committee Research Division, Research Information Note 196, 1, 1990a.

Pyatt, D. G., Long term prospects for forests on peatland, *Scott. For.,* 44(1), 19, 1990b.

Rosen, M., Forest Drainage Manual, Ontario Ministry of Natural Research, Science and Technocal Series 3, 1989, 1.

Ruseckas, J., Theoretical Background for Forest Hydromelioration, Leningrad Forestry Academy, Leningrad, 1991, 32. (In Russian)

Sabo, E. D., Hydrotechnical forest reclamation fund in the USSR, *Proc. 8th Peat Congr.,* Leningrad, USSR, 3, 58, 1988.

Seuna, P., Long-term influence of forestry drainage on the hydrology of an open bog in Finland, International Association of Hydrological Sciences, IAHS-AISH Publication 130, 141, 1980.

Stanek, W., A forest drainage experiment in northern Ontario, *Pulp Paper Mag. Can.,* 69, 58, 1968.

Starr, M. R. and Päivänen, J., The influence of peatland forest drainage on runoff peak flows, *Suo,* 32, 79, 1981.

Taylor, J. A., The peatlands of Great Britain and Ireland, in *Ecosystems of the World,* Vol. 4B, *Mires: Swamp, Bog, Fen and Moor,* Gore, A. J. P., Ed., Elsevier, Amsterdam, 1983, chap. 1.

Thompson, D. A., Forest drainage schemes, *For. Comm. Leaflet,* 72, 1, 1979.

Trettin, C. C., Application of water management techniques to hardwood silviculture, in *Proc. Conf. Northern Hardwood Resource: Management and Potential,* Mroz, G. D. and Reed, D. D., Eds., Michigan Technical University, School of Forestry and Wood Products, 1986, 183–192.

Trettin, C. C., Johnson, J. R., and Misiak, R.D., Hydrologic effects of a prescription drainage system on a forested wetland in northern Michigan, in *Peat and Peatlands — Diversification and Innovation,* Vol. 1, Jeglum, J. K. and Overend, R. P., Eds., Canadian Society of Peat and Peatlands, 1991, pp. 175–183.

Trottier, F., Prescriptions and programs for forest drainage in Quebec, in *Peat and Peatlands — Diversification and Innovation,* Vol. 1, Jeglum, J. K. and Overend, R. P., Eds., Canadian Society of Peat and Peatlands, 1991, pp. 207–212.

Valk, U., *Eesti Sood* (Estonian Peatlands), Tallinn, Valgus, 1988, 343.

Vasander, H., Plant biomass, its production and diversity on virgin and drained southern boreal mires, *Publ. Dept. Bot., Univ. Helsinki,* 18, 1, 1990.

Vasander, H., Laine, J., and Laiho, R., Effect of drainage on plant species diversity on forested mires, in *Northern Forested Wetlands: Ecology and Management,* Trettin, C. C. et al., Eds., Lewis Publishers, Boca Raton, FL, 1997.

Verry, E. S., The hydrology of wetlands and man's influence on it, *Publ. Acad. Finl.,* 5, 41, 1988.

Vompersky, S. E., Current status of forest drainage in the USSR and problems in research, in *Peat and Peatlands — Diversification and Innovation,* Vol. 1, Jeglum, J. K. and Overend, R. P., Eds., Canadian Society of Peat and Peatlands, 1991, pp. 13–18.

Wells, E. D., Effects of refertilization of an 18-year-old Japanese larch (*Larix leptolepis*) peatland plantation in western Newfoundland, Canada, in *Peat and Peatlands — Diversification and Innovation,* Vol. 1, Jeglum, J. K. and Overend, R. P., Eds., Canadian Society of Peat and Peatlands, 1991, pp. 129–138.

Zalitis, P., Yield capacity on drained forest sites in Latvia, in Annual reports summarized, Research and Production Association "Silava," Latvian Forestry Research Institute, Salaspils, 1990, pp. 4–5.

Zobel, M., Estonian mires and their utilization, *Suo,* 43(1), 25, 1992.

Zoltai, S. C. and Pollett, F. C., Wetlands in Canada: their classification, distribution, and use, in *Ecosystems of the World,* Vol. 4B, *Mires: Swamp, Bog, Fen and Moor. Regional Studies,* Gore, A. J. P., Ed., Elsevier, Amsterdam, 1983, chap. 8.

CHAPTER 4

Policy for Conservation of the Functions and Values of Forested Wetlands

Clayton D. A. Rubec

CONTENTS

1-56670-177-5/97/$0.00+$.50
© 1997 by CRC Press, Inc.

INTRODUCTION

In the last 200 years, about 14 percent (or 20 million ha) of the wetlands have disappeared from Canadian landscapes. Canada has a special position in the world in that about a quarter of the wetlands that now remain are in Canada. About 15 years ago Canadians saw a need for a wider variety of tools to focus on wetland conservation. Today there are many different approaches to that issue in Canada, one of which is policy.

The battle we face deals not only with wetlands as a threatened landscape in many areas of Canada but also with conserving overall biodiversity in these ecosystems. Wetland conservation is linked to many of the current environmental and socioeconomic issues in our nation. The range of factors that are involved and the agencies that must become involved in finding solutions to wetland decline are many, including fisheries, water, wildlife, parks, and forestry sectors, for example. In the last several decades we have seen major initiatives related to conservation of the functions of our wetlands. The protection of migratory birds and their habitats, for example, is being undertaken through initiatives such as the North American Waterfowl Management Plan and the Western Hemisphere Shorebird Reserves Network.

There has been increased recognition of the value of wetland ecosystems in recent years. Wetland loss due to agriculture, urbanization, industrial development, water management projects, and a variety of related activities has made wetland conservation an issue in many jurisdictions. Modification or loss of wetland ecosystems has been extensive in some regions. For example, over 70 percent of the wetlands in the southern portions of the central Prairies, southern Ontario, and the Fraser Lowland in British Columbia have been converted to other land uses. However, wetland disturbance has been minimal in lightly populated areas of Canada, such as the boreal zone, where most of our forested wetlands are found.

Figure 1 is a map focusing on where wetlands and their related biodiversity values are at greatest risk in Canada. This map was originally published by the North American Wetlands Conservation Council (Canada) in the *Wetland Evaluation Guide* developed by Bond et al. (1992). The dark (black) areas on the map depict the areas where land use pressures from agriculture and urban development on wetlands are at their greatest. The stippled (grey) areas are where agriculture and forestry as well as emerging issues such as hydroelectric development pose threats to wetland integrity in Canada. These areas also include coastal zones where there has been extensive development for ports of harbors. The lightest (white) areas have minor or no threats to wetland resources.

FORESTED WETLAND FUNCTIONS AND VALUES

Forested wetlands comprise up to 70 percent of the 127 million ha of wetlands in Canada. Much of this is dominated by peatland soils at least 40 cm deep. These forested peatlands provide many critical ecological functions, including:

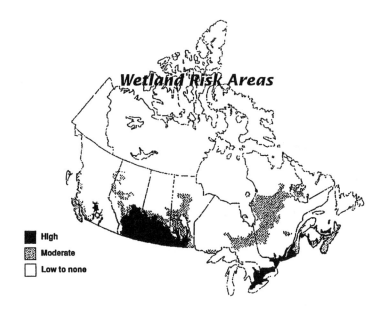

Figure 1   Wetland risk areas.

- biomass production and export,
- habitat for fish and wildlife,
- nutrient retention and removal,
- groundwater recharge and discharge,
- water quantity and quality stabilization,
- hunting and trapping resource areas,
- a source for forest products.

The economic values of forest products harvested from forested wetlands in Canada is estimated to exceed $500 million (Canadian) annually (Rubec et al., 1988). Extensive discussions of wetland and peatland functions are given in many other sources, such as Mitsch and Gosselink (1986) and Hook et al. (1988).

FORESTED WETLAND ISSUES IN CANADA

One of the economic sectors in Canada that has taken a proactive position on wetland conservation is our forest industry. Forestry in wetlands and peatlands is active across all areas of the nation. However, only about 25,000 ha of Canadian peatlands are now actually drained to facilitate forestry, and most of this is for experimental purposes only. Considering that over 450 million ha are forested in Canada (Government of Canada, 1991a), wetland drainage to support forestry operations is not at all well established in Canada. Most of the harvesting of timber from Canada's large expanse of forested peatlands is carried out in winter during frozen ground conditions, minimizing site impacts and facilitating use of appropriate machinery. While peatland drainage is used to enhance tree growth in several European countries, this is not the case in Canada at present. Pilot peatland forestry developments in Alberta, northern Ontario, eastern Quebec, and Newfoundland are being used to evaluate peatland drainage as a forest management tool in Canadian conditions.

Sheehy (1993) has reviewed the major impacts and issues facing wetlands and peatlands in the managed forests of Canada. He notes that summer harvesting can be very damaging through the direct impacts of machinery and inhibition of regeneration after harvesting. Secondary effects include rutting, ponding, and impediment of water flow, and erosion of soil and peat materials on wetlands. Both on- and off-site, additional impacts include water quality effects, degraded water quality of waters flowing into wetlands, and effects due to poor practices related to equipment storage and landing and hauling of timber. The forest industry and government have funded extensive research allowing implementation of many solutions to these problems. Reducing and avoiding impacts on wetlands are now commonplace practices throughout the forest industry in Canada.

However, the overall impacts of peatland forestry are not well documented. Osborne (1982) discusses potential impacts and presents a set of assessment guidelines developed by Environment Canada for peatland developments. A study by Clarke-Whistler et al. (1984) provides a summary of the technical literature on the topic. This study was based on the premise that an understanding of the structure and functions of the peatland ecosystem was needed to evaluate the significance of potential environmental impacts. The study concluded that potential aquatic impacts arising from peatland development relate to alteration of the hydrological regime and impairment of water quality, with subsequent direct or indirect effects on aquatic biota.

Wildlife Habitat Impacts

Conservation of forested wetlands for their wildlife habitat and other ecological values is an important issue (Rubec, 1991; Keys, 1992). Wetland loss has become acute in some regions of Canada, a matter now being addressed through extensive policy initiatives nationwide. A diverse range of development factors has resulted in this loss of wetlands. While agricultural development, particularly in the prairie regions of Canada, is the single greatest cause of wetland loss in Canada, such loss in the forested zones of the nation is still perceived as a minor issue, despite the fact that little wetland monitoring data yet exists for these areas of the nation. The relative impact of large-scale hydroelectric development is significant but not documented in Canadian literature, for example.

Forested wetland development and resource utilization have mainly affected the bog wetland class and have had less effect on other wetland classes, such as marshes and swamps. This reflects the distribution of wetlands in our nation wherein peatlands such as bogs and fens are mainly associated with the boreal wetland region (National Wetlands Working Group, 1988). This area is where forestry operations are focused in Canada. Canada's prairie and temperate wetland regions are dominated by swamps and marshes and are largely agricultural or urban landscapes. Many of the conservation issues of greatest concern that affect wildlife in Canadian wetlands, such as agricultural drainage or infilling, urbanization, and industrial development, have relatively little relationship to the forest industry. The swamp, marsh, and open water wetland classes are favored habitat for most waterfowl and a wide range of other wildlife species, due to the diverse range of vegetation and the common occurrence of open water. In contrast, peatlands, whether they are forested or open, tend to have a minimum of open water, a low diversity of vegetation, and limited cover for waterfowl or other bird nesting purposes. The number of waterfowl and wildlife species and the total wildlife populations in bogs, for example, are generally lower in comparison to other wetland classes or to mineral soil ecosystems. However, a few species of small mammals, such as muskrat and beaver, and game species such as caribou, moose, and deer, utilize peatland habitat. Other species use peatlands on a seasonal basis. Rare or endangered bird and mammal species that are known to utilize peatlands include the whooping crane, trumpeter swan, piping plover, and wood bison.

Impacts on Vegetation and Rare Plant Species

Peatlands include many species that are not common in mineral soil ecosystems; some are considered unusual and unique in parts of Canada. Many of these species, however, are widely distributed throughout Canada's boreal wetland regions. The pitcher plant, for example, while the provincial flower for the Province of Newfoundland and Labrador, is found across the entire boreal zone of Canada. Outside Edmonton, Alberta, in the Wagner Bog Natural Area, 16 species of provincially or nationally rare orchids have been identified. Much work needs to be done to document the richness and variety of rare and endangered flora of Canada's forested peatlands. Hence, vegetation conservation, especially protection of rare or endangered species, is an issue relating to forested wetland and peatland utilization.

Many forested peatlands may also be refuges for a wide range of other biological resources, including invertebrate species. The Biological Survey of Canada's National Peatland Entomology Project is leading to a better understanding of the distribution and composition of the biodiversity of peatlands beyond the more obvious plants and animals or birds.

Release of Carbon

Release of carbon gases due to forested peatland developments, such as drainage, is another environmental issue of concern. Draining of a peatland lowers the water table and accelerates the decomposition process. As a result, organic carbon stored in the peat is released to the atmosphere as carbon dioxide. Release of carbon gases to the atmosphere, which is primarily due to the combustion of fossil fuels (including coal, peat, and petroleum products), has been suggested as one cause of global warming (the "greenhouse" effect). The loss of peatland vegetation as a net carbon accumulator through photosynthesis as well as the role of peatland waters in the carbon cycle have also been linked to greenhouse gases emissions.

A study by Gorham (1991) assessed the impact of peatland development on the carbon cycle. The study indicates that on a global basis, combustion of peat as a fuel releases about three times the quantity of carbon as is released from drained peatlands. Drained peatlands for forestry operations represent a significant percentage of the total area of developed peatlands on a global basis, particularly in nations such as Finland. The study notes that the release of methane from undrained peatlands has a greater impact than the combined total impact from all peatlands areas drained for all applications. The current very low level of development of peatlands in Canada for forestry applications does not appear to have a significant impact on the global carbon cycle.

Water Quality and Quantity Impacts

The effect on downstream aquatic ecosystems of drainage waters from developed peatlands for forestry or peat harvesting has been viewed as a significant environmental issue (Keys, 1992). Physical parameters such as suspended solids are a concern, for example. Drainage of a bog can result in the exposure of peat particles that can be transported into the drainage system and leave the peatland. Sedimentation ponds are a standard inclusion in the design of peatland development sites in most nations. The design of the ponds must incorporate sufficient residence time to permit settling of solids during periods of peak rainfall.

Chemical parameters, such as pH and a range of elements, are also a consideration in the development of a peatland for forestry or peat harvesting operations. These factors tend to receive less emphasis because natural drainage waters from peatlands tend to already have a low pH. The establishment of an extensive network of drainage ditches enhances the opportunity for precipitation to be transported off a peatland.

Reforestation of Degraded Peatlands

Reforestation of peatlands used for peat harvesting is practiced in several European countries. This involves the use of the drainage systems left at the conclusion of peat harvesting. In some cases, a 30 to 50 cm (or more) layer of peat is left on the peatland for afforestation purposes. In other cases, deep ploughing is carried out to blend the basal peat with the underlying mineral soil. Fertilizer and lime may also be applied to enhance forest species growth or increase the pH. Similar techniques can be used to develop biomass production sites, where rapidly growing species such as willow (*Salix* spp.), alder (*Alnus* spp.), or cattail (*Typha* spp.) can be harvested as a fuel source. Very few sites in Canada have been reforested after peat operations so far.

THE CANADIAN APPROACH TO WETLAND CONSERVATION

There are five major aspects to the Canadian approach to wetland conservation, each of which is briefly discussed below. These aspects are:

- wetland securement and protection,
- wetland restoration and enhancement,
- designation of important wetland sites,
- modification of sectoral policies,
- development of wetland policies and/or legislation.

Wetland Securement and Protection

One of the national political objectives in Canada is protection of at least 12 percent of our landscapes. This objective reflects governmental and nongovernmental endorsement of the Canadian Wilderness Charter (Government of Canada, 1990). In Canada over half the efforts to secure land area for conservation objectives has been driven by wildlife habitat objectives, not for parks or public recreation objectives. "Secured area" in this context refers to lands and waters set aside by legal agreement for conservation purposes either by government or the private sector. It is inclusive of the spectrum from highly protected ecological reserves and parks to the land stewardship initiatives of many partner groups in Canada. A study was recently undertaken by Rick Bryson and Associates (1993) funded by Environment Canada and the Canadian Forest Service. This revealed that the Government of Canada to date has protected over 7.1 million ha of wetlands in its network of federal protected areas. The study included 180 federal sites: national wildlife areas, national parks, and migratory bird sanctuaries. Provincial and territorial government initiatives likely would match the federal leadership in this area, but hard statistics are difficult to compile. This would suggest that about 15 million ha of wetlands, or 12.5 percent of all our nation's wetlands, are under a protected or conservation management status in Canada today. This same study provided detailed information on the status of protected forest land in Canada's federal protected areas.

Wetland Restoration and Enhancement

Canadians are involved in many habitat restoration or enhancement programs, such as the North American Waterfowl Management Plan (NAWMP), in cooperation with the United States and Mexico, the Great Lakes Wetland Action Plan, Plan 2000 in the St. Lawrence River system in Quebec, and the Fraser River Action Plan in British Columbia. These are examples of large enhancement and restoration programs. Environment Canada has provided significant funding and support for these federal–provincial cooperative "flagship" programs involving wetland conservation. NAWMP is one of the largest land use programs in North America's history, having already expended in the area of $187 million (Canadian) in Canada over the 1987 to 1994 period. It has resulted in securement of over 830,000 ha of wetland and upland habitat across Canada so far.

The North American Waterfowl Management Plan, initiated in 1986, was developed jointly between Canada, the United States, and Mexico to reverse or modify activities that destroy or degrade waterfowl habitat, primarily wetlands. This multilateral initiative has proposed to invest over $1.5 billion in wetland conservation and management in Canada. The plan recognizes that, in the face of major alterations to the landscape by people, the continued maintenance and restoration of wetlands will be necessary to provide suitable habitat for waterfowl and many other wildlife species.

Designation of Important Wetland Sites

Canadians are also using site designation initiatives, such as the Ramsar Convention, to assist in securing wetlands. There are many additional mechanisms used to identify significant or sensitive wetland areas. One of the responsibilities of the Canadian Wildlife Service is tracking activities in Canada under the Convention on Wetlands of International Importance (the Ramsar Convention). Canada is one of the most active countries in the world in the use of Ramsar site designation to identify wetlands of importance. These sites are protected through the use of federal, provincial, territorial, and nongovernment legal mechanisms, as Ramsar designation per se provides no formal protection in Canada. There are 33 Ramsar sites in Canada covering over 13 million ha in total, individually ranging in size from 244 to over 6.2 million ha. Of the 750 Ramsar sites in the world, Canadian sites represent about 30 percent of the global total area designated under the Ramsar Convention. Hence, Ramsar is one mechanism that all the governments in Canada have used to identify and highlight sites that are of particular interest not only to Canadians but to the world.

Modification of Sectoral Policies

In the last several years we have seen major undertakings to modify existing economic and resource management policies and programs in sectors such as forestry and agriculture, to globally benefit wetland biodiversity. The implementation of GATT and NAFTA and international trade negotiations are seen as opportunities to substantially influence wetland conservation in Canada (Patterson and Rubec, 1993; Patterson, 1994).

Development of Wetland Policies and/or Legislation

The functions that wetlands provide are important to Canadians. Virtually all wetland conservation and management policies to date in Canada have focused to some degree or entirely on the functions that our wetland systems provide. The main areas have been water quality and flood damage reduction, resource harvest from wetland sites, and the recreational values of our wetlands. These are the functions that Canadians value the most and from which they derive the greatest amount of direct financial benefit, estimated to exceed $10 billion (Canadian) annually (Rubec et al., 1988).

In the mid-1980s, our federal and provincial governments, as well as the industry and private sectors, recognized that in addition to our initiatives for wetland protection, restoration, enhancement, and designation, they needed to do more. They felt there was a need to put wetland conservation at the forefront as a matter of public policy. In the 1986 to 1987 period, government and other sectors in Canada started to create wetland policy, relative to their own jurisdictions, industry groups, or private sector groups. However, they recognized that Canada could not do this as one single national initiative. It would have to be done sector by sector and jurisdiction by jurisdiction, within some kind of framework that all could agree upon. That led to a series of workshops from 1987 through 1990.

In early 1987, Environment Canada sponsored a Non-Government Organizations Workshop on Wetland Conservation Policy. This workshop developed a series of recommendations directed to all governments in Canada concerning the need for wetland policy. These recommendations were sent to all Environment and Natural Resource Ministers across the country. The Federal–Provincial Committee on Land Use in June 1987 produced a report entitled, *A Framework for Wetland Policy in Canada.* This report was endorsed by the full committee, and members agreed to encourage use of this framework, as appropriate, in their own jurisdictions. Also in early 1987, the Federal Interdepartmental Committee on Land identified the need to develop a wetlands policy statement to supplement the "wise land use" provisions of the Federal Policy on Land Use. The Federal Water Policy adopted in 1987 also identified wetlands conservation as a significant water resource issue.

Another major event in 1987 was the Third Meeting of the Contracting Parties to the Ramsar Convention, held in Regina, Saskatchewan. One result was the drafting of a document called *Wise Use Principles* for the wetland resources of the world. A commitment that Canada adopted when endorsing those principles was the development of wetland policies.

Subsequently, a major national policy conference, the Sustaining Wetlands Forum, was held in April 1990. This conference produced recommendations for action including a call for all jurisdictions in Canada to adopt mutually supporting wetland conservation and management policies. This conference included a keynote address by the Prime Minister of Canada who called for urgent action to conserve the nation's wetland resources. Subsequently, Canada's *Green Plan* announced the federal government's commitment to adopting a federal wetland policy.

THE FEDERAL POLICY ON WETLAND CONSERVATION

One of the original considerations in the development of the federal wetland policy was that it should be crafted to deliver Canadian commitments to the Ramsar Convention on Wetlands of International Importance wherever possible.

It was also apparent that greater influence on land use decisions by federal departments and agencies would assist in meeting Canada's commitments under the North American Waterfowl Management Plan. More recently, following Canada's endorsement of the International Convention on Biodiversity in 1992, it is expected that Canada's federal wetland policy will form a portion of implementation initiatives for this important and far-reaching international agreement.

The federal government initiated a consultation process toward implementing this federal wetland policy in late 1987. The development of the North American Waterfowl Management Plan and the Federal Water Policy were ongoing in that period. The Government of Canada consulted with its partners across Canada in developing the policy, and it was publicly announced in March of 1992. The Sustaining Wetlands Forum and the federal policy recognized that wetlands conservation is an economic opportunity for Canada that must involve the private sector. The federal initiative was designed to complement what other governments would do and the wide range of federal government activities regarding wetlands management and conservation. It also demonstrates leadership; Canada was the first government in the world to announce a wetland policy. Today, at least 30 other countries are developing national or federal level policies.

The Federal Policy on Wetland Conservation (FPWC) (Government of Canada 1991b) focuses on the sustainable, wise use of wetlands in Canada, consistent with the "Wise Use Principles" developed by the Ramsar Convention. The federal policy applies to all of its programs, projects, agencies, and policies. It will be implemented through existing programs and budgets. The Canadian Wildlife Service of Environment Canada acts as an interagency advisor, assisting departments in the design of mechanisms to implement the policy in their programs. Environment Canada is publishing a report, entitled *The Federal Policy on Wetland Conservation: Implementation Guide for Federal Managers,* to accompany this policy (Lynch-Stewart, 1994).

Goals of the Federal Wetland Policy

The stated objective of the Government on Canada as articulated in this new policy with respect to wetland conservation is:

to promote the conservation of Canada's wetlands to sustain their ecological and socio-economic functions, now and in the future.

In support of the above objective, the government of Canada, in cooperation with the governments of its ten provinces and two territories as well as the Canadian public, will strive to achieve the following goals:

- *maintenance* of the functions and values derived from wetlands throughout Canada;
- *no net loss* of wetland functions on federal lands and waters;
- *enhancement and rehabilitation* of wetlands in areas where the continuing loss or degradation of wetlands or their functions have reached critical levels;

- *recognition* of wetland functions in resource planning, management and economic decision-making with regard to all federal programs, policies and activities;
- *securement* of wetlands of significance to Canadians;
- *recognition of sound, sustainable management practices* in sectors such as forestry and agriculture that make a positive contribution to wetlands conservation while also achieving wise use of wetland resources;
- *utilization* of wetlands in a manner that enhances prospects for their sustained and productive use by future generations.

The FPWC is focused on areas of federal jurisdiction and management of wetlands under direct federal authority. As Canada is a federal state, wetlands in its ten provinces are generally under provincial regulation, except on federal lands such as national parks. However, in its two northern territories, most wetlands are under federal management. Hence, while the policy will apply directly to an estimated 29 percent of Canada's wetland base (its federally managed wetlands), it also will touch on how the federal government affects other wetlands through its federal programs, policies, and shared fiscal programs with the provinces and territories.

One of the goals of the federal policy is to ensure that its programs, policies, and expenditures do not result in a net loss of wetland functions. "No net loss of functions" includes balancing unavoidable losses of wetlands with mitigative action such as replacement so that further reductions to wetland functions may be prevented. In general, this means that, where development of particular wetlands in critical areas must proceed, wetland loss must be mitigated by replacement of wetland functions in close proximity. This may have a significant effect on federal expenditures related to federal–provincial forest development agreements. In implementing this concept in Canada, guidelines will call on the expertise and experience developed in "no net loss of habitat" applications in other jurisdictions. To this end the federal government in cooperation with the North American Wetlands Conservation Council (Canada) has published a report synthesizing North American experience and recommendations for implementing "no net loss" in Canada (Lynch-Stewart, 1992).

Policy Strategies for Wetland Conservation

The Federal Policy on Wetland Conservation outlines seven strategies to provide for the wise use and management of wetlands so that they can continue to provide a broad range of functions on a sustainable basis. These strategies are aimed at building on past achievements and working in concert with ongoing initiatives for wetland conservation, in particular the North American Waterfowl Management Plan. The policy promotes a nonregulatory, cooperative approach. The strategies set out a direction for putting the federal house in order, to manage federal wetlands, and to ensure delivery of effective wetlands science and public awareness actions both nationally and internationally. The strategies focus on:

1. Developing public awareness
2. Managing wetlands on federal lands and waters and in other federal programs
3. Promoting wetland conservation in federal protected areas
4. Enhancing cooperation with federal, provincial, territorial, and nongovernment partners
5. Conserving wetlands of significance to Canadians
6. Ensuring a sound scientific basis for policy
7. Promoting international actions

A NATIONAL WETLAND COUNCIL

To promote wetland program coordination in Canada, the Federal Minister of the Environment created the North American Wetlands Conservation Council (NAWCC) (Canada) in April 1990. This council, working closely with a parallel council in the United States and provincial and nongovernment partners, acts as the senior Canadian body for coordinating implementation of the North American Waterfowl Management Plan. The NAWCC (Canada) mandate also includes promotion of awareness of wetland science, management, and policy issues within Canada and coordination of Canadian involvement in international wetland conservation initiatives. It has also established a national Secretariat and is publishing a report series, entitled the "Sustaining Wetlands Issues Papers," on wetland science, management, and policy topics of national interest.

THE NATIONAL SCENE

Governments in most Canadian provinces have promoted a nonregulatory approach to wetland conservation. The introduction of nonregulatory wetland management and conservation policies is proceeding. Each of the provincial governments has developed a public review or consultation process for their wetland conservation strategies or policies. In two provinces, Prince Edward Island and New Brunswick, complementary regulatory procedures that result in consideration of wetlands in the environmental assessment process are also used.

Ontario has announced both policy and legislation through the provincial *Planning Act* in regard to wetlands and conservation management. Alberta has a policy in regard to the southern parts of the province, the southern "white zone," and is actively pursuing a policy that will affect what is called the forested "green zone" for the northern part of the province. That has been emulated in Saskatchewan, where wetland policy underwent public consultation for the southern parts of the province. Manitoba has established wetland policy with regard to wetlands, tied to water and sustainable development programs. In Atlantic Canada, wetland policy development is occurring in Nova Scotia and New Brunswick. Prince Edward Island has specific legislation with regard to wetlands protection through the 1989 *Environmental Protection Act*.

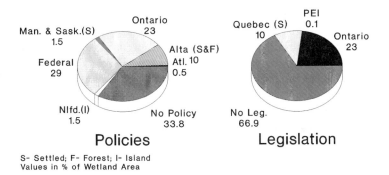

S- Settled; F- Forest; I- Island
Values in % of Wetland Area

Figure 2 Policy and legislation for wetlands in Canada.

In the private sector, several major industry groups that utilize peatland resources have brought forward industry-specific wetland management policies. The Canadian Sphagnum Peat Moss Association, whose member companies produce over $180 million (Canadian) in peat products each year, established a Peatland Restoration Policy focused on peatland reclamation and restoration after resource harvesting and cooperation with environmental interests and regulators. In 1992, the Canadian Pulp and Paper Association (CPPA) released a *Wetlands Policy Statement.* This statement lays out a series of sustainable development commitments with regard to use of wetland ecosystems by one of the nation's largest employers. The CPPA noted that it is committed to sustaining wetlands through integrated resource management and to mainte-nance of the ecological and socioeconomic functions of wetlands over the long term, and that it supports a coordinated, cooperative approach involving all stakeholders.

Looking across the nation on an area basis, Figure 2 summarizes the status of policy and legislative mechanisms for wetland conservation. Wetland manage-ment or conservation policy now applies to about two-thirds of the area of all the wetlands in Canada. On the legislative side, it reveals that less than one-third of this wetland area is covered by a legislative approach. Hence, the dominant approach so far in Canada has been the use of policy to achieve wetland conser-vation. In some provinces such as Ontario, a combination of policy within a legislative approach is underway.

CONCLUSIONS: POLICY IMPLEMENTATION THROUGH A COOPERATIVE APPROACH

Wetlands have become an important component in the development of con-servation strategies across Canada. Through the preparation of provincial, terri-torial, and federal wetland policies and programs, a common focus on the global themes of sustainable development and partnerships has emerged. Wetland and waterbird conservation are inextricably linked to these themes at the international,

national, regional, and local levels. In Canada, these linkages are achieved through the multitude of agreements and partnerships that complement jurisdictional arrangements and achieve cooperation in conservation.

Canada has established many commitments to wetland conservation that are far reaching nationally and internationally. The Federal Policy on Wetland Conservation is but one example of the federal government's interests in wetlands. The Government of Canada is also committed to furthering the strong cooperation inherent in the North American Waterfowl Management Plan with the United States and Mexico. On the global scene, Canada's long-standing support and interest in the Ramsar Convention and the principles of "wise use" have been well recognized.

REFERENCES

Bond, W. K., Cox, K. W., Heberlein, T., Manning, E. W., Witty, D. R., and Young, D. A., Wetland Evaluation Guide. Final Report of the Wetlands Are Not Wastelands Project, Sustaining Wetlands Issues Paper, No. 1992-1. North American Wetlands Conservation Council (Canada), Ottawa, Ontario, 1992, 121 pp.

Rick Bryson and Associates, Forest and Wetland Attributes for Highly Protected Conservation Areas in Canada, Contract Report, North American Wetlands Conservation Council (Canada), Ottawa, Ontario, 1993.

Clarke-Whistler, K., Snodgrass, W. J., McKee, P., and Rowsell, J. A., Development of an Innovative Approach to Assess the Ecological Impact of Peatland Development, Report No. NRCC 24129, National Research Council of Canada, Ottawa, Ontario, 1984, 204 pp.

Gorham, E., Northern peatlands: role in the carbon cycle and probable responses to climatic warming, Ecol. Appl., 1, 182–195, 1991.

Government of Canada, Canada's Green Plan, Ottawa, Ontario, 1990, 174 pp.

Government of Canada, Chapter 10 — Forestry, in The State of Canada's Environment, Environment Canada, Ottawa, Ontario, 1991a.

Government of Canada, The Federal Policy on Wetland Conservation, Environment Canada, Ottawa, Ontario, 1991b, 14 pp.

Hook, D. D., et al., Eds., The Ecology and Management of Wetlands, 2 Vols., Timber Press, Portland, Oregon, 1988.

Keys, D., Canadian Peat Harvesting and the Environment, Sustaining Wetlands Issues Paper, No. 1992-3, North American Wetlands Conservation Council (Canada), Ottawa, Ontario, 1992, 29 pp.

Lynch-Stewart, P., No Net Loss. Implementing "No Net Loss" Goals to Conserve Wetlands in Canada, Sustaining Wetlands Issues Paper, No. 1992-2, North American Wetlands Conservation Council (Canada), Ottawa, Ontario, 1992, 35 pp.

Lynch-Stewart, P. and Lee, G., Experience with implementation of the Federal Policy on Wetland Conservation, in Wetland Policy Implementation in Canada: Proceedings of a National Workshop, Rubec, C. D. A., compiler, Report No. 94-1, North American Wetlands Conservation Council (Canada), Ottawa, Ontario, 1994.

Mitsch, W. J. and Gosselink, J. G., Wetlands, Van Nostrand Reinhold, New York, 1986.

National Wetlands Working Group, Wetlands of Canada, Ecological Land Classification Series, No. 24, Environment Canada and Polyscience Publications Inc., Montreal, Quebec, 1988, 452 pp.

Osborne, J. M., Potential environmental impacts of peatland development, in Proceedings, Symposium '82 on Peat and Peatlands, Sheppard, J. D., Musial, J., and Tibbetts, T. E., Eds., Canadian Society for Peat and Peatlands, Dartmouth, Nova Scotia, 1982, pp. 198–219.

Patterson, J. H., Wetlands, agriculture and international trade interactions, in Wetland Policy Implementation in Canada: Proceedings of a National Workshop, Rubec, C. D. A., compiler, Report No. 94-1, North American Wetlands Conservation Council (Canada), Ottawa, Ontario, 1994.

Patterson, J. H. and Rubec, C. D. A., Agricultural trade and biodiversity: a win-win opportunity for environment and economy, *Waterfowl 2000*, 5(1), 1, 1993.

Rubec, C. D. A., Peat resources use in Canada: a national conservation issue, in Proceedings, International Symposium on Peat and Peatlands: The Resource and Its Utilization, Grubich, D. N. and Malterer, T. J., Eds., Duluth, MN, August 19–22, 1991, International Peat Society, 1991, pp. 390–408.

Rubec, C. D. A., Lynch-Stewart, P., Kessel-Taylor, I., and Wickware, G. M., Wetland utilization in Canada, Chapter 10, in *Wetlands of Canada*, Ecological Land Classification Series, No. 24, Environment Canada and Polyscience Publications Inc., Montreal, Quebec, 1988, 452 pp.

Sheehy, G., Conserving Wetlands in the Managed Forests, Sustaining Wetlands Issues Paper Series, No. 1993-2, North American Wetlands Conservation Council (Canada), Ottawa, Ontario, 1993, 37 pp.

CHAPTER **5**

A Comparison of Wetland Mapping Using SPOT Satellite Imagery and National Wetland Inventory Data for a Watershed in Northern Michigan

Trae A. Forgette and John A. Shuey

CONTENTS

INTRODUCTION

With the advent of powerful, yet relatively inexpensive personal computers, the use of nonpoint source pollution models to assess the effects of proposed land-use changes on local watersheds in rapidly developing areas has become feasible. For these models, numerous input layers are required over large areas, including land cover, soil hydrology, and slope. Recent studies have shown that satellite imagery can be used to facilitate the gathering of this spatial data for input into nonpoint source pollution models (Koches, 1992; Welch, 1987). Because wetlands are an important component in controlling water movement and nonpoint source pollution, an accurate, up-to-date inventory of wetlands within a watershed is also needed for the models. Field surveys, though highly accurate, require extensive resources and may be impractical for large-scale inventories. U.S. Fish and Wildlife Service National Wetlands Inventory (NWI) mapping, based on conventional aerial photography and manual photointerpretation techniques, has proven to be relatively reliable, and the information is available in digital format for many areas of the country. Similarly, the Michigan Resource Inventory System (MIRIS) land cover classifications are mapped using monochrome and infrared aerial photography. However, photointerpretation is time consuming and requires special skills and experience. Therefore, these maps are not updated on a frequent basis, and, in areas of rapid development, may not accurately reflect current conditions that are necessary for nonpoint source pollution model implementation by watershed managers.

Satellite imagery has emerged as a potential tool for large-scale wetland surveys for the following reasons: (1) data is collected in digital format, which facilitates its analysis by Geographic Information Systems (GIS), and avoids the need for extensive digitizing; (2) sensors include the near infrared wavelength band for mapping vegetation, wetlands, and land use/land cover; (3) satellites provide a synoptic overview of large regions; (4) data is orthorectifed and geocoded; (5) images are acquired on a continuous basis; and (6) costs are relatively low. However, the success of satellite imagery to map wetland types accurately has been mixed and appears to be site-specific (Koches, 1992; Podolsky and Conkling, 1991; The Federal Geographic Data Committee, 1992; Welch, 1987).

SPOT (Systeme Probatoire d'Observaton de la Terre) satellite imagery was used to map wetland areas in a small watershed in northern Michigan. Results were compared to NWI and MIRIS wetland maps to determine if the wetland areas within the watershed could be more refined through remote sensing.

METHODS

Project Area

The Mitchell Creek Watershed is located in Grand Traverse County near Traverse City, Michigan (Figure 1). The watershed encompasses approximately 3812 ha (38.1 km^2), and drains into the southern tip of East Grand Traverse Bay in northern Lake Michigan (Niehaus et al., 1991). The watershed is fairly steep and drains south to north by several intermittent streams (Fulcher, 1991). The soils are primarily sand, loamy sand, gravelly sandy loam, and sandy loam with some muck soils in low-lying areas (Soil Conservation Service, 1991). Because these soils have high infiltration rates, runoff potential is relatively low when the land is left undeveloped. The greatest threat to water quality in this watershed is nonpoint source pollution. Wetlands within this watershed have been shown to

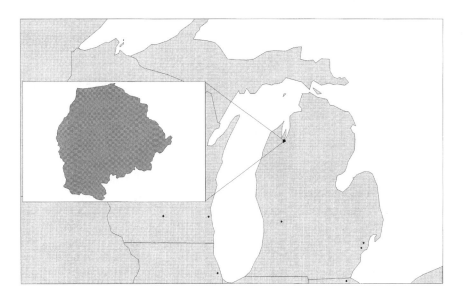

Figure 1 Study area located in Grand Traverse County, MI.

play an important role in reducing nonpoint source pollution by acting as nutrient sinks during runoff events (Niehaus et al., 1991).

Satellite Imagery and Analysis Software

SPOT is a series of French satellites (SPOT 1 and SPOT 2) that use three electro-optical sensors covering the green (0.50 to 0.59 µm), red (0.1 to 0.68 µm), and near infrared (0.79 to 0.89 µm) bands. SPOT 2 imagery is processed daily and is constantly being acquired over the United States. For this project, two multispectral SPOT images encompassing the Mayfield 7.5-min quadrangle were selected from a series of SPOT satellite images taken from 1987 through 1992. The selected images contained little atmospheric and ground distortion (i.e., cloud cover and snow, respectively) and were taken in early spring (May 1988) and summer (June 1992). These two images allowed the comparison of leaf-off and leaf-on conditions.

The computer platform used for classifying the images was a desktop PC, with a 486 DX processor, running at a clock speed of 33 MHz, and operating in the DOS environment. Software used for image processing, classification, and statistical comparison was IDRISI (The IDRISI Project, Clark University). This software was selected because it is a powerful, yet inexpensive raster (grid-based) analysis package.

Wetland Map Comparison

General land-use areas were identified from a 7.5-min (1:24,000 scale) topographic map (Mayfield quadrangle), Michigan Department of Natural Resources aerial photography, and field observation. Using the IDRISI software, signature files (i.e., training sites) were derived for each of several land-use and land-cover types and assigned to each of the three bands from both the spring and summer images.

Maximum likelihood and minimum distance to means classifications were conducted based on 14 signature files and six spectral bands, three from the spring image combined with three from the summer image. The maximum likelihood classification technique is based on the probability density function associated with a particular training site's signature. Pixels are then assigned to the most likely class based on a comparison of the probability that the class belonged to each of the signatures being considered. The minimum distance to means classification technique is based on the mean reflectance on each band for a signature. Pixels are then assigned to the class with the mean closest to the value for that pixel.

The images were reclassified several times to combine similar classifications (e.g., water and shallow water). A series of filtering techniques (mean, median, and mode) were used to "smooth" the data through image generalization, random pixel noise removal, and data gap-filling. Similar cover types were grouped for each image to show open water, wetland, upland, urban, and agriculture cover types (Figures 2 and 3). The two images were then combined using a maximum

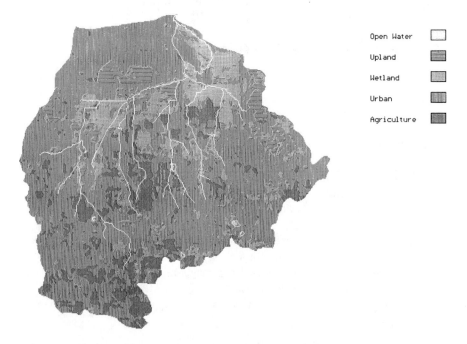

Open Water

Upland

Wetland

Urban

Agriculture

Figure 2   Maximum likelihood classification of the Mitchell Creek watershed in Grand Traverse County, MI, using multispectral SPOT satellite data.

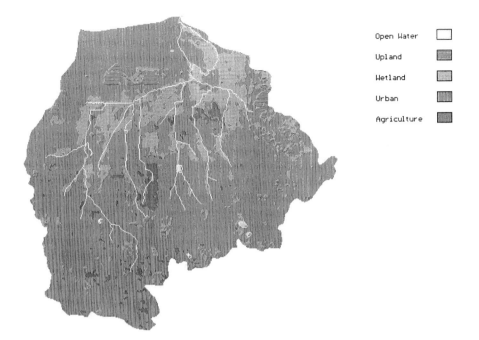

Open Water

Upland

Wetland

Urban

Agriculture

Figure 3   Minimum distance classification of the Mitchell Creek watershed in Grand
Traverse County, MI, using multispectral SPOT satellite data.

overlay technique to create a composite land use map of the watershed (Figure
4). This map was used for the remainder of the comparisons.

Wetland areas determined using the SPOT classifications were compared with
two existing wetland maps of the same quadrangle: National Wetlands Inventory
(NWI) and Michigan Resource Information System (MIRIS) data. A digitized soil
map of the watershed was obtained from the Soil Conservation Service to compare
areas with hydric soils in the watershed to the wetland areas of the SPOT, NWI,
and MIRIS wetland maps. A test of the significance of spatial relationships
between the SPOT, NWI, and MIRIS wetland distribution maps was performed
using the Chi Square test of independence (Barber, 1988; Remillard and Welch,
1993). This nonparametric test was used to gain an understanding of relationships
that may exist between two classifications. IDRISI software uses pixel cell frequency data arranged in contingency tables to perform a Chi Square test of
independence between images. IDRISI was also used to test the strength of
association between maps by measuring Cramer's V and the Kappa Index of
Agreement (KIA). These measures attempt to standardize Chi Square, with the
resulting coefficients ranging from 0, indicating no correlation between the
classes, to 1, indicating a perfect correlation (Carstensen, 1987; The IDRISI
Project, 1992; Remillard and Welch, 1993; Rosenfield and Fitzpatrick-Lins, 1986).

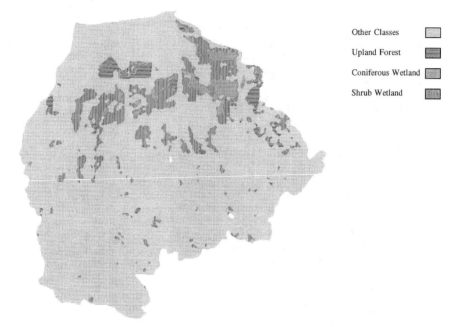

Other Classes

Upland Forest

Coniferous Wetland

Shrub Wetland

Figure 4   Forest and wetland types classified from multispectral SPOT satellite data. This image was created using a maximum overlay technique of original images created by maximum likelihood and minimum distance classification routines.

RESULTS OF COMPARISON AND DISCUSSION

The SPOT, NWI, and MIRIS wetland maps correlate well with the hydric soil data for the watershed, with a positive association occurring on all three maps. For this analysis, the null hypothesis of the Chi Square test of independence was that the wetland classification cross-classified with hydric soils data were independent of one another, and differences were due to random error. The calculated Chi Square for each map was larger than the critical value for each map, suggesting that a significant statistical association between the maps existed. Cramer's V values of 0.572, 0.683, and 0.710 were calculated for the SPOT, NWI, and MIRIS wetland maps, respectively (Table 1). The corresponding KIA values were similar, with values of 0.554, 0.661, and 0.697 calculated. Although exhibiting a relatively strong association with hydric soils (Cramer's V = 0.572), the SPOT wetland map was not as closely associated as the NWI nor the MIRIS wetland maps.

A direct comparison of the SPOT wetland map with the NWI and MIRIS wetland maps also indicated a relatively strong association. Calculated values of Cramer's V and KIA show the relatively strong relationship between the SPOT wetlands and NWI (Cramer's V = 0.581 and KIA = 0.581) and SPOT wetlands and MIRIS (Cramer's V = 0.636 and KIA = 0.635) (Table 2). A comparison of

Table 1   A Comparison Between Hydric Soils and SPOT, NWI, and
MIRIS Classified Wetlands Located Within the Mitchell Creek
Watershed, Grand Traverse County, Michigan

Comparison	d.f.	Chi Square value ($\chi^2_{0.001,1}$)[a]	Cramer's V[b]	Overal Kappa[c]
SPOT vs. hydric soils	1	116,264 (10.8)	0.572	0.554
NWI vs. hydric soils	1	165,684 (10.8)	0.683	0.661
MIRIS vs. hydric soils	1	178,919 (10.8)	0.710	0.697

[a] Chi Square critical value calculated for one degree of freedom and P = 0.001.
[b] Cramer's V = $\sqrt{\chi^2/(N \cdot \min(r-1, c-1))}$; where N = total number of observations = 355,407; r = number of rows; c = number of columns.
[c] Overall Kappa based on the Kappa Index of Agreement for wetland and nonwetland areas.

the NWI and MIRIS wetland maps resulted in Cramer's V and KIA values of 0.765 and 0.763, respectively. These higher values of association are to be expected, since landcover for both NWI and MIRIS classifications is derived from aerial photography.

Areas of discrepancy between the SPOT-derived wetland areas and those found in the NWI and MIRIS data sets warrant further field investigation and verification. For each of the two comparisons, four large areas of disagreement occurred within the watershed and are identified in Figures 5 and 6. Disassociation at sites 3 and 4 occurred in both the NWI and MIRIS comparisons. Site 3 was found to be classified as coniferous wetland using SPOT data if the green spectral band of the summer image was not used in the initial classification of the raw satellite image. However, overlays of the hydric soil data suggest that this area may not be wetland due to the apparent lack of hydric soils. Additional field verifications would be necessary to resolve this discrepancy.

Table 2   A Comparison Between a SPOT-Derived Wetland
Map and NWI and MIRIS Classified Wetland Maps
for the Mitchell Creek Watershed, Grand
Traverse County, Michigan

Comparison	d.f.	Chi Square value ($\chi^2_{0.001,1}$)[a]	Cramer's V[b]	Overall Kappa[c]
SPOT vs. NWI	1	119,768 (10.8)	0.581	0.581
SPOT vs. MIRIS	1	143,909 (10.8)	0.636	0.635
NWI vs. MIRIS	1	208,071 (10.8)	0.765	0.763

[a] Chi Square critical value calculated for one degree of freedom and P = 0.001.
[b] Cramer's V = $\sqrt{\chi^2/(N \cdot \min(r-1, c-1))}$; where N = total number of observations = 355,407; r = number of rows; c = number of columns.
[c] Overall Kappa based on the Kappa Index of Agreement for wetland and nonwetland areas.

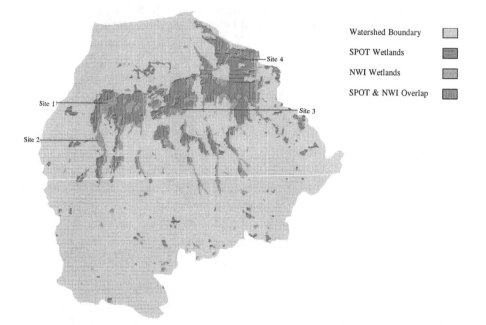

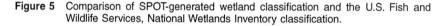

Figure 5 Comparison of SPOT-generated wetland classification and the U.S. Fish and Wildlife Services, National Wetlands Inventory classification.

Site 4, identified as wetland using SPOT data, is actually a duneridge-swale complex, where a majority of the vegetation consists of mixed coniferous wetland species intermingled with a distinct series of long, thin ridges of upland species. Neither the NWI nor MIRIS classifications classified this wetland-upland mosaic to be wetland, although this area should be considered a wetland–upland complex and has been described as an important nutrient sink for the watershed (Niehaus et al., 1991).

Without extensive field verification, no clear explanations for the wetland classification differences for Sites 1 and 2 and the many smaller areas of discrepancy in the NWI and MIRIS comparisons can be made. There are several possible explanations for these differences. First, many of these areas were along Mitchell Creek and were classified by SPOT data to be largely deciduous forest. This may in part be a function of spectral resolution problems resulting from different land covers with similar radiance being placed in the same classification unit. This can result in drier forested and scrub–shrub wetlands being classified as upland forest (e.g., alder along the stream classified as young aspen). This difference in wetland area may also be due to the spatial resolution limitations of present SPOT multispectral data, which is presently confined to a 20×20 m^2 pixel, which represents approximately 1 acre. However, in practice, it takes a box with 3 pure pixels on a side to consistently identify an object. If the object does not have a square shape, or if adjacent land cover has a similar radiance, the number of pixels needed for accurate classification rises (The Federal Geographic Data

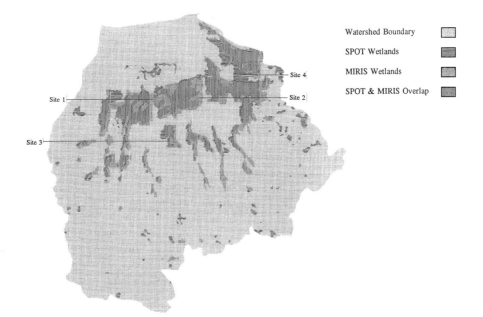

Figure 6   Comparison of SPOT-generated wetland classification and the Michigan Resource Inventory System classification.

Committee, 1992). This may explain the problems of the SPOT classification in areas along Mitchell Creek, where long, thin wetlands occur. Additional satellite scenes from time periods of known wetland inundation and optimum wetland identification conditions may have resolved these problems. Furthermore, the training fields or signature files used for processing the raw satellite images were, by design, limited to well-known field sites, and likely did not capture the entire array of spectral diversity present in these wetland types.

CONCLUSIONS

The results of this investigation indicate that using SPOT multispectral data is useful in generating cost- and time-effective composite wetland maps over large areas that are comparable to wetland maps created by more time-consuming traditional methods. In addition, SPOT-generated wetland maps can be cost-effectively updated on a regular basis to reflect real-time changes that are occurring within a study area. However, there are likely to be subtle differences between SPOT-derived wetland maps and photo-interpreted wetland maps. Accurate determinations of reflectance values from known wetland vegetation would allow researchers to "fine-tune" the cluster analyses and would likely improve the classification. Given the ease with which up-to-date wetland maps can be created using SPOT multispectral images, this method should have wide applicability with land use planners and managers in the near future.

ACKNOWLEDGMENTS

The authors wish to thank Jim Lively of the Northwest Michigan Council of Governments for providing the NWI, MIRIS, and soils data files used for our analyses. In addition, we wish to thank the Grand Traverse County Drain Commission, the State of Michigan, Great Lakes Environmental staff members, and Battelle Memorial Institute's Internal Research and Development Program for supporting this project.

REFERENCES

Barber, G. M., *Elementary Statistics for Geographers,* Guilford Press, New York, 1988.

Carstensen, L. W., A measure of similarity for cellular maps, *Am. Cartogr.,* 14 (4), 345, 1986.

Fulcher, J., Mitchell Creek Hydrologic Investigation, Michigan Department of Natural Resources unpublished report, 1991.

Koches, J. K., The Use of Geographic Information Systems and Remote Sensing Techniques in the Analysis of the Grand River Watershed, Publication No. MR-92-7, Water Resources Institute, Grand Valley State University, 1992.

Niehaus, S., Shuey, J., Roache, T., Harris, C., and DeGraeve, G. M., Mitchell Creek watershed non-point source pollution study, final report to the Grand Traverse County Drain Commission, 1991.

Podolsky, R. and Conkling, P., Satellite search aids wetlands visualization, *GIS World,* 12, 80, 1991.

Remillard, M. M. and Welch, R. A., GIS technologies for aquatic macrophyte studies: modeling applications, *Landscape Ecol.,* 8 (3), 163, 1993.

Rosenfield, G. H. and Fitzpatrick-Lins, K., A coefficient of agreement as a measure of thematic classification accuracy, *Photogrammetric Eng. Remote Sensing,* 52 (2), 223, 1986.

Soil Conservation Service, Soil Survey: Grand Traverse County Michigan, Series 1958, No. 34, U.S. Department of Agriculture, 1991.

The Federal Geographic Data Committee, Wetlands Subcommittee, Application of Satellite Data for Mapping and Monitoring Wetlands, 1992.

The IDRISI Project, IDRISI Technical Reference, Clark University, March 1992.

Welch, R. A., Integration of photogrammetric, remote sensing and database technologies for mapping applications, *Photogrammetric Rec.,* 12 (70), 409, 1987.

Ecology and Vegetation

CHAPTER 6

Stand Structure, Dynamics, and Diversity of Virgin Forests on Northern Peatlands

Yrjö Norokorpi, Erkki Lähde, Olavi Laiho, and Timo Saksa

CONTENTS

INTRODUCTION

The past decade has witnessed a kindling of interest toward the application of nature-oriented methods in the treating and regeneration of forests as alternatives to heavy-handed forest treatment methods (Lähde et al., 1991, 1992; Hagner, 1992). Particular concern has been expressed over the extensive clear fellings in the ecologically sensitive northern areas where the multiple-use function of forests also receives particular emphasis (Bäckström, 1984). The development of nature-oriented methods requires information on the structure and development of forests in nature (Lähde et al., 1991). Little attention has been directed at this topic.

The 1980s witnessed the development of a strengthened need also to protect the remnants of old virgin forests. As well as being based on the need to protect forest ecosystems as such, their protection includes factors such as acting as comparative areas for research focusing on managed forests and monitoring of the environment, and as refuges for endangered animals and plants (Bråkenhielm,

1-56670-177-5/97/$0.00+$.50

1982). The object itself — old virgin forests — continues to lack an adequate, distinct definition. Consequently, this has led to a situation of increasing need for research in the field of succession in virgin forests.

The need for protection of endangered species, increasing emphasis on forest uses other than wood production and on natural regeneration, and the regulation of atmospheric carbon through forests have, in the course of the recent years, made us aware of the need to maintain the natural biodiversity of our forests. This is one of the principles approved as one of the foremost objectives in forestry through international conventions, both at UNCED's meeting in Rio de Janeiro in 1992 and at the Ministerial Conference on the Protection of Forests in Europe held in 1993 in Helsinki (Ministerial Conference..., 1993).

Biological diversity (or biodiversity) is a term used when referring to the general diversity of and variation in the flora, fauna, and micro-organisms and their habitats. When applied to the forest ecosystem, the growing stock of trees plays a central role in the formation of the habitat. Biodiversity of the growing stock means species abundance in a contiguous stand and intraspecies variation in size, age, and genetic composition. The spatial arrangement of the trees shows variation both vertically and horizontally. Natural stands and stands very close to being natural best fulfill the requirements of biodiversity (Norokorpi et al., 1994).

Highly representative collections of material depicting the structure of virgin forests on mineral soils have been published recently in Finland (Lähde et al., 1991; Norokorpi et al., 1994). They are based on the national forest survey conducted in the 1920s. Mostly they were all-sized mixed woods, with the stem distribution curves resembling an inverted letter J; i.e., there was an abundance of small-diameter trees and the number of trees per diameter class diminished on moving into classes of bigger diameters. Strictly defined, even-sized forests (range of DBH no more than 15 cm) amounted to only 1 percent of the total containing the advanced forests (age more than 40 years) and to 6 percent when including also the young ones. There were, however, some stands that were all-sized in structure, but whose stem distribution resembled a normal distribution.

Gustavsen and Päivänen (1986) used the data collected in conjunction with the third national forest inventory (conducted in Finland in the years 1951 to 1953) to study the growing stocks on Finland's virgin peatlands. In particular, they examined stand volume and increment by site type and described stand structures according to the mean percentage of stem distributions without differentiating between tree species and age classes. The stem distributions in peatland stands were also, to a marked extent, all-sized.

Previous studies have been conducted in Finland with the aim of determining development series for what have been referred to as "natural normal" stands. To this end, measurements have focused on those parts of stands that have appeared to be as even-sized as possible, and attention has been directed at the way in which the dominant storey has developed (e.g., Ilvessalo, 1920, 1937).

The aim of this study is to describe the natural structure, dynamics, and diversity of the growing stock on virgin, undrained peatlands in the boreal forest zone in Finland. The material is based on the first national forest inventory conducted during the years 1921 to 1924, when two-thirds of Finland's forests were still in a natural state. The stand structure is described in terms of species-specific stem diameter distributions by dominant tree age class in order to formulate the succession and dynamics of the growing stock by tree species-dominated site type group in different regions of Finland. Species composition and size variation of trees are used as diversity characteristics of forest stands by development stage.

MATERIALS AND METHODS

The virgin forests included in the material of the first national inventory in Finland (NFI I, data collected in 1921 to 1924) divided into four geographical regions (Figure 1). Region A (South Finland) consists of the southern boreal vegetation zone and a southern part of the middle-boreal zone, to which Region B belongs. Regions C and D form the northern boreal zone. Forested peatlands

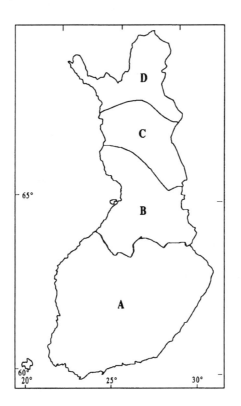

Figure 1   The geographical regions of Finland: A = South Finland, B = Southern part of North Finland, C = Central part of North Finland, and D = Northern part of North Finland.

had been classified into two site-type groups in the following way (Ilvessalo, 1927): (1) spruce mires (dominant tree species Norway spruce [*Picea abies* (L) Karsten] and downy birch [*Betula pubescens* Ehrh.]) and (2) pine mires (dominant tree species Scots pine [*Pinus sylvestris* L.]). Site types of low productivity (volume growth less than 1 m³ ha⁻¹ yr⁻¹ during a 100-year rotation) were excluded from this study.

A further means used in dividing the material was to apply age classes of dominant trees and growing stock volumes as follows:

Region	Young	Middle-aged	Old			
			Age class of dominant trees, years			
A	1. 1–40	2. 41–80	3. >80			
B	1. 1–60	2. 61–100	3. >100			
C	1. 1–60	2. 61–120	3. >120			
D	1. 1–80	2. >80	3. Undefined			
			Growing stock volume class, m³ ha⁻¹			
A	1. ≤20	2. 21–40	3. 41–80	4. 81–120	5. 121–160	6. >160
B	1. ≤20	2. 21–40	3. 41–80	4. 81–120	5. >120	
C	1. ≤20	2. 21–40	3. 41–80	4. >80		
D	1. ≤20	2. 21–40	3. 41–80	4. >80		

The data were collected using systematic line surveys with sample plots 10 × 50 m in size being measured at certain intervals. Each sample plot represented a uniform stand. In the event that a systematically placed sample plot fell on the boundary between different stands, it was relocated so as to be entirely within a single uniform stand (Ilvessalo, 1927). All trees with height exceeding breast height (measured from estimated stump height) were tallied and measured. Stem distributions were presented by species or groups of tree species: (1) Scots pine, (2) Norway spruce, (3) Birch (*Betula* sp.), (4) Alder (*Alnus* sp.), (5) Aspen (*Populus tremula* L.), and (for diameters above 20 cm for all broadleaves combined) in 5-cm diameter classes: 1 to 5, 5 to 10, ... > 35 cm. Detailed descriptions of the measurement procedure are available in contemporary publications (Heikinheimo, 1924; Ilvessalo, 1927). The following terms were applied for the various DBH class combinations: small-sized trees = 1 to 10 cm, middle-sized trees = 10 to 20 cm and large-sized trees ≥20 cm.

The stand structure based on DBH distribution was classified in accordance with Smith (1962) and Daniel et al. (1979) as modified by Lähde et al. (1991) and Norokorpi et al. (1994):

1. Small-sized stands: trees present only in the three smallest diameter classes (DBH ≤ 15 cm)
   a.   Stem distribution resembling a normal distribution
   b.   Stem distribution resembling an inverted letter J
   c.   Other structure

2. Advanced stands: trees present also or only in classes other than the three
   smallest diameter classes
   a. Even-sized: trees present in no more than three successive diameter classes
      i. Even-sized, resembling a normal distribution
      ii. Other even-sized
   b. All-sized: trees present in at least four diameter classes
      i. All-sized, storeyed. At least one of the second, third, or fourth diameter
         classes without trees
      ii. All-sized, regular stem distribution resembling an inverted letter J and
          trees present in at least the four smallest DBH classes
      iii. All-sized, stem distribution approximately normal, and other structures

The target diameter (or mean diameter corresponding to rotation) was set at
30 cm. Even-sized stands were defined to be stands in which the DBH varied no
more than by 15 cm. All other structures were defined as all-sized.

RESULTS

There were a total of 520 peatland sample plots; of these, 214 (41 percent)
were located in spruce mires and 306 (59 percent) in pine mires. Peatland stands
amounted to about a quarter of all virgin forests. They were most frequent in the
southern part of North Finland (35 percent) and least frequent in South Finland
and the northern part of North Finland (19 percent). The relative proportion of
pine mires increased in the south-to-north direction.

The proportion of old stands (in terms of their dominant storey) in spruce
mires was at least 50 percent in all the regions (Table 1); at its maximum, this
was observed in the northern boreal zone (Regions C and D). In pine mires, the
proportion of old stands increased steeply from South Finland (17 percent) to the
northern part of North Finland (87 percent) while the proportion of young stands
showed a concurrent decrease from 35 to 4 percent.

The mean total stem tally amounted to ca. 3000 stems/ha on South Finland's
peatlands. It was reduced to half of this level in the northern boreal region's
spruce mires and to one-quarter in pine mires (Table 1). There were decreasing
numbers of stems from young to old age classes in most regions, except for pine
in the central part of North Finland (Region C), in which there were a more or
less similar number for all age classes. Stem volume decreased from south to
north by about 50 percent, and tended to increase with age within each region
(Table 1). However, for pine in North Finland (Region D), the youngest age class
had the most volume.

The stands formed mixedwoods in all stages of development, with tree species
composition manifesting wide variation. In the spruce mires, the southern regions
were dominated by spruce (in terms of numbers), while the northern regions were
dominated by broadleaved species. The proportion of spruce generally increased
as the stands aged, while the proportions of broadleaved species decreased. In

Table 1 Stem Number, Volume, Distribution of Tree Species (% of Stem Number), and Occurrence of Broadleaved Species (% of Plots) by Age Class and Region on Peatlands

Region	Age class	Number of plots	x̄	Stems/ha range	V, m³/ha	Distribution of tree species, % stems			Occurrence of broadleaves, % plots		
						Pine	Spruce	Broadleaf	Birch	Aspen	Alder
				Spruce mires							
A	1–40	15	4562	1770–9880	27	13	31	56	80	13	13
	41–80	27	3764	610–11760	87	6	54	40	92	11	7
	>80	46	2068	600–4670	118	6	70	24	91	13	2
	x̄	89	3012	600–11760	93	7	59	34	89	12	6
B	1–60	12	2773	1290–6200	41	8	30	62	91	18	4
	61–100	22	2181	1050–4060	58	4	46	50	95	6	
	>100	32	1647	630–4080	75	2	72	26	68	9	1
	x̄	68	2001	630–6200	63	4	56	40	82		
C	1–60	0									
	61–120	7	1922	1030–2500	55	4	36	60	100		
	>120	22	1542	200–4200	65	1	54	45	90		9
	Undef.	17	1397	840–2330	50	2	38	60	94		
	x̄	46	1546	200–4200	58	2	45	53	91		7
D	1–80	0									
	>80	2	1245	1190–1300	36	0	53	47	100		
	Undef.	9	1627	810–3340	54	1	28	71	100		
	x̄	11	1558	810–3340	51	1	32	67	100		

Pine mires

A	1–40	27	3644	540–9060	25	40	10	50	88		19
	41–80	38	2651	940–6280	40	51	10	39	89	3	
	>80	13	2140	90–3820	89	51	29	20	84	1	6
	x̄	79	2882	90–9060	44	48	13	39	87		4
B	1–60	24	2640	280–23200	27	64	1	35	62	4	
	61–100	23	1756	220–5190	31	52	27	21	60		
	>100	61	982	170–2830	41	60	25	15	52	1	
	x̄	112	1518	170–23200	35	56	21	21	56		1
C	1–60	2	980	750–1210	15	17	26	57	100		
	61–120	10	1127	280–2130	30	46	26	28	60		
	>120	26	1092	160–3130	37	46	33	21	76		
	Undef.	22	1214	80–2410	27	20	37	43	90		
	x̄	60	1139	80–3130	31	35	33	32	80		
D	1–80	2	2385	1680–3090	33	15	20	65	100		
	>80	5	492	340–830	27	26	38	36	100		
	Undef.	48	710	100–1880	19	27	15	58	91		
	x̄	55	751	100–3090	20	26	17	57	93		

the pine mires, pine dominated in Regions A and B; in Region C, pine, spruce, and birch were equally represented; and in Region D, downy birch exceeded 50 percent. Among the broadleaved species, birch was dominant in all regions. Aspen and alder occurred in minor proportions in the two southern regions only, except for a limited occurrence of alder in spruce mires, Region C. In general, the broadleaved species are shortlived and tend to decrease in numbers as they mature and die; in contrast, spruce is longer lived and increased in relative numbers in older stands, especially in the southern regions.

The stem diameter distributions were clearly skewed to the right (i.e., resembled an inverted letter J) on all sites and in all regions; there were clearly more small-sized trees (DBH ≤ 10 cm) than large-sized trees (DBH ≥ 20 cm). The degree of skewness was reduced as the age of the dominant trees increased and the further north a particular stand was located; this was also accompanied by a reduction in the overall stem tally (Figure 2). In Regions A and B, the number of small-sized trees decreased by about 50 percent when comparing the young stands to middle-aged stands and again by about 50 percent when comparing middle-aged stands to old stands (Table 1). Further north, in Regions C and D, there were few stands of the young classes of spruce and pine, and valid comparisons could not be made (Table 1). The lack of young stands may reflect a tendency of stands to develop an all-sized structure. However, in terms of size, there was a relative abundance of small-sized broadleaved species and pine even in old stands. This indicates that they have the potential to regenerate as an understorey in various stages of stand development.

An increase in stand volume was associated with an increase in the stem tally, with the bigger diameter classes exhibiting the relatively higher increases. This reduced the degree of skewness in the stem diameter distribution. As stem diameter increased, the proportion of spruce increased in the spruce mires and that of pine increased in the pine mires (Figure 3).

On average, even-sized stands accounted for 14 percent of the total number (spruce mires 8 percent and pine mires 18 percent). The majority of the stands (86 percent) were all-sized to various extents (Table 2). All-sized regular stands amounted to 63 percent of the total (spruce mires 74 percent and pine mires 55 percent). The mean percentage for all-sized stands that were storeyed was 9 (spruce mires 2 percent and pine mires 14 percent). The proportion of all-sized regular stands decreased in the south-to-north direction and fertile-to-poor sites. The proportion of even-sized stands increased as the age of the dominant storey volume increased. Structurally, the modal class of the stem diameter distribution usually occurred in the smallest DBH class, especially within the South Finland region.

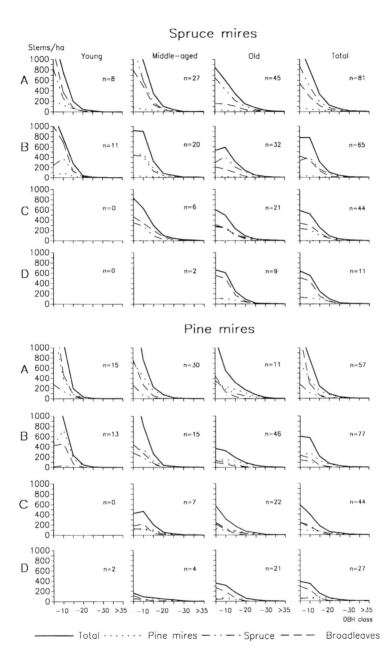

Figure 2   Average diameter class distribution of tree species and in total by age class of dominant trees and region in spruce and pine mires. If the number of plots (n) was less than three, the data were not given.

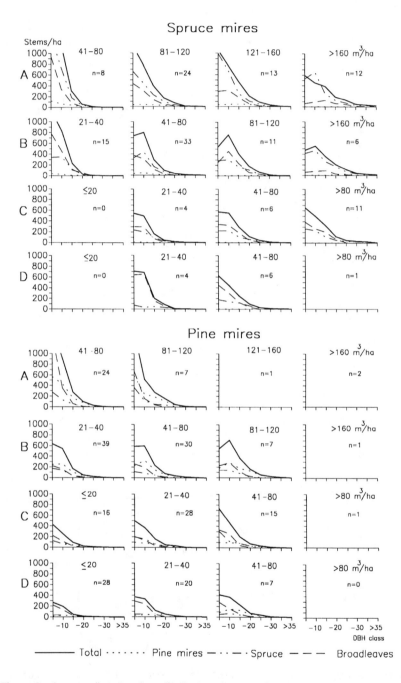

Figure 3   Average diameter class distribution of tree species and in total by volume class and region in spruce and pine mires.

Table 2 Distribution of Stand Structure Classes (%)

Region	Age class	Spruce mires							Pine mires						
				Stand structure class							Stand structure class				
		N	a+b+c	A1	A2	B1	B2	B3	N	a+b+c	A1	A2	B1	B2	B3
A	1–40	15	46			7	47		27	55			4	37	4
	41–80	27	7				86	7	38	16				84	
	>80	46	2				83	15	13	8		8		61	23
	Total	89	11			1	78	10	79	28		1	1	64	6
B	1–60	12	25			8	59	8	24	33				67	
	61–100	22	18			5	59	18	23	17			9	70	4
	>100	32				3	66	31	61	6			26	42	26
	Total	68	10			4	64	22	112	15			16	53	16
C	1–60	0							2					100	
	61–120	7					100		10	20				60	20
	>120	22				5	72	23	26	4			4	69	23
	Undef.	17					88	12	22	5			14	81	
	Total	46				2	83	15	60	7			7	73	13
D	1–80	2					100		2				50	50	
	>80	9	11				56	33	5				20	20	60
	Undef.								48	25			33	27	15
	Total	11	9				64	27	55	22			33	27	18

Note: Structure classes: Small-sized (DBH <15 cm): a = resembling even-sized, b = resembling all-sized, c = others; Advanced (DBH >15 cm): A1 = even-sized, resembling normal distribution, A2 = even-sized, other structures, B1 = all-sized, storeyed, B2 = all-sized, regular, B3 = all-sized, stem distribution approximately normal and other structures.

DISCUSSION

The material used in this study is a representative systematic sample of virgin forests in Finland from a period when most of the country's forests were still undisturbed by man or were seminatural (Heikinheimo, 1924). It also provides us with significant indications as to the structure of forests in the southern, middle, and northern boreal zones when viewed from a wide perspective. Growing stock data on current virgin forests in Finland are available only selectively, and these pertain to segments of strict nature reserves and national parks. However, the value of the measurement data from Finland's first national forest inventory is reduced by the fact that the inventory took place at the end of a cooler-than-average climatic period (Sirén, 1961). It would appear that the data underestimate the yield and regeneration capacity of the growing stock, especially in the case of the northern parts of the country (Paavilainen and Tiihonen, 1988).

This study revealed that regular all-sizedness is the dominant structure in almost all stages of stand development. These results are in accordance with previous studies involving peatlands (Heikurainen, 1971; Gustavsen and Päivänen, 1986; Hökkä and Laine, 1988). Broadleaved species (i.e., *Betula* sp., *Alnus* sp., and *Populus* sp.) were present in all strata. Their proportions usually grow with increasing site fertility. The abundance and number of broadleaved species are good indicators of diversity of the tree layer (Hunter, 1990).

The change in species composition mostly conformed to the generally recognized trends in succession: almost without exception, the proportion of broadleaved species decreased and that of spruce increased on all site types and in all regions as the age of the dominant trees increased. The proportion of broadleaved species decreased most of all in South Finland and least of all in the northern part of North Finland. The steepness of this change appears to have been fairly directly related to the density and height of the growing stock (i.e., to increasing shading). The decrease in the proportion of broadleaved species in South Finland was also the result of the almost total absence of the relatively short-lived alder from old forests. Very little alder was present in the northern parts of the country.

Detailed descriptions have been given of the structure and dynamics of only virgin spruce stands in northern Finland. Stand development following forest fires on moist mineral soil sites has led to the emergence of all-aged mixed birch–spruce stands (Sirén, 1955). Stand structure has remained much the same, even during the second tree generation (i.e., the secondary forest stage) even though the proportion of birch has diminished. The age difference between the oldest and the youngest trees may be as much as 400 years (Norokorpi, 1979, 1992). A typical feature of old spruce forests is that trees die and fall over singly and in small groups. The openings thus created, especially depressions left by uprooted trees and in the immediate vicinity of fallen trees, enable trees belonging to the understorey to continue growing (Norokorpi, 1992). Thus, regulated by ecological conditions, forests avoid degeneration by regenerating through small-scale disturbance spots (Runkle, 1982). With the mosaic of forest compartments of varying sizes and the fragmentation of site types as the basis, forests in the coniferous zone often develop into diverse forests of varying structure (Pobedin-ski, 1988; Leemans, 1991).

It is generally accepted that spruce is the most shade-tolerant of the main Nordic tree species, birch is less tolerant, and pine is the least. With these species as the basis, it has been estimated that it is primarily spruce stands that have the capacity to develop into stands with all-sized regular structure (Mikola, 1984). Spruce is known to be able to recuperate even from underneath a dense overstorey (Cajander, 1934). Birch is also able to develop as understoreys, even when dominated by spruce (Lähde et al., 1991, 1992; Laiho, 1992). Pine understoreys develop best underneath a pine stand; an overstorey of birch is also suitable (Laiho, 1992). The material used in this study supports the aforementioned understanding of the occurrence of pine and birch understoreys on peatland sites as well. The successful development of pine as an understorey on virgin and drained peatlands has been previously reported by Heikurainen (1971), Hånell (1984), Gustavsen and Päivänen (1986), Hökkä and Laine (1988), and Hökkä et al. (1991).

The proportion of all-sized storeyed stands averaged 9 percent. In reality most of the undergrowth was located outside those stands. All-sized regular stands contained undergrowth in the form of lower storey or small-sized trees. When the undergrowth included in the all-sized stands of other structures is added to this, one can conclude that there was abundant undergrowth in nearly all stands. The regeneration potential of naturally developed forests on peatlands is thus very high.

In the development of silvicultural regimes emulating natural succession and stand structure, the findings of this study indicate that we could endeavor to retain our forests as all-sized mixed stands or steer their development in this direction.

REFERENCES

Bäckström, P.-O., Ed., *Skogsföryngring i fjällnärä skogar. Forskarrapport,* Sveriges Lant-bruksuniversitet, Skogsvetenskapliga fakulteten, Umeå, 1984, 86 pp.

Bråkenhielm, S. Ed., Urskogar, Inventering av urskogsartade områden i Sverige, Del 1, Allmän del (Summary: Virgin forests — inventory of old natural forests in Sweden), *Naturvårdsverket, Rapport,* 1507, 108, 1982.

Cajander, E. K., Kuusen taimistojen vapauttamisen jälkeisestä pituuskasvusta. Referat: Über den Höhenzuwachs der Fichtenpflanzenbestände nach der Befreiung, *Commun. Inst. Forestalis Fenniae,* 19 (5), 59, 1934.

Daniel, V., Helms, J. A., and Baker, F. S., *Principles of Silviculture,* 2nd ed., McGraw-Hill, New York, 1979, 500 pp.

Gustavsen, H. G. and Päivänen. J., Luonnontilaisten soiden puustot kasvullisella metsä-maalla 1950-luvun alussa (Summary: Tree stands of virgin forested mires in the early 1950s in Finland), *Folia Forestalia,* 673, 27, 1986.

Hagner, M. Ed., Silvicultural Alternatives. Proceedings from an internordic workshop, June 22–25 1992, *Swed. Univ. Agric. Sci., Dept. Silviculture Rep.,* 35, 214, 1992.

Hånell, B., Skogsdikningsboniteten hos Sveriges torvmarker, *Rapporter i skogsekologi och skoglig marklära,* 50, 125, 1984.

Heikinheimo, O., Suomen metsien metsänhoidollinen tila, *Commun. Inst. Forestalis Fenniae,* 9 (4), 12, 1924.

Heikurainen, L., Virgin peatland forests in Finland, *Acta Agralia Fennica,* 123, 11, 1971.

Hökkä, H. and Laine, J., Suopuustojen rakenteen kehitys ojituksen jälkeen (Summary: Post-drainage development of structural characters in peatland forest stands), *Silva Fennica,* 22, 45, 1988.

Hökkä, H., Piiroinen, M.-L., and Penttilä, T., Läpimittajakauman ennustaminen Weibulljakaumalla Pohjois-Suomen mänty-ja koivuvaltaisissa ojitusaluemetsiköissä (Summary: The estimation of basal area-dbh distribution using the Weibull-function for drained pine- and birch-dominated and mixed peatland stands in North Finland), *Folia Forestalia,* 781, 22, 1991.

Hunter, M. L., Jr., *Wildlife, Forests, and Forestry. Principles of Managing Forests for Biological Diversity,* Prentice Hall, Englewood Cliffs, NJ, 1990, 370 pp.

Ilvessalo, Y., Kasvu- ja tuottotaulut Suomen eteläpuoliskon mänty-, kuusi- ja koivumetsille (Referat: Ertragstafeln für die Kiefern-, Fichten- und Birkenbestände in der Südhälfte von Finnland), *Acta Forestalia Fennica,* 15, 94, 1920.

Ilvessalo, Y., Suomen metsät. Tulokset vuosina 1921–1924 suoritetusta valtakunnan metsien arvioimisesta (Summary: The forests of Suomi [Finland]. Results of the general survey of the forests of the country carried out during the years 1921–1924), *Commun. Inst. Forestalis Fenniae,* 11, 241, 1927.

Ilvessalo, Y., Perä-Pohjolan luonnonnormaalien metsiköiden kasvu ja kehitys (Summary: Growth of natural normal stands in Central North Suomi [Finland]), *Commun. Inst. Forestalis Fenniae,* 24 (2), 162, 1937.

Lähde, E., Laiho, O., Norokorpi, Y., and Saksa, T., The structure of advanced virgin forests in Finland, *Scand. J. For. Res.,* 6 (2), 527, 1991.

Lähde, E., Laiho, O., Norokorpi, Y., and Saksa, T., Stand structure of thinning and mature conifer-dominated forests in boreal zone, in Silvicultural alternatives. Proceedings from an internordic workshop June 22–25, 1992, Hagner, M., Ed., *Swed. Univ. Agric. Sci., Dept. Silviculture Rep.,* 35, 58, 1992.

Laiho, O., Understoreys in the forests of Finland, in Silvicultural alternatives. Proceedings from an internordic workshop, June 22–25 1992, Hagner, M., Ed., *Swed. Univ. Agric. Sci., Dept. Silviculture Rep.,* 35, 100, 1992.

Leemans, R., Canopy gaps and the establishment patterns of spruce in two old growth coniferous forests in central Sweden, *Vegetatio,* 93, 157, 1991.

Mikola, P., Harsintametsätalous (Summary: Selection system), *Silva Fennica,* 18, 293, 1984.

Ministerial Conference on the Protection of Forests in Europe, 16–17 June 1993 in Helsinki, Ministry of Agriculture and Forestry, Helsinki, ISBN 951-47-8283-6, 1993, 56 pp.

Norokorpi, Y., Old Norway spruce stands, amount of decay and decay-causing microbes in northern Finland, *Commun. Inst. Forestalis Fenniae,* 97 (6), 77, 1979.

Norokorpi, Y., Natural structure and development of forests as a basis for alternative silvicultural methods in northern Finland, in Silvicultural alternatives. Proceedings from an internordic workshop June 22–25, 1992, Hagner, M., Ed., *Swed. Univ. Agric. Sci., Dept. Silviculture Rep.,* 35, 78, 1992.

Norokorpi, Y., Lähde, E., Laiho, O., and Saksa, T., Luonnontilaisten metsien rakenne ja monimuotoisuus Suomessa (Summary: Stand structure and diversity of virgin forests in Finland), Finnish Forest Research Institute, *Research papers,* 495, 54, 1994.

Paavilainen, E. and Tiihonen, P., Suomen suometsät vuosina 1951–1984 (Summary: Peatland forests in Finland in 1951–1984), *Folia Forestalia,* 714, 29, 1988.

Pobedinski, A. V., Comparative evaluation of even-aged and uneven-aged stands (in Russian), *Lesnoe Khozyaistvo,* 2, 40, 1988.

Runkle, J. R., Patterns of disturbance in some old-growth mesic forests of eastern North America, *Ecology,* 63, 1533, 1982.

Sirén, G., The development of spruce forest on raw humus sites in northern Finland and its ecology, *Acta Forestalia Fennica,* 62 (4), 408, 1955.

Sirén, G., Skogsgränstallen som indikator för klimatfluktnationerna i norra Fennoskandien under historisk tid, *Commun. Inst. For. Fenn.,* 54 (2), 66, 1961.

Smith, D. M., *The Practice of Silviculture,* John Wiley & Sons, New York, 1962, 577 pp.

CHAPTER 7

Relationships Between Groundwater Characteristics, Vegetation, and Peatland Type in the Hiawatha National Forest, Michigan

Gregory M. Kudray and Margaret R. Gale

CONTENTS

INTRODUCTION

Peatlands are an important group of wetlands characterized by the presence of an organic soil, typically to some minimum depth. This group is roughly correspondent to European "mires" (Moore, 1984) and to the Histosol soil order in the USDA Soil Taxonomy system (Soil Survey Staff, 1994). Two main types of peatlands are often differentiated (Gore, 1983), ombrotrophic peatlands, or bogs, that are mainly isolated from groundwater influence, and minerotrophic peatlands, or fens, that are recipient to groundwater that has flowed through mineral soil or bedrock. Additionally, minerotrophic peatlands may be subdivided into types that lack a significant tree cover, fens, and forested peatlands or swamps (Heinselman, 1963). Mader (1991) has suggested that swamps can be considered fens with trees. In a northern Michigan study, Schwintzer and Tomberlin (1982) identified two main groups, (1) bogs and (2) fens and swamps, on the basis of

1-56670-177-5/97/$0.00+$.50
© 1997 by CRC Press, Inc.

shallow groundwater chemistry, but recognized that strong vegetation differences exist between fens and swamps. Schwintzer (1981) suggested that bogs, fens, and swamps should be considered separate wetland types based on vegetation and water chemistry.

While the terms *bog, fen,* and *swamp* are used widely as broad concepts by both layman and scientist, there remains enough confusion about their specific definition that the United States National wetland classification avoids the terms entirely (Cowardin et al., 1979). However, the Canadian National wetland classification system embraces the terms, seeking to refine their definitions in a Canadian context (Zoltai, 1988). In this study the two most common bases for peatland classification, vegetation and groundwater chemistry, will be examined in the context of the bog, fen, swamp concept. An initial classification of plots into bog, fen, and swamp types using vegetation characteristics will be compared against a discriminant analysis based on groundwater variables. Additionally, correlations between groundwater characteristics and the bryophyte and vascular plant communities will be interpreted to examine vegetation–groundwater relationships across a broad range of peatland types.

MATERIALS AND METHODS

The western half of the Hiawatha National Forest is located in the Upper Peninsula of Michigan, U.S. (Figure 1) with an area of about 200,000 ha. The landscape is variable; Albert et al. (1986) identified five ecodistricts, and 17 land type associations have been mapped by the Forest Service (Ball and Padley, 1993). A total of 61 peatland plots, stratified throughout the major land type associations and ecodistricts, were sampled during the 1992 and 1993 field season. Peatlands were defined by the presence of at least 40 cm of peat (Zoltai, 1988), which generally corresponds to the Histosol soil order definition (Soil Survey Staff, 1994).

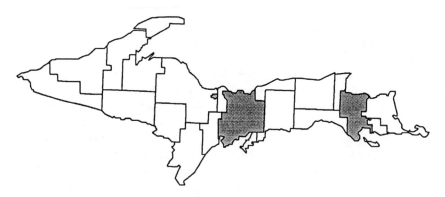

Figure 1   Location of the Hiawatha National Forest, Upper Peninsula of Michigan, U.S. Sample plots were located in the western unit.

Circular 200 m² plots were established at random distances along predetermined transects. At each plot all vascular and bryophyte plant species were identified; plots were visited twice during the growing season to facilitate identification of early or late blooming species. Perforated plastic pipe was installed at each plot and referenced to ground level to establish the location of the groundwater table. Each year during the third week of July, all of that year's plots were revisited to determine the location of the groundwater table, to collect water samples, and to measure groundwater temperature in an excavated pit. Electrical conductivity and pH were determined on refrigerated samples after the end of the sampling week. Water samples were treated with acid in the field and then shipped in refrigerated coolers to the Tennessee Valley Authority (TVA) Environmental Chemistry laboratory at Chattanooga, TN, for analysis. Water samples collected for the determination of total phosphorus (P) and total Kjeldahl nitrogen (N) were preserved with H_2SO_4 and analyzed with a colorimeter. Water samples collected for the determination of total potassium (K) were preserved with HNO_3 and analyzed on an atomic absorption spectrophotometer. Water samples collected for the determination of aluminum (Al), iron (Fe), manganese (Mn), calcium (Ca), zinc (Zn), and magnesium (Mg) (all total elemental analysis) were preserved with HNO_3 and analyzed on an inductively coupled argon spectrophotometer. Water samples collected for the determination of total organic carbon (C) were preserved with H_2SO_4 and analyzed using the combustion infrared method. Groundwater electrical conductivity was corrected (K_{corr}) for the influence of hydrogen ions (Sjörs, 1950). Preliminary data analysis indicated that P, N, K, Al, Fe, Mn, Ca, Zn, and Mg were not normally distributed and variances were not homogeneous. A log transformation was applied to meet these assumptions for further statistical analysis.

Individual plots were identified as bog, fen, or swamp based on plot vegetation and physiognomy. Swamps were dominated by trees or shrubs (Mitsch and Gosselink, 1993). Shrub thicket swamps were differentiated from fens by an abundant tall (>2.0 m) shrub cover (Jeglum et al., 1974; Swanson and Grigal, 1989; Jeglum, 1991). Treed swamps had a cover of at least 15 percent in trees >5 m tall. Tree cover in bogs and fens was sparse and stunted if present. Fens were dominated by *sphagnum*, sedge, and low shrub. Bogs were separated from fens by the absence of minerotrophic vegetation indicators. Common indicator species that were found in fens but not in bogs included *Alnus rugosa, Betula pumila, Carex stricta, Menyanthes trifoliata,* and *Potentilla palustris.*

Detrended correspondence analysis (Hill, 1979) was used to examine the vegetation data. Detrended correspondence analysis (DCA) provides an ordination of plots; each plot is assigned a value along four ordination axes that allows correlations with environmental factors or a visual representation. A DCA ordination was performed with each of the following data sets: vascular plants only, bryophytes only, and vascular plants and bryophytes combined. The vegetation data was presence/absence; no cover values were used. In each ordination, species that occurred in only one plot were eliminated from the vegetation data set. DCA

Table 1   Discriminant Analysis Classification Matrix
Based on Groundwater Variables That Were
Significant at the 5 Percent Level for Entry
into the Discriminant Analysis Model. The
Variables pH, Ca, Groundwater Temperature,
and Mn Were Used in the Model

	Predicted				
	Bog	Fen	Swamp	Plots	Accuracy
			Actual		
Bog	8	0	1	9	89%
Fen	1	23	6	30	77%
Swamp	2	2	18	22	82%
Total				61	
Overall					80%

default settings regarding axes rescaling and detrending were used. Plot scores from DCA axes were correlated to water chemistry variables using Spearman's rank order correlation.

Linear discriminant analysis with prior probabilities was used to examine the ability of the groundwater variables to differentiate the bog, fen, and swamp types. Variables were entered stepwise into the discriminant equation based on F-value (5 percent significance level); only those variables meeting this criterion were retained in the final model.

RESULTS AND DISCUSSION

Only pH, Ca, groundwater temperature, and Mn (in order of decreasing F-value) were significant at the 5 percent level for entry into the discriminant analysis model. A discriminant analysis using these variables was able to predict bog, fen, and swamp membership with an accuracy of 80 percent (Table 1). This same level of accuracy was achieved if Mn was not used. While only pH, Ca, groundwater temperature, and Mn were significant variables in the discriminant analysis, other variables, notably Mg and K_{corr}, were significantly different between groups (Table 2) but were not entered into the discriminant model due to strong correlations with variables already entered. Prediction accuracy was highest in the order, bogs > swamps > fens. The greatest amount of misclassification was between the fen and swamp type. Some richer fens were classified as swamps, while some more acidic swamps were classified as either fens or bogs.

The influence of mineral rich groundwater on peatland floristics and classification has long been recognized (Shotyk, 1988) and it would be expected that such groundwater variables as pH, Ca, Mg, and K_{corr} would show the largest differences between peatland types (Table 2). The entry of groundwater temperature as a significant discriminating variable is more surprising. Mean groundwater temperature is lowest in swamps and highest in fens (Table 2), which probably reflects hydrologic factors such as depth to water table and rate of

Table 2   Mean Values and Standard Deviations for Groundwater Variables

	Bog	Fen	Swamp
pH	4.38 (0.28)	6.06 (0.65)	6.07 (0.99)
F = 18.95; P = 0.00			
Ca (mg/l)	8.43 (6.54)	40.67 (43.65)	89.96 (59.46)
F = 18.53; P = 0.00			
Temperature (Celsius)	14.02 (1.26)	15.71 (3.06)	13.64 (2.37)
F = 4.3; P = 0.018			
Mn (μg/l)	71.00 (61.21)	726.53 (1386.91)	1624.27 (2501.66)
F = 8.3; P = 0.007			
K (mg/l)	3.11 (1.64)	1.81 (1.22)	2.18 (1.59)
F = 1.73; P = 0.186			
K_{corr}	37.56 (7.63)	126.2 (116.53)	179.94 (120.29)
F = 13.04; P = 0.00			
Al (μg/l)	3500.00 (2630.00)	5030.67 (6252.20)	15068.64 (16454.18)
F = 4.91; P = 0.011			
N (mg/l)	36.77 (32.07)	18.70 (24.34)	32.28 (41.83)
F = 1.79; P = 0.175			
Water table depth (cm)	7.81 (12.11)	0.61 (13.02)	7.85 (7.80)
F = 3.14; P = 0.051			
Mg (mg/l)	1.49 (0.71)	5.80 (5.61)	11.84 (7.83)
F = 17.43; P = 0.000			
Zn (μg/l)	114.44 (81.41)	110.70 (126.02)	276.36 (285.48)
F = 3.66; P = 0.032			
P (mg/l)	1.75 (1.24)	0.99 (1.04)	2.00 (2.90)
F = 2.13; P = 0.128			
Fe (μg/l)	4765.56 (3171.74)	43153.33 (67561.67)	87206.82 (114105.8)
F = 4.85; P = 0.011			
C (mg/l)	60.00 (22.96)	51.47 (21.49)	67.23 (41.14)
F = 0.70; P = 0.498			

Note: Variables are listed in order of entry into the discriminant analysis model. F-values are based on log-transformed values (pH, groundwater temperature, and depth to groundwater were not transformed).

groundwater discharge, as well as vegetation influences in the form of canopy shading and insulation by *sphagnum* mosses. The average water table was closest to the surface in fens and lower in bogs and swamps (Table 2). While this is only one midseason measurement, it suggests that the difference between open and treed peatlands may not be as strongly related to differences in water table depth as it is to the degree of minerotrophic groundwater influence. Swamp groundwater had the highest levels of Ca, Mn, Al, Mg, Zn, P, and Fe (Table 2). Average Ca levels were higher in swamps than fens. Schwintzer (1981) had similar Ca levels in both fens and swamps. Bogs were lowest in Ca, Mn, Al, Mg, and Fe, in agreement with their hydrologically isolated landscape position. Average N, P, and K levels were quite high in bogs as compared to the other types. The low productivity of bogs may be related more to nutrient availability or low Ca and Mg levels than to total N, P, and K content in the groundwater. Levels of groundwater pH, Ca, and K_{corr} are slightly higher for bogs than has been reported for other studies in the region (Schwintzer, 1981; Glaser et al., 1981; Glaser, 1983; Jeglum, 1991). This may reflect the underlying limestone bedrock in much

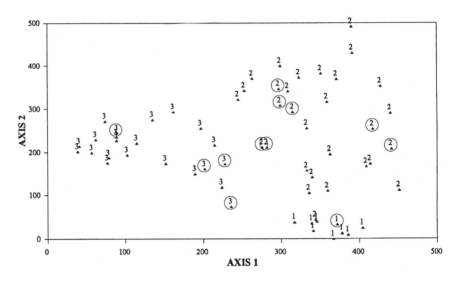

Figure 2   Detrended correspondence analysis ordination of vascular plants and bryophytes. Circled plots are plots misclassified on the basis of water chemistry discriminant analysis using the variables pH, Ca, groundwater temperature, and Mn. 1 = bog, 2 = fen, 3 = swamp.

of the study area and the relatively shallow peat accumulations at some of the sampled bogs.

A DCA ordination of vascular and bryophyte vegetation combined (Figure 2) shows a pattern similar to the groundwater chemistry discriminant analysis. Plots misclassified on the basis of water chemistry tend to be borderline between wetland types on the DCA ordination. A classification of wooded peatland vegetation by Jeglum (1988) did not form primary divisions corresponding to bogs, fens, and swamps, but grouped poor swamps with poor fens and bogs. The water chemistry misclassification of two of the more acidic swamp plots into the bog group parallels Jeglum's (1988) vegetation classification. These plots are more similar, in vegetation and water chemistry, to bogs or poor fens than to other richer swamps. Swamps misclassified as fens had vegetation similarities to fens. One interesting exception was a thicket swamp that was strongly predicted to be a fen on the basis of water chemistry although the vegetation was similar to other swamps. Several misclassified plots represent an intermediate group of peatlands, with a treestory typically dominated by *Picea mariana* and *Larix laricina*, that, when further subdivided, could be classified as poor swamps or treed fens (Jeglum 1988, 1991). These plots can be contrasted with the main group (Figure 2) of shrub thickets and swamps that share a richer vegetation and water chemistry. The richer forested swamps are usually dominated by a tree story of *Thuja occidentalis* with a more diverse ground flora. Separating the continuum of peatland types into two groups (Gore 1983), bogs and fens (including swamps), includes a very broad range of vegetation and groundwater characteristics within the fen class. A further division, in agreement with Schwintzer (1981), recognizing a swamp type, appears justified based on vegetation and groundwater chem-

Table 3   Spearman Rank–Correlation Coefficients Between
Groundwater Variables and Plot Scores from Axes
1 and 2 of Detrended Correspondence Analysis
Ordinations Using Vascular Plants Only and
Bryophytes Only (Presence/Absence)

	Vascular plants only		Bryophytes only	
	Axis 1	Axis 2	Axis 1	Axis 2
pH	−0.20	−0.53[b]	−0.51[b]	0.46[b]
Ca	−0.54[b]	−0.51[b]	−0.70[b]	0.31[a]
Temperature	0.44[b]	−0.06	0.28[a]	0.21
Mn	−0.38[b]	−0.20	−0.52[b]	0.08
K	−0.05	0.17	0.04	−0.17
K_{corr}	−0.39[b]	−0.41[b]	−0.57[b]	0.36[b]
Al	−0.35[b]	−0.01	−0.20	−0.25
N	−0.24	0.11	−0.09	−0.13
Water table depth	−0.41[b]	0.14	−0.21	−0.31[a]
Mg	−0.53[b]	−0.46[b]	−0.67[b]	0.29[a]
Zn	−0.21	0.11	−0.11	−0.16
P	−0.24	0.07	−0.12	−0.14
Fe	−0.16	−0.11	−0.16	0.07
C	−0.01	0.08	0.09	−0.33[a]

[a] Significant at 5%.
[b] Significant at 1%.

istry. Another group, consisting of poorer swamps and treed fens that are inter-mediate in vegetation and often misclassified by the groundwater chemistry discriminant analysis, would represent the next possible division based on this data.

The relationship between vegetation composition and groundwater character-istics can be inferred from the correlation of the DCA stand scores with the water variables (Table 3). A DCA ordination using only vascular plants was most strongly correlated with the variables (1) Ca, (2) Mg, and (3) groundwater tem-perature for axis 1. Axis 2 was correlated with (1) pH, (2) Ca, and (3) Mg. Jeglum (1991) reported high correlations of Ca and pH with plot scores from a vegetation ordination of wooded peatlands. When bryophytes only were considered, axis 1 correlations were in the order (1) Ca, (2) Mg, and (3) K_{corr}. Bryophyte axis 2 correlations were ranked (1) pH, (2) K_{corr}, and (3) C. The vegetation–groundwater relationship reflects the strong influence of mineral enrichment in structuring vegetation composition in peatlands.

REFERENCES

Albert, D. A., Denton, S. R., and Barnes, B. V., Regional Landscape Ecosystems of Michigan, School of Natural Resources, University of Michigan, Ann Arbor, MI, 1986.

Ball, J. and Padley, E., Delineation of Landscape Ecological Units in the Eastern Upper Peninsula of Michigan (draft 12/13/93), Eastern Upper Peninsula Interagency Group, 1993, 15 pp.

Cowardin, L. M., Carter V., Golet F. C., and Laroe E. T., Classification of Wetlands and Deepwater Habitats of the United States, U.S. Fish and Wildlife Service, United States Department of the Interior, Report No. FWS/OBS-79/31, 1979, 103 pp.

Glaser, P. H., Vegetation patterns in the North Black River peatland, northern Minnesota, *Can. J. Bot.,* 61, 2085, 1983.

Glaser, P. H., Wheeler, G. A., Gorham E., and Wright, H. E., The patterned mires of the Red Lake peatland, northern Minnesota: vegetation, water chemistry and landforms, *J. Ecol.,* 69, 575, 1981.

Gore, A. J. P., Introduction, *Ecosystems of the World,* Vol. 4A, *Mires: Swamp, Bog, Fen and Moor,* Gore, A. J. P., Ed., General Studies, Elsevier, Amsterdam, 1983, chap. 1.

Heinselman, M. L., Forest sites, bog processes, and peatland types in the glacial Lake Agassiz region, Minnesota, *Ecol. Monogr.,* 33, 327, 1963.

Hill, M. O., DECORANA – A FORTRAN program for detrended correspondence analysis and reciprocal averaging, *Ecology and Systematics,* Cornell University, Ithaca, NY, 52 pp.

Jeglum, J. K., Haavisto V. F., and Boissonneau A. N., Toward a Wetland Classification for Ontario, Department of the Environment, Canadian Forestry Service, Sault Ste. Marie, Ont. Information Report O-X-215, 1974, 54 pp.

Jeglum, J. K., The use of TWINSPAN, tabular analysis and Finnish/Swedish concepts in classifying wooded peatlands in Ontario, in Symposium '87: Wetlands/Peatlands, Edmonton, Alberta, Rubec, C. D. A. and Overend, R. D., compilers, 1988.

Jeglum, J. K., Definition of trophic classes in wooded peatlands by means of vegetation types and plant indicators, *Ann. Bot. Fennici,* 28, 175, 1991.

Mader, S. F., Forested Wetlands Classification and Mapping: A Literature Review, National Council of the Paper Industry for Air and Stream Improvement, Technical Bulletin No. 606, New York, 1991, 99 pp.

Mitsch, W. J. and Gosselink, J. G., *Wetlands,* Second Edition, Van Nostrand Reinhold, New York, 1993.

Moore, P. D., The classification of mires: an introduction, in *European Mires,* Moore, P. D., Ed., Academic Press, London, 1984, chap. 1.

Schwintzer, C. R., Vegetation and nutrient status of northern Michigan bogs and conifer swamps with a comparison to fens, *Can. J. Bot.,* 59, 842, 1981.

Schwintzer, C. R. and Tomberlin, T. J., Chemical and physical characteristics of shallow groundwaters in northern Michigan bogs, swamps, and fens, *Am. J. Bot.,* 69, 1231, 1982.

Shotyk, W., European contributions to the geochemistry of peatland waters, 1890–1940, in Symposium '87: Wetlands/Peatlands, Edmonton, Alberta, Rubec, C. D. A. and Overend, R. D., compilers, 1987.

Sjörs, H., On the relation between vegetation and electrolytes in north Swedish mire waters, *Oikos,* 2, 241, 1950.

Soil Survey Staff, Keys to Soil Taxonomy, Sixth Edition, Soil Conservation Service, United States Department of Agriculture, Washington, D.C., 1994.

Swanson, D. K. and Grigal, D. F., Vegetation indicators of organic soil properties in Minnesota, *Soil Sci. Soc. Am. J.,* 53, 491, 1989.

Zoltai, S. C., Wetland environments and classification, in *Wetlands of Canada,* National Wetlands Working Group, Ecological Land Classification Series No. 24, Environment Canada, Ottawa, Ontario, and Polyscience Publications Inc., Montreal, Quebec, 1988, chap. 1.

CHAPTER 8

Late Holocene Paleoecological Reconstruction of a Coastal Peat Bog along the St. Lawrence Maritime Estuary, Quebec

Michelle Garneau

CONTENTS

INTRODUCTION

In Quebec, there is a paucity of studies dealing with detailed and local paleoecological reconstructions with the use of different integrated techniques (pollen and spores as well as macrofossils analysis). Detailed analysis of three cores collected in a coastal peat bog has been carried out using the integrated scientific approach of the Hugo de Vries Laboratory from the University of Amsterdam (van Geel, 1972, 1978; Pals et al., 1980; Bakker and van Smeerdijk, 1982; van Geel et al., 1981, 1983, 1984, 1986, 1989; Kuhry, 1988; Ran, 1990).

1-56670-177-5/97/$0.00+$.50
© 1997 by CRC Press, Inc.

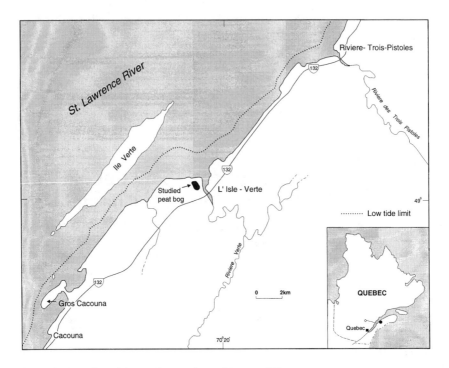

Figure 1   Location of the study area (48°00′N, 69°22′W).

The studied peat bog covers a 2 km² surface and is located at the edge of a *Spartina* coastal marsh at l'Isle-Verte on the south shore of the St. Lawrence maritime estuary, Quebec (Figure 1). The paleoecological reconstruction was prompted by the particular geomorphic context of the coastal plain, which led to the development of a peat bog at the edge of the marsh. Detailed analysis of three cores (depth between 120 to 130 cm) allowed the local reconstruction of the coastal area from the lowering of the sea level that followed the Mitis transgressive episode (3000 to 2500 BP; Dionne, 1988).

The three main objectives of this research were (1) to establish the detailed stratigraphic record of the coastal area from the onset of the intertidal sedimentation until today, (2) to document and understand the gradual transformation of the coastal marsh into a peat bog, and (3) to ascertain the correspondence between the topographic sequence and the chronostratigraphic sequence (Garneau, 1993).

METHODS

Most of the modern results are derived from an integrated ecological survey. The methodological approach made possible the relations between vegetation units and geomorphic and pedological units (Bertrand et al., 1983; Bertrand, 1984; Garneau, 1984). The results of the surface ecological survey guided the choice of the sampling sites of the three cores. Three types of bog environments were distinguished from a combination of homogenous vegetation cover and sediment composition; all three types were sampled. The cores were collected from the modern surface down to the basal silt unit that records the beginning of the intertidal sedimentation. The cores were collected with a Coûteaux corer (diameter of 10 cm; Coûteaux, 1962). Between 72 to 75 samples (thickness of 1 cm) were analyzed in each core, and integrated microfossil (pollen, spores, algae, rhizopods, etc.) and macrofossil (plant and animal macrofossils) analyses were conducted.

For pollen analysis, 1 cm³ was treated following Faegri and Iversen technique (1989). Each sample was contaminated with *Lycopodium* spikes (Stockmarr, 1971), treated with hot 10 percent KOH, 10 percent HCl, sieved and acetolyzed. Palynomorphs were separated by a mixture of bromoform and alcohol (specific gravity of 2). For macrofossil analysis, constant specified volume of 3 cm³ was used. The material was heated in a 5 percent KOH solution to disperse the particles, washed and sieved through a 180 μm screen (Garneau, 1987, 1992, 1993). Macrofossils remaining on the screen were hand picked under a stereo-scopic microscope (magnitude 10 to 40×). Identification of some plant parts (e.g., plant epidermis) was done under a photonic microscope (magnitude 250 to 400×). Nine organic samples (three levels for each core) were submitted for radiocarbon dating. Samples were cleaned of their roots and sent for conventional dating.

The results from the paleoecological reconstruction of the three cores were classified into two main zones (and different subzones) associated to two distinct past ecosystems. The recognition of these two zones was based on the biostrati-graphic composition of the sediment as well as on the assemblages similarities that characterized each core. The zonation was guided by the objective of local paleoecological reconstruction and hence, was not included in successional series.

RESULTS

Topographic Sequence

Figure 2 presents the ecogeomorphic context of the coastal plain and the location of the three cored sites. The vegetation is characterized by different communities parallel to the shore that correspond to different temporal percent-ages of tidal submergence during the year. The combination of local topography and tidal submergence explains the sediment distribution, the associated pedogenic processes, and the vegetation dynamics.

The lower section of the intertidal marsh (low marsh) is distinguished by high rates of annual submersion (>60 to 5 percent yearly). In the lowest part, the very high rates of submersion (>60 to 25 percent yearly) explain the absence of vascular vegetation and the abundance of algae attached to pebbles deposited over the basal silt unit under the influence of ice-drift processes. Higher in altitude, the algae community is replaced by a *Spartina alterniflora* community. In the highest sections of the low marsh, the lower rates of submersion (25 to 5 percent yearly) favor an increase in interspecific competition and hence, a greater floristic

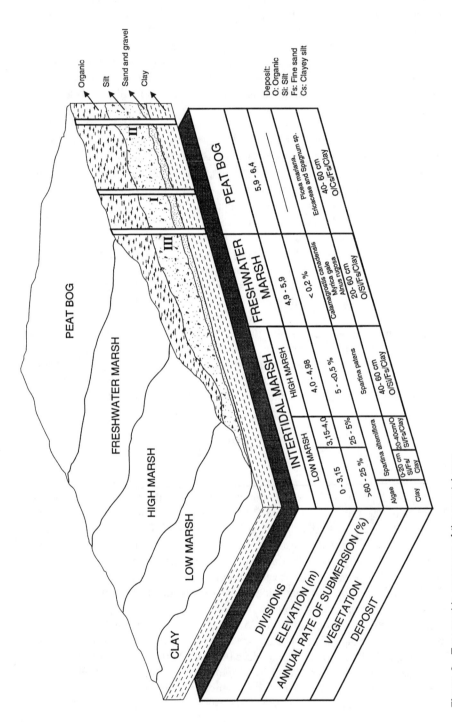

Figure 2 Topographic sequence of the coastal area.

Table 1   Radiocarbon Dates from the Three Cores

Core	Labo. #	^{14}C BP	Material	Depth (cm)
I	UL-293	180 ± 50	*Sphagnum* sp.	25
	UL-296	500 ± 60	Wood	60
	UL-292	1060 ± 60	Debris of *Spartina* sp.	90
II	Beta-26497	480 ± 70	Wood	27
	Beta-26498	1250 ± 80	Coal	50
	Beta-26499	1430 ± 90	Debris of *Spartina* sp.	75
III	Beta-25721	170 ± 70	*Sphagnum* sp.	18
	Beta-25722	420 ± 60	Wood	65
	Beta-25724	1020 ± 60	Debris of *Spartina* sp.	115

diversity, as evidenced by the presence of *Salicornia europaea, Spergularia marina, Plantago maritima,* and *Triglochin maritima* with *Spartina alterniflora.*

Above 4 m of altitude (high marsh), the *Spartina alterniflora* community is replaced by the *Spartina patens* community that is occasionally inundated (5 to <0.5 percent yearly). This community is distinguished by the presence of several halophytic species (*Spartina patens, Festuca rubra, Puccinellia langeana, Glaux maritima, Potentilla egedii, Limonium nashii, Stellaria humifusa*), which are added to the ones of the previous community in the topographic sequence.

The limit of the intertidal section is marked by a notable decrease of halophytes and an increase of hygrophytes. The freshwater marsh succeeds to this section and is characterized by accidental or rare frequency of submersion (<0, 2 percent yearly). The vegetation is mainly represented by grasses (*Spartina pectinata, Calamagrostis canadensis*), sedges (*Carex paleacea, Carex mackenziei, Scirpus maritimus*) and a few shrubs such as *Spiraea alba* and *Myrica gale.* The highest section of the freshwater marsh is colonized by a dense shrubland of *Alnus rugosa.* As opposed to topographic sequences that grade into forested mesic communities along the St. Lawrence coastline (Couillard and Grondin, 1986), the topographic sequence of the Isle-Verte Bay ends in a forested peat bog dominated by *Picea mariana* and *Sphagnum* sp.

Radiocarbon Dating

Nine radiocarbon dates (three for each core) have been obtained on the three cores. Table 1 presents the results for each core in a chronological order.

Chronostratigraphic Sequence

Two sedimentary environments, identified as zone 1 and zone 2, have been recognized in the three cores; these are associated with two distinct ecosystems (Figure 3). Zone 1 (coastal ecosystem) is characterized by a high mineral content (at the base, 90 percent of the total volume of the sample) decreasing gradually toward the top. The pollen content reveals local presence of several herb species, which is confirmed by macrofossil analysis: Gramineae, Chenopodiaceae, Caryophyllaceae, and *Plantago* sp. Some other fossils are restricted to zone 1: Foraminifereae — *Elphidium* sp., Rivulariaceae cyanobacteria — *Gloeotrichia* sp.,

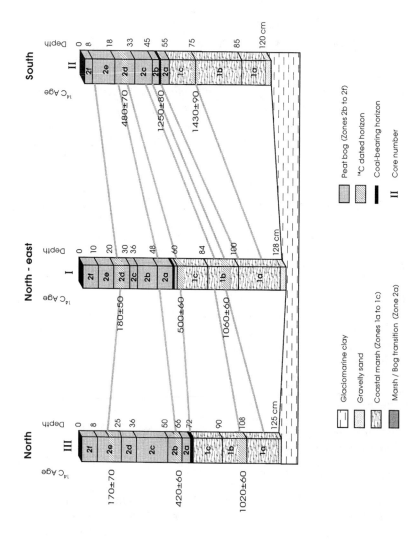

and their akinetes, Zygomicotina chlamydospores — *Glomus* sp., as well as Bryozoan statoblasts and Hydroid branches.

A thin charcoal layer with *Gelasinospora* ascospores (carbonicoulous and lignicolous ascomycete) marks the transition between zone 1 and zone 2 (subzone 2a). Pollen concentration as well as decomposition rate of the organic matrix are high at this level. Zone 2 (peat ecosystem) is characterized by an organic matrix whose decomposition varies with depth and therefore with the different vegetation assemblages. Cores I and III are mainly composed of bryophytes, while core II is characterized by a woody peat deposit.

Pollen spectra of zone 2 reveal a greater species diversity than in zone 1. At the base of zone 2, saline species are replaced by Gramineae and Compositae tubuliflorae species. Mosses as *Calliergon stramineum*, *Drepanocladus fluitans*, and *Sphagnum fimbriatum* are abundant and indicate a fen community (subzone 2b). The assemblage is also dominated by *Alnus rugosa* with sedges and other *Sphagnum* sect. Acutifolia. The *Alnus* shrubland was followed by a coniferous forest of *Picea mariana* and *Larix laricina* (subzone 2c). An outstanding change through subzone 2d at a time between 400 and 200 BP is a decrease in various woody remnants and an increase of Cyperaceae, Graminae, Liliaceae, and *Typha latifolia*. The top of the sequence is distinguished by cultivated and ruderal species (subzone 2f).

DISCUSSION

The two main sedimentary units identified in the three cores are associated with two distinct environments or ecosystems. Zone 1 (coastal ecosystem) reveals the constant presence of *Spartina* in the mineral sequence. *Spartina* is the main species that characterizes the present day vegetation in saline coastal marshes along the St. Lawrence (Gauthier et al., 1980). The organic (peat) sequence (zone 2: peat ecosystem) corresponds to different vegetation assemblages that evolved with time and under different local and/or regional processes (dryness, rise of moisture) into a *Picea mariana* and *Sphagnum* forested peat bog.

The mineral unit assemblages in the three cores (zone 1) are equivalent to the present coastal topographic sequence and reveal no major sea level fluctuation during this period of sedimentation (approximately from 2500 to 500 BP). The sedimentation rate was high in core II (1.48 mm/year) compared to the two other cores (core I: 0.57 mm/year and core III: 0.8 mm/year) and explains the low content of organic material retraced in the sediment. When compared with the two other cores, the high sedimentation rate of core II combined with its total depth suggests that the site was probably located on a paleo-channel of Rivière Verte. This particular geomorphic context (Figure 4) may also explain the bio- and chronostratigraphic differences between site II and sites I and III.

At approximately 2500 BP (2400 ± 60 BP; UQ-560) (Bertrand, 1984), iso-static rebound and sedimentation resulted in a decreasing marine influence and a progressive adaptation along the coastal plain to brackish (zone 1c) and later, to freshwater conditions (zone 2a). The gradual transformation of the intertidal

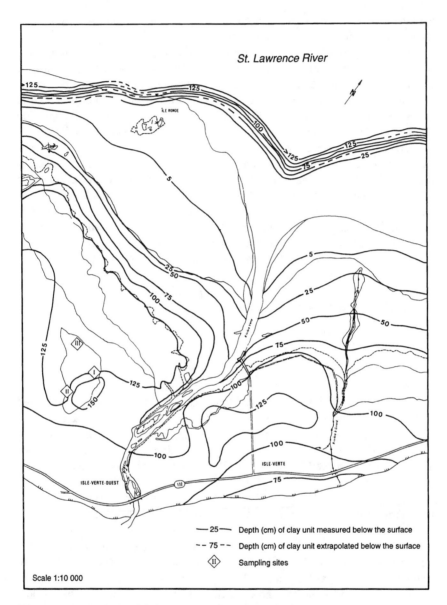

Figure 4 Depth of the clay unit measured from the surface.

marsh seems to have taken place in the same order as present day successions along the shoreline. Plant communities of *Spartina pectinata* and *Calamagrostis canadensis* (zone 2a) were replaced by *Alnus rugosa* (zone 2b) communities between approximately 1000 and 500 BP. Well-drained conditions have lately been registered at the three core locations (zone 2c: rise in the volume of hyphae as well as fern spores and pollen and macrofossils of shrubs and trees) as a probable consequence of the local accumulation of *Alnus* remnants that impeded fluctuations of the water table. Between 400 and 200 BP (subzone 2d), a major change in palynomorphs and plant macrofossils assemblages through the three peat sequences indicates deterioration of drainage conditions. These changes caused the opening of the forest cover registered in subzone 2c (composed mainly of *Picea*, *Larix,* and *Betula*), which led to the formation of a *Typha latifolia* swamp. There are no sufficient results yet that can explain this transformation. The local change in surface drainage conditions could be associated with different reasons (local, regional, or extra-regional) that may have influenced the delicate equilibrium of the peatland ecosystem (Miller and Futyma, 1987). Reasons such as an increase in moisture or precipitation in the region or the persistence of a snow cover on the surface of the bog can be considered. Even though not yet proved by stratigraphic record in the region because of the paucity of researches along the Saint Lawrence River, a slight rise of the sea level associated with climatic deterioration could also be possible. A rise of sea level may have caused, by hydrostratigraphic pressure, a rise of the water table without direct submergence of the peat bog site. The interpretation of a climatic deterioration can also be supported by indirect evidence associated to this period, such as the abandonment of coastal sites by native people and the increase of ice drift processes in modern, lowlying marshes along the estuary (Chapdelaine et al., 1991; Garneau, 1993).

CONCLUSION

Local and detailed study of the Isle-Verte peat bog and the integration of different analytical techniques confirm the capability of detecting century-scale environmental changes associated with allochthonous factors that can modify the internal dynamism of peat ecosystems. The major environmental change detected in the peat bog between 400 and 200 BP indicates local drainage deterioration. This can be explained by different local or regional factors, but no other detailed studies have been realized yet in the region. Lake deposits in the north central United States and Canada analyzed from Swain (1978), Bernabo (1981), Gajewski (1987) and Campbell (1992) have registered a rise in atmospheric moisture and a lowering of annual temperatures at that period. Therefore, correspondence between topographical sequences and time sequences is not necessarily direct, as previously stated in traditional ecology. Major shifts, as registered in vegetation and animal assemblages of subzones 2d of the three cores of the Isle-Verte peat bog, indicate that allochthonous and/or autogenous factors can influence the internal dynamism of an ecosystem. Further studies on these correspondences should be pursued in the future to recognize the different factors that can influence and modify the time/space dynamics of these fragile ecosystems.

REFERENCES

Bakker, M. and Van Smeerdijk, D. G., A palaeoecological study of a Late Holocene section from "Het Ilperveld," western Netherlands, *Rev. Palaeobot. Palynol.*, 36, 95, 1982.

Bernabo, J. C., Quantitative estimates of temperature changes over the last 2700 years in Michigan based on pollen data, *Quaternary Res.*, 15, 143, 1981.

Bertrand, P., Le secteur côtier Cacouna-Isle-Verte: géomorphologie et classification des sols, Masters thesis, Department of Geography, Laval University, 1984, 259 pp. (unpublished).

Bertrand, P., Garneau, M., and Jurdant, M., Carte écologique du secteur côtier Cacouna-Isle-Verte et Notice Explicative, Canadian Wildlife Service, Environment Canada, 1983, 86 pp.

Campbell, I. D., Climate, People and Trees: The Little Ice Age in Southern Ontario, Canada, PhD thesis, Department of Botany, University of Toronto, 1992.

Chapdelaine, C., Tremblay, R., Chalifoux, E., Bourget, S., and Burke, A., Le système adaptatif des Iroquoiens dans la région de Québec. Final report 1990–1991, Department of Anthropology, University of Montréal, 1991, 55 pp.

Couillard, L. and Grondin, P., La végétation des milieux humides du Québec. Les Publications du Québec, Gouvernement du Québec, 1986, 400 pp.

Coûteaux, M., Notes sur le prélèvement et la préparation de certains sédiments, *Pollen Spores*, 4 (2), 317, 1962.

Dionne, J. C., Holocene relative sea-level fluctuations in the St. Lawrence estuary, Quebec, Canada, *Quaternary Res.*, 29, 233, 1988.

Faegri, K. and Iversen, J., *Textbook of Pollen Analysis,* 4th ed., John Wiley & Sons, New York, 1989, 328 pp.

Gajewski, K., Climatic impacts on the vegetation of eastern North America during the past 2000 years, *Vegetatio*, 68, 179, 1987.

Garneau, M., Cartographie et phyto-écologie du territoire Cacouna-Isle-Verte, Masters thesis, Department of Geography, Laval University, 1984, 388 pp. (unpublished).

Garneau, M., Reconstitution paléoécologique d'une tourbière littorale de l'estuaire du Saint-Laurent: analyse macrofossile et sporopollinique, *Géographie physique Quaternaire,* 41 (1), 333, 1987.

Garneau, M., Analyses macrofossiles d'un dépôt de tourbe dans la région de Hot Weather Creek, Péninsule de Fosheim, Ile d'Ellesmere, Territoires du Nord-Ouest, *Géographie physique Quaternaire,* 46 (3), 285, 1992.

Garneau, M., Reconstitution paléoécologique d'une tourbière en position littorale sur la rive sud de l'estuaire du Saint-Laurent, Isle-Verte, Québec, Canada, Ph.D. thesis, University of Amsterdam, Amsterdam, The Netherlands, 1993, 218 pp.

Gauthier, J., Lehoux, D., and Rosa, J., Les marécages intertidaux dans l'estuaire du Saint-Laurent, Canadian Wildlife Service, Environment Canada, 1980, 93 pp.

Kuhry, P., Palaeobotanical-palaeoecological studies of tropical high Andean peat bog sections (Cordillera Oriental, Columbia), *Dissertationes Botanicae,* 116. Edition J. Cramer, Berlin, 1988, 241 pp.

Miller, N. G. and Futyma, R. P., Paleohydrological implications of holocene peatland development in northern Michigan, *Quaternary Res.,* 27, 297, 1987.

Pals, J. P., Van Geel, B., and Delfos, A., Palaeoecological studies in the Klokkeweel bog near Hoogkarspel (Prov. of Noord-Holland), *Rev. Palaeobot. Palynol.,* 30, 371, 1980.

Ran, E. T. H., Dynamics of vegetation and environment during the Middle Pleniglacial in the Dinkle valley (The Netherlands), *Mededelingen Rijks Geologische Dienst,* 44–3, 141, 1990.

Stockmarr, J., Tablets with spores used in absolute pollen analysis, *Pollen Spores,* 13 (4), 615, 1971.

Swain, A. M., Environmental changes during the past 2000 years in North Central Wisconsin: analysis of pollen, charcoal, and seeds from varved lake sediments, *Quaternary Res.,* 10, 55, 1978.

Van Geel, B., Palynology of a section from the raised bog "Wietmarscher Moor" with special reference to fungal remains, *Acta Bot. Neerlandica,* 21 (3), 261, 1972.

Van Geel, B., A palaeoecological study of Holocene peat bog sections in Germany and in The Netherlands, based on the analysis of pollen, spores and macro- and microscopic remains of fungi, algae, cormophytes and animals, *Rev. Palaeobot. Palynol.,* 25, 1, 1978.

Van Geel, B., Bohncke, S. J. P., and Dee, H., A palaeoecological study of an upper Late Glacial and Holocene sequence from "de Borchert," the Netherlands, *Rev. Palaeobot. Palynol.,* 31, 367, 1981.

Van Geel, B., Coope, G. R., and Van Der Hammen, T., Palaeoecology and stratigraphy of the Late Glacial type section at Usselo (the Netherlands), *Rev. Palaeobot. Palynol.,* 60, 25, 1989.

Van Geel, B., De Lange, L., and Wiegers, J., Reconstruction and interpretation of the local vegetation succession of a Late Glacial deposit from Usselo (the Netherlands), based on the analysis of micro- and macrofossils, *Acta Bot. Neerlandica,* 33 (4), 535, 1984.

Van Geel, B., Hallewas, D. P., and Pals, J. P., A Late Holocene deposit under the Westfriese Zeedijk near Enkhuizen (Prov. of Noord-Holland, the Netherlands): palaeoecological and archaeological aspects, *Rev. Palaeobot. Palynol.,* 38, 269, 1983.

Van Geel, B., Klink, A. G., Pals, J. P., and Wiegers, J., An upper Emsian lake deposit from Twente, Eastern Netherlands, *Rev. Palaeobot. Palynol.,* 46, 31, 1986.

CHAPTER 9

Changes in Species Diversity in Peatlands Drained for Forestry

Harri Vasander, Raija Laiho, and Jukka Laine

CONTENTS

INTRODUCTION

In Finland there are about 10 million ha of peatlands, which is some 30 percent of the area of the country. The two main peatland complex types encountered in Finland are aapa mires and ombrotrophic (raised) bogs. The bog area can be divided into concentric and eccentric bogs, while the aapa mire area is divided into the southern, main, and northern complex types (Ruuhijärvi, 1983; Eurola et al., 1984). Cajander (1913) in his classic work distinguished more than 80 mire site types which is about the number of botanically distinguished mire site types even nowadays in Finland (Eurola et al., 1984). In practical peatland forestry more than 30 undrained mire site types are distinguished (Laine and Vasander, 1990).

1-56670-177-5/97/$0.00+$.50

Approximately 14 million ha of peatlands have been drained for forestry worldwide. Finland, with a little more than 5 million drained ha (= more than 50 percent of the peatlands in the country), accounts for about one-third of this area (Paavilainen and Päivänen, 1995). Drainage increases wood production (e.g., Keltikangas et al., 1986 and the references therein), which is the economic reason for forest drainage in Finland (e.g., Heikurainen, 1980).

Early research on drained peatlands focused on the succession of mire plant communities to forest vegetation following peatland drainage (Keltikangas, 1945; Nieppola, 1986). General trends in secondary postdrainage succession were clarified in the 1950s by Sarasto (1957, 1961), who distinguished three separate phases in postdrainage succession: the attack, fighting, and compensation phases. During the 1980s, attempts were made to compare ground vegetation biomass and productivity on undrained and drained sites in Fennoscandia. It was found that after drainage the biomass and production of tree and field layers usually increase while those of the moss layer decrease. These changes increase the biomass and production after drainage, except on unfertilized raised bogs, where the small increase in tree and field layers cannot compensate for the decrease in the moss layer (Vasander, 1982; Elina et al., 1984; Reinikainen et al., 1984).

Changes in species diversity after drainage have seldom been studied (cf. Vasander, 1984, 1987), although Sarasto (1951) presented data on the plant species richness of drained sites. He found the number of plant species was higher on nutrient-rich sites than on nutrient-poor sites, both in the undrained and drained state. After drainage, the number of plant species was highest in the attack phase.

The aim of this paper is to analyze local and regional effects of forest drainage on the species diversity of drained peatlands. The review is based on published and partly reanalyzed data.

MATERIAL AND METHODS

Material 1

Data was gathered from southern Finland and the Karelian region of Russia. Biomass samples were collected from 80 sites, representing 51 mire site types from ombrotrophy to eutrophy, and from densely wooded spruce mires to treeless fens. Of these, 59 sites (39 site types) were on undrained peatlands; the remainder had been drained $\leq$ 60 years. The data was used to compare the species diversity–biomass relationships in the field and moss layers between undrained and drained peatlands (all drained sites pooled). For more details, see Vasander (1987).

Vegetation coverage was measured and biomass harvested from 10 to 30 systematically laid-out quadrats of 1 m^2. Biomass samples were taken in the middle of these quadrats from an area of 0.5, 0.25, or 0.0625 m^2. If the sites consisted of different subcommunities (e.g., hummocks and hollows), these subcommunities were sampled separately. When calculating total values for these sites, the subcommunity values were weighted with their relative proportion of occurrence (Heikurainen, 1951).

Material 2

Data were collected at 60 sites drained about 25 years ago. According to the Finnish classification system (Laine and Vasander, 1990; Laiho and Laine, 1994; Paavilainen and Päivänen, 1995) the sites were herb-rich sedge birch-pine fen (RhSR), tall-sedge pine fen (VSR), cottongrass-sedge pine fen (TSR), low-sedge *Sphagnum papillosum* pine fen (LkR), cottongrass pine bog (TR), and ridge-hollow pine bog (KeR). These site types form a continuum from mesotrophic (RhSR) to oligotrophic and, ultimately, ombrotrophic (TR, KeR). The major factors determining the phase of the secondary plant succession at a particular site are nutrient level and hydrological conditions, as controlled by ditch depth and ditch spacing. For more details, see Laine and Vanha-Majamaa (1992).

Plant cover was estimated using 1 m^2 quadrats, with a sampling design of 20 per site. In addition, 26 environmental variables were measured at each site. Of these, only four (volume of the tree stand, median water level, total N content and C to N ratio of the 0 to 20 cm surface peat) were used in analyses of plant community data to avoid multicollinearity problems (Laine and Vanha-Majamaa, 1992).

Material 3

The data consist of mires, both undrained and drained, which belong to the same series of site types as the previous data: RhSR, VSR, TSR, LkR and TR. Altogether, measurements were made at 82 sites on 43 mires. The sites were of different ages after drainage, with the oldest having been drained for 55 years. Some of the sites had been fertilized. As the drainage age of the sites varied, the phase of secondary succession at a site within a particular site type was largely related to the time elapsed since drainage. For more details, see Laiho and Laine (1994) and Laine et al. (1995).

The coverage of each plant species present was determined using coverage classes, as the vegetation was mapped in the area of the whole 10 × 30 m sample plot. Altogether, 31 environmental variables were measured in the field or in the laboratory from peat samples taken from the site. Of these, five (drainage age, tree stand volume, total C contents, total N contents, and pH of the 0 to 20 cm surface peat) were used in the analysis.

Numerical Analyses

The diversity in Material 1 was measured using the Shannon index ($H' = -\Sigma p_i \ln p_i$, where p_i is the proportion of the total biomass contributed by each species) (Shannon and Weaver, 1948). In Materials 2 and 3, the inverse of the Simpson diversity index ($N_2 = 1/\Sigma p_i^2$, where p_i is the proportion of the total coverage contributed by each species) (Hill, 1973) was used. Both of these diversity indices measure the alpha-diversity (within-community diversity) of the sites. We used the beta- and gamma-diversity concepts (Whittaker, 1972; Figure 1) for the effect of species change on diversity. Beta-diversity in general is defined

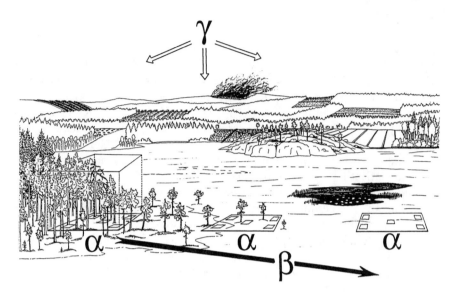

Figure 1   Diversity concepts. Alpha-diversity describes within-community (habitat) diversity. Beta-diversity describes the change of diversity between communities. Gamma-diversity refers to the regional or landscape level. Drawing by Hannu Nousiainen.

as the change in species composition along the environmental gradients within a habitat hyperspace (Whittaker, 1972). We determined beta-diversity as the difference (or separation in ordination diagrams) between the plant communities within a set of communities sampled from various peatlands within a particular geographical area. Gamma-diversity is an extension of the alpha-diversity to a regional (or landscape) level (Figure 1). Determined in this way, there is no need to analyze the distances between individual sites and the types of possible barrier (e.g., rivers, highways, etc.) existing between sites.

Canonical correspondence analysis (CCA) (ter Braak, 1986, 1987) was used in analyzing materials 2 and 3, since it allows direct comparison of the environmental information with the plant community data (Jongman et al., 1987) and provides a calculation of the Simpson index (ter Braak, 1990).

RESULTS

In Material 1, the alpha-diversity (H′) in the field layer was lower on drained sites, but in the moss layer and combined understory, no clear difference could be found between undrained and drained sites. A weak negative relationship was found between biomass and diversity in the moss and field layers (Figure 2; Table 1).

Secondary succession after drainage was clearly discernible in Material 2. The first axis in the CCA ordination is a "forest vegetation succession" gradient, which strongly correlates with such variables as median water level, nutrient level, and stand volume (Figure 3; Table 2). The sites where secondary succession was only

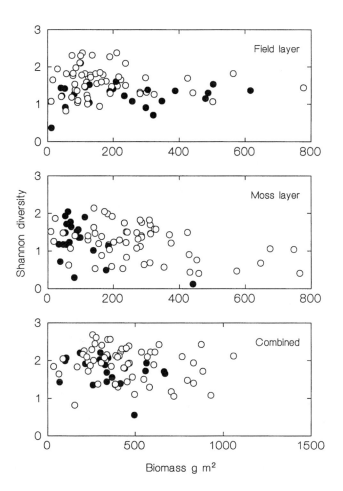

Figure 2   The relationship between biomass and Shannon index (H′) in the field layer, moss layer, and combined understory in Material 1. Dots represent drained mires and open circles undrained ones. Correlation coefficients in different vegetation layers for undrained, drained, and combined sites are given in Table 1.

Table 1   **Spearman Rank Correlation Values Between Biomass and Shannon Index in the Field, Moss, and Combined Understory Vegetation in Material 1**

Layer	r_u	r_d	r_c
Field	−0.015	0.073	−0.076
Moss	−0.301	−0.256	−0.296
Combined	−0.141	−0.317	−0.137

Note: The values are given separately for undrained ($n = 59$, r_u), drained ($n = 21$, r_d), and combined ($n = 80$, r_c) sites.

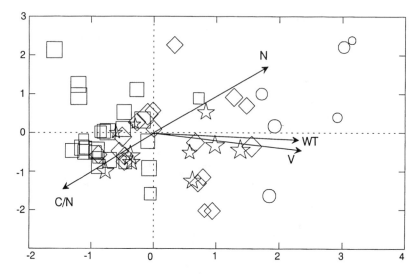

Figure 3   The CCA ordination of Material 2. Symbols: RhSR = circles, VSR = diamonds,
TSR = stars, LkR, TR, KeR = squares. The size of the symbol is the square root
of the species diversity value (the inverse of the Simpson diversity index); the
larger the symbol, the higher the diversity. Vectors: V = the volume of the tree
stand, WT = median water level, N = total nitrogen content of the surface peat
(0 to 20 cm), C/N = carbon–nitrogen ratio of the surface peat. The eigenvalue
for Axis 1 is 0.273 and for Axis 2 0.114.

in the initial stages (LkR, TR, KeR) had higher diversity values than the sites
where secondary succession had proceeded further (VSR, RhSR, Figure 3).

In Material 3, the first ordination axis again formed a clear "forest vegetation
succession" gradient, with undrained sites on the left and drained sites on the
right. In this material, this axis was correlated with drainage age (Age vector in
Figure 4), tree stand volume (V in Figure 4), and total C content in peat, which
is used as a measure of the degree of decomposition. The second axis is a nutrient
level–pH gradient (Table 2). The diversity was in general higher on the more
nutrient-rich sites along the drainage age gradient (RhSR, VSR) (Figure 4).

The change in species diversity along the drainage age gradient was rather
small within all site types. No difference could be found in ANOVA when the
diversities of different drainage age groups were compared. The highest diversity
values were found on fairly young drained sites. On the more nutrient-rich sites,
the total change in the vegetation composition was bigger (grey arrows added in
Figure 4 show the average extent of the vegetation succession for each site type).
The mesotrophic species of undrained RhSR sites (e.g., *Sphagnum subsecundum*,
Potentilla palustris) had been replaced by mesic forest species (e.g., *Trientalis
europaea, Dryopteris carthusiana, Rubus idaeus*). Similarly, the oligotrophic
species (e.g., *Sphagnum papillosum, Carex lasiocarpa, C. rostrata*) had been
replaced by forest species such as *Pleurozium schreberi, Melampyrum pratense,
Vaccinium myrtillus,* and *V. vitis-idaea*. Detailed vegetation changes are given in
Laine et al. (1995).

Table 2   Intraset Correlations in CCA for the First Two Axes in Materials 2 and 3

Variable	Axis 1	Axis 2
Material 2		
Median water level	0.846	−0.069
Tree stand volume	0.861	−0.135
C/N of surface peat	−0.513	−0.457
N cont. of surface peat	0.682	0.534
Material 3		
Drainage age	0.680	−0.244
Tree stand volume	0.860	0.044
C cont. of surface peat	0.536	0.179
N cont. of surface peat	0.469	0.496
pH of surface peat	0.043	−0.314

The beta-diversity describing the differences between plant communities (site types) decreased along the drainage age gradient, as the hydrological conditions became more uniform and forest species were able to invade the sites. This is shown in Figure 4 as a smaller separation between the older drained sites representing different site types (arrow heads nearer each other than arrow tails).

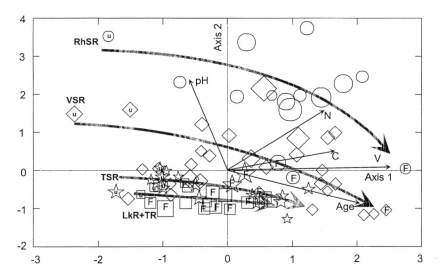

Figure 4   The CCA ordination of Material 3. Symbols: RhSR = circles, VSR = diamonds, TSR = stars, LkR, TR = squares. The size of the symbol is the square root of the species diversity value (the inverse of the Simpson diversity index); the larger the symbol, the higher the diversity. The sites marked with "u" are undrained and those marked with "F" have been fertilized. Vectors: V = the total volume of the tree stand, Age = years since drainage, C = total carbon content of the surface peat (0 to 20 cm), N = total nitrogen content of the surface peat, pH = the water pH of the surface peat. General trends of the secondary succession of individual site types are marked with grey arrows. The eigenvalue for Axis 1 is 0.365 and for Axis 2 0.257.

DISCUSSION

In Fennoscandia, where the bedrock is mainly granite lacking carbonates, mire site types with high nutrient levels are usually also wet. Nutrients are transported to the site in moving water. Wet conditions also allow nitrogen-fixing microbes to increase the amount of nitrogen in the peat. Considerable microhabitat variation created by water flows increases plant species diversity on more nutrient-rich sites (e.g., Karlin and Bliss, 1984).

Drainage for forestry and the consequent water level drawdown cause many changes in site properties, such as the oxidation of the inorganic complexes (e.g., Fe, S, and Al) and organic compounds (DeVries and Breeuwsma, 1987), increasing the acidity of the aerated surface peat and diminishing the original difference between nutrient-rich (originally less acid) and nutrient-poor (originally more acid) sites (Lukkala, 1929). The nutrient input via water inflow is cut off by ditches, which diminishes the nutrient availability for the plants, especially on minerotrophic mires. On the other hand, the increased aeration of the surface peat allows for an accelerated rate of nutrient mineralization (e.g., Lieffers, 1988; Freeman et al., 1993). The subsidence of the mire surface also brings new nutrient reserves from deeper peat layers into the plant root zone (Laiho and Laine, 1994). Even if the availability of nutrients remains on the same level, various species groups react differently to the changed hydrochemistry. Tree stands particularly benefit from this change because of an increased oxygen supply for roots in the aerated surface peat (Boggie, 1977). The increased tree layer biomass is an important ecological factor for the other vegetation layers through shading and as a long-term nutrient sink (Reinikainen et al., 1984; Laine and Vanha-Majamaa, 1992).

These changes initiate a secondary succession in the vegetation. Succession rates are faster in wetter and more nutrient-rich sites. Species adapted to wet (flark level) conditions are the first to disappear after drainage. This was noted also in Material 3, where mesotrophic flark level species decreased in abundance quite quickly and eventually disappeared from the sites after drainage. Hummock level species, adapted to drier conditions, and shade-tolerant species may even benefit from the changed environment (Eurola et al., 1984; Laine et al., 1995). Spore-germinating mosses may quickly invade the "empty" space left behind by the vanishing mire bryophytes. Invasion of vascular plants from the surrounding forests is a slower process. This is reflected in the difference between the development of field and moss layer diversities after drainage in Material 1.

In the long run, the change brought about by drainage does not always significantly affect the alpha-diversity. After the initial increase (the attack phase in Sarasto, 1951), which may be attributed to the simultaneous occurrence of original mire species, colonizing forest species and fugitive species, the species diversity declines again when the forest species take dominance. Hotanen and Vasander (1992) reanalyzed Sarasto's (1961) data with multivariate methods and found that the postdrainage succession could be seen as a gradual shift of the site type successional phases along the DCA axes. The total variation and diversity of the vegetation of different site types decreased with the postdrainage succession as the hydrology became more uniform.

If all plant communities of a peatland are drained, the beta-diversity declines, as can be deduced from the results of Material 3. This reflects a general trend in disturbed ecosystems: biomass, production, and dominance increase while diversity decreases (e.g., Bakelaar and Odum, 1978). However, if only a part of a peatland is drained, the invading forest vegetation may even increase the beta-diversity within the peatland.

The effect of long-term water-level drawdown after drainage on regional gamma-diversity may be expected to be negative. There are no species known to be dependent on drained mires. The species pool on drained mires consists of a mixture of mire and forest species, the composition of which depends on the successional stage of the site. Mire species are gradually substituted by forest species, which already dominate in the surrounding forests.

Analogous results have also been obtained from studies of birds (Väisänen and Rauhala, 1983), spiders and soil fauna (Koponen, 1979, 1985; Markkula, 1986) on drained peatlands in Finland.

REFERENCES

Bakelaar, R. G. and Odum, E. P., Community and population level responses to fertilization in an old-field ecosystem, *Ecology,* 59, 660, 1978.

Boggie, R., Water-table depth and oxygen content of deep peat in relation to root growth of Pinus contorta, *Plant Soil,* 48, 447, 1977.

Cajander, A. K., Studien über die Moore Finnlands, *Acta For. Fenn.,* 2, 1, 1913.

De Vries, W. and Breeuwsma, A., The relation between soil acidification and element cycling, *Water Air Soil Pollut.,* 35, 293, 1987.

Elina, G. A., Kuznecov, O. L., and Maksimov, A. I., *Strukturno-funkcional'naja organizacija i dinamika bolotnyh ekosistem Karelii,* Nauka, Leningrad, 1984, 128 pp.

Eurola, S., Hicks, S., and Kaakinen, E., Key to Finnish mire types, in *European Mires,* Moore, P. D., Ed., Academic Press, London, 1984, chap. 2.

Freeman, C., Lock, M. A., and Reynolds, B., Climatic change and the release of immobilized nutrients from Welsh riparian wetland soils, *Ecol. Eng.,* 2, 367, 1993.

Heikurainen, L., Eräs suokasvillisuuden analysoimismenetelmä (Referat: Ein Verfahren zur Analisierung der Moorvegetation), *Silva Fennica,* 70, 1, 1951.

Heikurainen, L., Input and output in Finnish forest drainage activity, in Proc. 6th Int. Peat Congress, Duluth, MN, 1980, 398.

Hill, M. O., Diversity and evenness: a unifying notation and its consequences, *Ecology,* 54, 427, 1973.

Hotanen, J.-P. and Vasander, H., Post-drainage development of vegetation in southern Finnish peatlands studied by numerical analysis (in Finnish with English summary), *Suo,* 43, 1, 1992.

Jongman, R., ter Braak, C., and van Tongeren, O., *Data Analysis in Community and Landscape Ecology,* Pudoc, Wageningen, 1987, 299 pp.

Karlin, E. F. and Bliss, L. C., Variation in substrate chemistry along microtopographical and water-chemistry gradients in peatlands, *Can. J. Bot.,* 62, 142, 1984.

Keltikangas, M., Laine, J., Puttonen, P., and Seppälä, K., Vuosina 1930–1978 metsäojitetut suot: ojitusalueiden inventoinnin tuloksia (Summary: Peatlands drained for forestry during 1930–1978: results from field surveys of drained areas), Acta For. Fenn., 193, 1, 1986.

Keltikangas, V., Ojitettujen soiden viljavuus eli puuntuottokyky metsätyyppiteorian valossa (Summary: The fertility of drained bogs as shown by their tree producing capacity, considered in relation to forest types), Acta For. Fenn., 53, 1, 1945.

Koponen, S., Differences of Spider Fauna in Natural and Man-Made Habitats in a Raised Bog, The National Swedish Environment Protection Board Report PM 1151, 104, 1979.

Koponen, S., Soiden hämähäkkilajiston muutoksista (Abstract: On changes in the spider fauna of bogs), Memo. Soc. Fauna Flora Fenn., 61, 19, 1985.

Laiho, R. and Laine, J., Nitrogen and phosphorus stores in peatlands drained for forestry in Finland, Scand. J. For. Res., 9, 251, 1994.

Laine, J. and Vanha-Majamaa, I., Vegetation ecology along a trophic gradient on drained pine mires in southern Finland, Annales Bot. Fenn., 29, 213, 1992.

Laine, J. and Vasander, H., Suotyypit, Kirjayhtymä, Helsinki, 1990, 80 pp.

Laine, J., Vasander, H., and Laiho, R., Long-term effects of water level drawdown on the vegetation of drained pine mires in southern Finland, J. Appl. Ecol., 32, 785, 1995.

Lieffers, V. J., Sphagnum and cellulose decomposition in drained and natural areas of an Alberta peatland, Can. J. Soil Sci., 68, 755, 1988.

Lukkala, O. J., Ueber den Aziditätsgrad der Moore und die Wirkung der Entwässerung auf denselben, Commun. Inst. Quaestionum For. Finl., 13, 1, 1929.

Markkula, I., Comparison of the communities of the oribatids (Acari: Cryptostigmata) of virgin and forest-ameliorated pine bogs, Ann. Zool. Fenn., 23, 33, 1986.

Nieppola, J., Cajanderin metsätyyppiteoria. Kirjallisuuteen perustuva tarkastelu (Summary: Cajander's theory of forest site types. Literature review), Folia For., 654, 1, 1986.

Paavilainen, E. and Päivänen, J., Peatland Forestry. Ecology and Principles, Springer Verlag, Berlin-Heidelberg, 1995, 248.

Reinikainen, A., Vasander, H., and Lindholm, T., Plant biomass and primary production of southern boreal mire-ecosystems in Finland, in Proc. 7th Int. Peat Congress 4. Dublin, Ireland, 1984, 1.

Ruuhijärvi, R., The Finnish mire types and their distribution, in Ecosystems of the World, Vol. 4B, Mires: Swamp, Bog, Fen and Moor. Regional Studies, Gore, A. J. P., Ed., Elsevier, Amsterdam, 1983, chap. 3.

Sarasto, J., Metsäojituksen vaikutuksesta eräiden rämeiden pintakasvillisuuteen (Summary: On the influence of forest ditching on the surficial vegetation of some hummocky peat moors), Suo, 2, 57, 1951.

Sarasto, J., Metsän kasvattamiseksi ojitettujen soiden aluskasvillisuuden rakenteesta ja kehityksestä Suomen eteläpuoliskossa (Referat: Über Struktur und Entwicklung der Bodenvegetation auf für Walderziehung entwässerten Mooren in der südlichen Hälfte Finnlands), Acta For. Fenn., 65, 1, 1957.

Sarasto, J., Über die Klassifizierung der für Walderziehung entwässerten Moore, Acta For. Fenn., 74, 1, 1961.

Shannon, C. E. and Weaver, W., The Mathematical Theory of Communication, University of Illinois Press, Urbana, IL, 1948, 117 pp.

ter Braak, C. J. F., Canonical correspondence analysis: a new eigenvector technique for multivariate direct gradient analysis, Ecology, 67, 1167, 1986 pp.

ter Braak, C. J. F., CANOCO — A FORTRAN Program for Canonical Community Ordination by [Partial] [Detrended] [Canonical] Correspondence Analysis, Principal Component Analysis and Redundancy Analysis (Version 2.0), TNO Institute of Applied Computer Science, Wageningen, 1987, 95.

ter Braak, C. J. F., Update Notes: CANOCO Version 3.10, Agricultural Mathematics Group, Wageningen, 1990, 35 pp.

Väisänen, R. A. and Rauhala, P., Succession of landbird communities on large areas of peatland drained for forestry, *Ann. Zool. Fenn.,* 20, 115, 1983.

Vasander, H., Plant biomass and production in virgin, drained and fertilized sites in a raised bog in southern Finland, *Ann. Bot. Fenn.,* 19, 103, 1982.

Vasander, H., Effect of forest amelioration on diversity in an ombrotrophic bog, *Ann. Bot. Fenn.,* 21, 7, 1984.

Vasander, H., Diversity of understorey biomass in virgin and in drained and fertilized southern boreal mires in eastern Fennoskandia, *Ann. Bot. Fenn.,* 24, 137, 1987.

Whittaker, R. H., Evolution and measurement of species diversity, *Taxon,* 21, 213, 1972.

CHAPTER 10

Undergrowth as a Regeneration Potential on Finnish Peatlands

Olavi Laiho, Erkki Lähde, Yrjö Norokorpi, and Timo Saksa

CONTENTS

INTRODUCTION

Norway spruce (*Picea abies* [L.] Karsten), Scots pine (*Pinus sylvestris* L.), downy birch (*Betula pubescens* Ehrh.), and silver birch (*B. pendula* Roth.) are the main tree species in Finland. Next in frequency are gray alder (*Alnus incana* [L.] Moench) and aspen (*Populus tremula* L.). All these species are native to Finland and form undergrowth. Spruce, as a climax species, forms vigorous undergrowth on peatland and upland sites of medium and high fertility. Pine is typically an undergrowth species of poorer sites. Birch undergrowth occurs mainly on spruce sites and prefers a sparse overstory (Laiho, 1992). Downy birch is particularly common on peatlands.

Since 1921, Finland's forests have been inventoried eight times. Throughout the period of these national forest inventories (NFI), undergrowth has constituted a proportion of 2 to 3 percent of the area of forest land (Ilvessalo, 1942, 1956; Kuusela and Salminen, 1983; Salminen, 1993). This low proportion could be only of minor importance as regeneration potential. However, the amount of undergrowth depends on how it is defined.

1-56670-177-5/97/$0.00+$.50

In the inventories, emphasis has been on the visual appraisal of well-spaced undergrowth capable of further development. This useful undergrowth must have met numerous requirements, e.g., sufficient density, commercially valuable tree species, high probability of surviving a liberation cutting, height less than half of the dominant trees, and two-storied stand structure (L. Ilvessalo, 1929; Y. Ilvessalo, 1951; Kuusela and Salminen, 1969; Laiho, 1992; Salminen, 1993). The application of these requirements must have omitted a high proportion of undergrowth from the inventory data.

The abundance of undergrowth in Finnish forests is supported by a study concerning so-called natural normal forests (Ilvessalo, 1920). Although its purpose was to deal with pure even-aged stands, 39 percent of the sample plots had to be accepted with distinct undergrowth. Undergrowth has also caused confusion in the development of the NFI age class distribution. When the first NFI was conducted in 1921 to 1924, 10-year-old forest in South Finland amounted to 7.5 percent, but 40 years later, 50-year-old forest covered 27.9 percent of the forest land (Kuusela, 1972). This increase in area has its explanation in undergrowth. A high proportion of this age class had been hiding as undergrowth and was liberated in felling operations during the intervening years, oftentimes at the age of 20 to 40 years. For such a great increase in acreage, the coverage of undergrowth must have been considerable.

Undergrowth has not been utilized actively. Its value has generally been rated low and its future outcome uncertain, because of the risk of damage when the dominant story is removed. Studies provide unambiguous support for the high value of undergrowth as the starting point for a new tree crop (Seppälä and Keltikangas, 1978; Isomäki, 1979). Consequently, more detailed knowledge on undergrowth is urgently needed.

The undergrowth present in peatland stands is especially abundant (Heikurainen, 1971; Hånell, 1984; Kaunisto and Päivänen, 1985). Often such undergrowth is very dense. Instead of being a separate story, most peatland stands are uneven-aged and structurally all-sized (Laiho et al., 1994b; Norokorpi et al., 1994). There is an abundance of small trees, and the number of stems diminishes fairly regularly as tree diameter increases (Gustavsen and Päivänen, 1986). Drainage seems to further strengthen this kind of structure by increasing the number of seedlings (Seppälä and Keltikangas, 1978; Keltikangas et al., 1986; Hökkä and Laine, 1988).

This study deals with the occurrence and characteristics of undergrowth on different peatland sites and, in comparison, on upland sites as well. It is based on analyses of the stem diameter distribution of sample plots provided by national forest inventories.

MATERIALS AND METHODS

The study is mainly based on the third (1951 to 1953) NFI. This inventory was the last one to be completed before large-scale undergrowth clearing was commenced in stands treated with low thinning. Thus, the stand structure at that time resembled naturally developed forests. Another positive feature was the use of fixed area samples plots as opposed to point-sampled plots of subsequent inventories (Kuusela and Salminen, 1969). Fixed area plots enable the plot-specific assessment employed in this study (Laiho et al., 1994a). The point sampled plots of the seventh NFI (1977 to 1984) were used, therefore, only on a stratum basis.

In the third NFI, sample plots were located at 1-km intervals along the survey line (Ilvessalo, 1951). Each sample plot represented a uniform stand; i.e., it was located entirely within one stand. Thus, the structure on any plot could not be an admixture of different stands. Peatland sites were divided into spruce mires, pine mires, and drained peatlands. Upland sites were included as a comparison. This study was restricted to advanced stands (young and advanced thinning stands, mature stands, and stands whose development class had not been determined) in South Finland with a growing stock of at least 40 m³ ha⁻¹. This volume limitation reduced the number of sample plots by 60 percent in the third NFI, and by 40 percent in the seventh NFI. The final material included 911 fixed area sample plots and 4501 point-sampled plots, respectively (Figure 1).

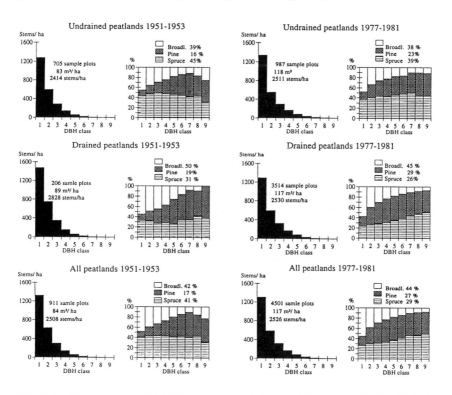

Figure 1 Occurrence of advanced stands on undrained and drained peatlands with their average volume, stand density, and stem distribution by tree species in southern Finland.

In the third NFI, circular plots of 0.1 ha were used for trees with a DBH of at least 10 cm. Smaller trees were tallied in a concentric circle 0.01 ha in size. The smallest trees tallied were 2 cm in DBH. Tallying was done by tree species but without classifying trees into stories. For the purpose of this study, DBH classes, originally 2 cm wide, were combined into nine 4-cm classes (2 to 6, ... > 34 cm). The stem distribution thus formed was used in the determination of individual stand structure for each sample plot.

Stand structure was determined in the same way as in previous studies (Lähde et al., 1991, 1992; Laiho et al., 1994b). It was based on the range and shape of the diameter distribution, as obtained from the inventory plots, with emphasis on small trees. The following is a description of the classification.

1. Even-sized stands; trees present in less than four consecutive diameter classes (even-sized in strict sense), or stem tally resembling normal distribution (even-sized in broad sense; i.e., Gibbs, 1978).
2. Storied stands: at least one empty intermediate diameter class within the range of 6 to 18 cm.
3. Regularly all-sized stands: stem distribution resembling an inverted letter J; trees present in at least the four smallest diameter classes, with the mode in the first or second class.
4. Other all-sized stands.

Because there were only a small number of the above-mentioned strictly even-sized stands, they were combined with the broadly even-sized stands. It is noted that the category of broadly even-sized (Gibbs, 1978) could also be regarded as a special case of all-sized, because the trees were present over a range of at least four diameter classes.

Stands with a separate understory formed only part of the group of storied stands. It also included stands where the stories almost or slightly overlapped. They had a twin-peaked stem diameter distribution. Often the understory peak is formed by another tree species than the dominant story. These twin-peaked sample plots classified in this study into the broadly even-sized or regularly all-sized stands. Regularly all-sized stands in their typical form have no distinguishable stories.

The practice is to use height as an undergrowth classification criterion. The lower story (corresponding to the understory) is then delimited either in relation to top height (< 50 percent; Assmann, 1970) or in absolute measure (< 10 m; Kern, 1966). In the inventories used in this study, DBH, not height, was measured from each tree. Therefore, DBH was used instead of height. Owing to the relatively small diameters in peatland stands, only trees with DBH < 6 cm were included into the undergrowth. Because the data used in this study lacks a great proportion of undergrowth (trees < 2 cm in DBH and seedlings), a density of 1000 stems per heactare (2 to 6 cm) was considered as a fully stocked undergrowth. In young seedling stands, a density of 900 to 1400 stems per hectare is regarded as acceptable in Finnish forestry practice.

RESULTS

A typical feature of peatland forest stands was their close spacing (Figure 1). The stand density was 53 percent higher than that of stands on upland sites. However, the stand volume was 27 percent lower. Small trees (DBH 2 to 6 cm; the smallest measured) were very abundant, with a proportion of 53 percent of the stem number (11 percent units more than in stands on upland sites). Small broadleaved trees were particularly numerous. This was characteristic for all the site classes.

Storied stands amounted to 5 percent of the sample plots. Small trees averaged 868 stems/ha. The treeless intermediate class usually was the second (6 to 10 cm). On 30 percent of the sample plots, the understory reached 1000 stems per hectare and on 57 percent attained 500 stems per hectare.

Regularly all-sized stands included 84 percent of the sample plots (Figure 2), 21 percent units more than upland sites. This structure class was characterized by its high stem tally, 2705 stems/ha, a value twice as much as that of storied stands. The total number of small trees was correspondingly 29-fold. These small trees (2 to 6 cm), the lower story, averaged 1453 stems per hectare. Of the sample plots, 58 percent in regularly all-sized stands had at least 1000 stems per hectare of this undergrowth, and 85 percent had at least 500 stems per hectare.

Even-sized stands numbered 8 percent of the sample plots. Eleven percent of these had a twin-peaked stem diameter distribution. Small trees averaged 671 stems per hectare. Structure class other all-sized stands lacked small trees completely.

Considering stem diameter distribution, differences in stand structure between peatland site groups were negligible. Mixed stands prevailed on all peatland sites, with two thirds of the sample plots being composed of both conifers and of broadleaves. In poor productive pine mires, the number of small pine trees was low, but the proportion of pine increased with increasing diameter as the proportion of spruce and broadleaves decreased. Small spruce trees were most abundant on productive spruce mires, and broadleaved trees more common on poor productive spruce mires and productive pine mires. On drained peatlands, broadleaved trees were the most numerous of all the site classes. Eighty-four percent of broadleaved trees were downy birch.

In the 1951 to 1953 NFI, 20 percent of the advanced stands on peatland were drained for forestry (Figure 1). Thirty years later the ratio was reversed, with 80 percent drained. On both occasions and regardless of drainage, stem diameter distributions were almost identical. Changes in tree species composition and in stand volume occurred, however.

DISCUSSION

In advanced stands, small (2 to 6 cm) trees, and even slightly bigger ones, belong to the undergrowth. Of course, undergrowth also includes trees with DBH

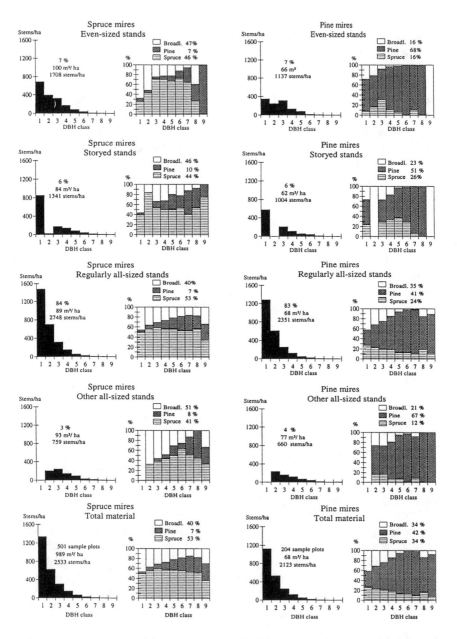

Figure 2   Occurrence (%) of stand structures with their average volume, stand density, and stem distribution by tree species in advanced stands on undrained peatland in southern Finland, 1951–1953.

< 2 cm and seedlings that have not yet reached breast height. Unfortunately, they were determined only as crown coverage in the third NFI (1951 to 1953). However, the number of trees with DBH 0 to 2 cm can be assumed to have been similar to that of the fourth NFI 10 years later, i.e., about 500 stems per hectare (Ilvessalo, 1962). The amount of seedlings under breast height varies from stand to stand, but is on average high (Sarvas, 1944; Räsänen et al., 1985; Hagner, 1992; Lähde, 1992). Therefore, even 500 stems/ha of small trees (2 to 6 cm) may be considered as a sufficiently stocked undergrowth, if one takes into account the smaller sizes of trees (0 to 2 cm) and seedlings that were probably present. Of course, in a more detailed evaluation, spatial distribution of trees should be known. It was not determined on these sample plots.

According to the results of the third NFI conducted in 1951 to 1953, useful pine understory was recorded on 0.4 percent and spruce understory on 1.2 percent of the area of productive peatland forests in South Finland (Ilvessalo, 1956). At that time, understory consisting of broadleaves was deemed useless. The determination of the useful understory had been assessed visually per stand. In this study the stem distribution of sample plots located in the same stands was examined. The proportion of sample plots with a separate understory (storied stands) amounted to 5 percent. Were the criteria of the visual appraisal applied to these sample plots, the said proportion would have been reduced to < 1 percent.

In reality, however, most of the undergrowth was located outside storied stands. Regularly all-sized stands represented 84 percent of the study material. All of these stands included undergrowth, usually in abundance. When the understory in even-sized stands is added to this, the proportion of stands containing undergrowth rises over 90 percent.

Fully stocked undergrowth with at least 1000 stems per hectare of small trees was present on 51 percent of the sample plots with sufficient (at least 500 stems/ha) undergrowth on 75 percent. On upland sites, these proportions were 25 and 47 percent, respectively. The proportions of useful undergrowth, 1.6 percent for peatland and 3.2 percent for upland sites, as indicated in the inventory results (Ilvessalo, 1956), are clearly underestimates. The main reason for this underestimate is the acceptance of only storied stands with clearly separate undergrowth in the visual assessment of understory (Ilvessalo, 1951). Stands on peatland are generally relatively uneven-aged (Heikurainen, 1971). This was confirmed by the high proportion of regularly all-sized stands in this material.

Growth in South Finland's peatland forests increased by 64 percent between the 1950s and the 1980s (Ilvessalo, 1956; Paavilainen and Tiihonen, 1988; Kuusela and Salminen, 1991). The increase for upland sites was 25 percent. A noticeable proportion of the growth increase in peatland stands is attributable to drainage for forestry (Paavilainen and Tiihonen, 1988). However, this positive reaction to drainage is not visible in this material because drainage has been applied to nutrient-poor sites with low stocking and in some cases to mires void of trees. Drainage may have a negative influence on future regeneration if a layer of raw humus is formed over the peat (Kaunisto and Päivänen, 1985).

It has proven difficult to assess the ability of stunted undergrowth to recover. The development of undergrowth under a dense canopy is slow; a tree 50 years old may not necessarily have reached breast height (Pöntynen, 1929; Sirén, 1951). Even the weakest undergrowth, however, is capable of recovering if liberated properly. Usually recuperation takes a few years (Cajander, 1934). Serious growth stagnation, advanced age, and large size do, however, hinder recuperation (Vaartaja, 1951). Once recuperated, stunted undergrowth will grow at the same rate as free grown seedlings of equivalent size (Näslund, 1944; Sarvas, 1951; Hatcher, 1967; Vuokila, 1970; Indermühle, 1978; Nilson and Haveraaen, 1983; Klensmeden, 1984).

The regeneration potential of peatland forests in the early 1950s was very high. Thirty years later, the situation was essentially the same (Lähde et al., 1992; Laiho et al., 1994b). Results of the eighth NFI (Salminen, 1993) prove that this character of South Finnish forests has prevailed. Actually, the value of the regeneration potential may increase because even sparse undergrowth does not mean significant losses in yield (Gustavsen, 1992). In addition, silver birch and even downy birch are now also commercially acceptable tree species. The full utilization of the undergrowth minimizes the need for artificial regeneration, shortens the rotation, and increases the internal biodiversity of stands.

REFERENCES

Assmann, E., *Principles of Forest Yield Study,* McGraw-Hill, New York, 1970, 502 pp.

Cajander, E. K., Kuusen taimistojen vapauttamisen jalkeisesta pituuskasvusta (Referat: Über den Höhenzuwachs der Fichtenpflanzenbestände nach der Befreiung), *Commun. Inst. For. Fenn.,* 19.5, 59, 1934.

Gibbs, C. B., Uneven-aged silviculture and management? Even aged silviculture and management? Definitions and differences, in Uneven-aged Silviculture and Management in the United States, USDA Forest Service, Timber Management Research, Washington, D.C., General Technical Report, WO-24, 1978, 18 pp.

Gustavsen, H. G. and Päivänen, J., Luonnontilaisten soiden puustot kasvullisella metsämaalla 1950-luvun alussa (Summary: Tree stands of virgin forested mires in the early 1950's in Finland), *Folia For.,* 673, 27, 1986.

Gustavsen, H. G., Vähäpuustoisten männiköiden ja kuusikoiden kehitys (Summary: The development of understocked pine and spruce stands), *Folia For.,* 796, 29, 1992.

Hagner, M., Biologiskt och ekonomiskt resultat i fältförsök med plockhuggning kombinerad med plantering (Summary: Biological and economical results from experiments with selective felling combined with enrichment planting), *Sveriges lantbruksuniversität, Institutionen för skogsskötsel, Arbetsrapporter,* 63, 129, 1992.

Hånell, B., Skogsdikningsboniteten hos Sveriges torvmarker, *Rapporter i skogsekologi och skoglig marklära,* 50, 125, 1984.

Hatcher, R. J., Balsam fir advance growth after cutting in Quebec, *For. Chron.,* 40, 86, 1967.

Heikurainen, L., Virgin peatland forests in Finland, *Acta Agralia Fenn.,* 123, 11, 1971.

Hökkä, H. and Laine, J., Suopuustojen rakenteen kehitys ojituksen jälkeen (Summary: Post-drainage development of structural characters in peatland forest stands), *Silva Fenn.,* 22, 45, 1988.

Ilvessalo, L., Puuluokitus ja harvennusasteikko (Summary: A tree classification and thinning system), *Acta For. Fenn.,* 34.38, 15, 1929.

Ilvessalo, Y., Tutkimuksia metsätyyppien taksatoorisesta merkityksestä nojautuen etupäässä kotimaiseen kasvutaulujen laatimistyöhön (Referat: Untersuchungen über die taxatorische Bedeutung der Waldtypen, hauptsächlich auf den Arbeiten für die Aufstellung der neuen Ertragstafeln Finnlands fussend), *Acta For. Fenn.,* 15, 157, 1920.

Ilvessalo, Y., Suomen metsävarat ja metsien tila, II Valtakunnan metsien arviointi (Summary: The forest resources and the condition of forests of Finland, The second national forest survey), *Commun. Inst. For. Fenn.,* 30, 446, 1942.

Ilvessalo, Y., III valtakunnan metsien arviointi, Suunnitelma ja maastotyön ohjeet (Summary: Third national forest survey in Finland, Plan and instructions for field work), *Commun. Inst. For. Fenn.,* 393, 67, 1951.

Ilvessalo, Y., Suomen metsät vuosista 1921–1924 vuosiin 1951–1953, Kolmeen valtakunnan metsien inventointiin perustuva tutkimus (Summary: The forests of Finland 1921–1924 to 1951–1953, the survey based on three national forest inventories), *Commun. Inst. For. Fenn.,* 47.1, 227, 1956.

Ilvessalo, Y., IV. Valtakunnan metsien inventointi, I. Maan eteläpuoliskon vesistöalueryhmät (Summary: The Fourth national forest inventory, I. Southern water area), *Commun. Inst. For. Fenn.,* 56.1, 112, 1962.

Indermühle, M. P., Struktur-, Alters-, und Zuwachsuntersuchungen in einem Fichten-Plenterwald der subalpinen Stufe, *Beiheft zu den Zeitschriften des schweizerischen Forstvereins,* 60, 98, 1978.

Isomäki, A., Kuusialikasvoksen vaikutus männikön kasvuun, tuotokseen ja tuottoon (Summary: The effect of spruce undergrowth on the increment, yield and returns of a pine stand), *Folia For.,* 392, 13, 1979.

Kaunisto, S. and Päivänen, J., Metsänuudistaminen ja metsittäminen ojitetuilla turvemailla, Kirjallisuuteen perustuva tarkastelu (Summary: Forest regeneration and afforestation on drained peatlands, A literature review), *Folia For.,* 625, 75, 1985.

Keltikangas, M., Laine, J., Puttonen, P., and Seppälä, K., Vuosina 1930–1978 metsäojitetut suot: ojitusalueiden inventoinnin tuloksia (Summary: Peatlands drained for forestry during 1930–1978: results from field surveys of drained areas), *Acta For. Fenn.,* 193, 94, 1986.

Kern, K. G., *Wachstum und Umweltfaktorem im Schlag- und Plenternwald, Vergleichende ertragskundlich-ökologische Untersuchungen in zwei Ta-Fi-(Bu)-Plenternwaldbeständen und zwei benachbarten Fi-Schlagwäldern des Hochschwarzwaldes,* Bayerischer Landwirtschaftsverlag, München-Basel-Wien, 1966, 232 pp.

Klensmeden, V., Stamvis blädning — Nagra studier åförsöksytor I Dalarna, Examensarbete 1984 86, Institutionen för skogsskötsel, Sveriges lantbruksuniversität, Umepåtoån, 1984, 38 pp.

Kuusela, K., Suomen metsävarat ja metsien omistus 1964–70 sekä niiden kehittyminen 1920–1970 (Summary: Forest resources and ownership in Finland 1964–70 and their development in 1920–70), *Commun. Inst. For. Fenn.,* 76.5, 126, 1972.

Kuusela, K. and Salminen, S., The 5th national forest inventory in Finland, General design, instructions for field work and data processing, *Commun. Inst. For. Fenn.,* 69.3, 72, 1969.

Kuusela, K. and Salminen, S., Metsävarat Etelä-Suomen kuuden pohjoisimman piirimetsälautakunnan alueella 1979–1982 sekäa koko Etelä-Suomessa 1977–1982 (Summary: Forest resources in the six northernmost Forestry Board Districts of South Finland, 1979–1982, and in the whole of South Finland, 1977–1982), *Folia For.,* 568, 79, 1983.

Kuusela, K. and Salminen, S., Suomen metsävarat 1977–1984 ja niiden kehittyminen 1952–1980 (Summary: Forest resources of Finland in 1977–1984 and their development in 1952–1980), *Acta For. Fenn.*, 220, 1991, 84.

Lähde, E., Regeneration potential of all-sized spruce-dominated stands, in Silvicultural Alternatives, Proceedings from an internordic workship June 22–25 1992, Hagner, M., Ed., *Swed. Univ. Agric. Sci. Dep. Silviculture Rep.*, 35, 111, 1992.

Lähde, E., Laiho, O., Norokorpi, Y., and Saksa, T., The structure of advanced virgin forests in Finland, *Scand. J. For. Res.*, 6, 527, 1991.

Lähde, E., Laiho, O., Norokorpi, Y., and Saksa, T., Stand structure of thinning and mature conifer-dominated forests in boreal zone, in Silvicultural Alternatives, Proceedings from an internordic workshop June 22–25 1992, Hagner, M., Ed., *Swed. Univ. Agric. Sci. Dep. Silviculture Rep.*, 35, 58, 1992.

Laiho, O., Understoreys in the forests of Finland, in Silvicultural Alternatives, Proceedings from an internordic workshop June 22–25 1992, Hagner, M., Ed., *Swed. Univ. Agric. Sci. Dep. Silviculture Rep.*, 35, 100, 1992.

Laiho, O., Lähde, E., Norokorpi, Y., and Saksa, T., Determining stem distribution by relascope and fixed sample plots in diverse stands. Forests, environment and new technology in northern Europe, Proceedings of the international conference, held in Petrozavdsk, the Republic of Karelia, Russia, September 1993, *Univ. Joensuu Res. Notes*, 17, 70, 1994a.

Laiho, O., Lähde, E., Norokorpi, Y., and Saksa, T., Varttuneiden metsiköiden rakenne 1950–luvun alussa (Summary: Stand structure of advanced forests in early 1950's in Finland), *Finnish For. Res. Inst. Res. Pap.*, 495, 90, 1994b.

Näslund, M., Den gamla norrländska granskogens reaktionsförmåga efter genomhyggning (Referat: Die Reaktionsfähigkeit des alten norrländishen Fichtenwaldes nach Durchhauung), *Meddelanden från statens skogs försöksanstalt*, 33, 194, 1944.

Nilsen, P. and Havaraaen, O., Årringbredder hos gjenstäende traer etter hogst I eldre granskog, *Norsk institutt för skogforskning Rapport*, 9, 13, 1983.

Norokorpi, Y., Lähde, E., Laiho, O., and Saksa, T., Luonnontilaisten metsien rakenne ja monimuotoisuus Suomessa (Summary: Stand structure and diversity of virgin forests in Finland), *Finnish For. Res. Inst. Res. Pap.*, 495, 54, 1994.

Paavilainen, E. and Tiihonen, P., Suomen suometsät vuosina 1951–1984 (Summary: Peatland forests in Finland in 1951–1984), *Folia For.*, 714, 29, 1988.

Pöntynen, V., Tutkimuksia kuusen esiintymisestä alikasvoksena Raja-Karjalan valtiomailla (Referat: Untersuchungen über das Vorkommen der Fichte (Picea excelsa) als Unterwuchs in den finnischen Staatswäldern von Grenz-Karelien, *Acta For. Fenn.*, 35.1, 235, 1929.

Räsänen, P. K., Pohtila, E., Laitinen, E., Peltonen, A., and Rautiainen, O., Metsien uudistaminen kuuden eteläisimmän piirimetsälautakunnan alueella, Vuosien 1978–1979 inventointitulokset (Summary: Forest regeneration in the six southernmost forestry board districts of Finland, Results from the inventories 1978–1979), *Folia For.*, 637, 30, 1985.

Salminen, S., Eteläisimmän Suomen metsävarat 1986–1988 (Summary: Forest resources of Southernmost Finland, 1986–1988), *Folia For.*, 825, 111, 1993.

Sarvas, R., Tukkipuun harsintojen vaikutus Etelä-Suomen yksityismetsiin (Referat: Einwirkung der Sägestammplenterungen auf die Privatwälder Südfinnlands), *Commun. Inst. For. Fenn.*, 33.1, 268, 1944.

Sarvas, R., Tutkimuksia puolukkatyypin kuusikoista (Summary: Investigations into the spruce stands of Vaccinium type), *Commun. Inst. For. Fenn.*, 39.1, 82, 1951.

Seppälä, K. and Keltikangas, M., Alikasvostaimikot Pohjanmaan ojitusaluiden hieskoi-
vikoissa (Summary: Occurrence of understorey seedlings in drained Betula pubescens
stands in Ostrobothnia), *Suo,* 29, 11, 1978.

Sirén, G., Alikasvoskuusten biologiaa (Summary: On the biology of undergrown spruce),
Acta For. Fenn., 58.2, 82, 1951.

Vaartaja, O., Alikasvosasemasta vapautettujen männyntaimistojen toipumisesta ja merkityksestä metsänhoidossa (Summary: On the recovery of released pine advanced
growth and its silvicultural importance), *Acta For. Fenn.,* 59.3, 133, 1951.

Vuokila, Y., Harsintaperiaate kasvatushakkuissa (Summary: Selection from above in intermediate cuttings), *Acta For. Fenn.,* 110, 45, 1970.

CHAPTER **11**

Plant Reintroduction on a Harvested Peat Bog

François Quinty and Line Rochefort

CONTENTS

INTRODUCTION

Peat harvesting for horticultural and other purposes is an important industry affecting peat bogs in southern Canada. The methods being used to extract the peat greatly affect the environment, leaving bare peat surfaces that can scarcely be recolonized by bog plants (Elling and Knighton, 1984). These abandoned fields rarely return through secondary succession to the original moss-dominated eco-system. Birches or ericaceous shrubs have often been the only species found, even 10 years after harvesting stopped, and the moss layer typical of bogs has usually been absent on abandoned surfaces (Spencer et al., 1993). In order to restore those peatlands, efforts must focus on the reintroduction of *Sphagnum* species that are responsible for specific conditions found in bogs, such as acidity and peat accumulation (Clymo, 1984). The general objective of our research program is to find methods to reintroduce peatland vegetation, in order to initiate a process that will eventually lead abandoned fields back to functional peatland ecosystems.

Small-scale greenhouse experiments (Campeau and Rochefort, in press) have shown that the spreading of *Sphagnum* fragments (diaspores) on bare peat sur-faces can lead to the reestablishment of a moss layer within a few months. Thus, collecting living plants on the surface layer of a natural bog (often referred to as the top spit layer), chipping this material, and spreading it on a harvested field seems to be a promising approach to restoring vegetation on abandoned sites. Greenhouse and field conditions, however, are different, and any plant material spread on bare peat surfaces faces adverse environmental conditions. As peatland restoration research is a new field in North America, it is assumed, based on the Dutch literature, that the most important problem that *Sphagna* and other plant diaspores face in colonizing bare peat surfaces is a water deficit (Schouwenaars, 1988a). The drainage necessary for harvesting purposes results in a drop of the water table generally far below 40 cm, which is considered a minimum for *Sphagnum* regeneration (Schouwenaars, 1988b). Moreover, it has been observed that dry conditions at the ground surface can lead to the formation of a crust and thus isolate plant diaspores from soil moisture (Spencer et al., 1993). Thus, rewetting harvested peat bogs by blocking drainage ditches is the first step before reintroducing plant diaspores. But even then, field observations show that dry conditions persist at the soil surface and that other interventions are needed to provide plant diaspores with favorable conditions.

In this study, one objective was to reduce evaporation from bare peat surfaces to encourage *Sphagnum* establishment, and currently the importance of reduced evaporation in the hydrological budget of harvested peatlands is being assessed (Ingram, 1983). As evaporation is closely related to wind and soil surface tem-perature (Enz et al., 1988), we tested two different treatments aiming to reduce this drying effect: windbreaks and ground cover (mulching). Windbreaks reduce evaporation by decreasing air flow (Rosenberg, 1976). Their efficiency is related to the porosity of the barrier in such a way that solid and very loose windbreaks are less effective than intermediate ones (Heisler and DeWalle, 1988). The hor-izontal extent of wind protection is a function of windbreak height, and it is accepted that the area of influence is ten times its height (Dickey, 1988). Mulching reduces evaporation by decreasing wind speed and soil surface temperature (Enz et al., 1988). Though crop residues such as straw are widely used, paper and plastic cover are among other types of ground cover that proved to be efficient in the matter (Rosenberg et al., 1983). Microtopography also helps plant diaspores establishment. Tillage of harvested peat bogs to loosen the soil surface can provide better aeration for vascular plants and more sheltered microsites for *Sphagnum* and moss diaspores (Salonen and Laaksonen, 1994).

Small-scale experiments successfully used material collected by hand in the natural bog (Campeau and Rochefort, in press). But to scale up those experiments, methods to collect large amounts of living plant material in the natural bog and to spread it on the harvested site by mechanical means must be found. Thus, the specific objective of this study is to estimate the potential of the reintroduced plant material to survive, grow, and multiply when evaporation is reduced by windbreaks, mulching, and the creation of a microtopogaphy. A second goal is to test the ability of machinery commonly available to do some of the work involved in the restoration process.

STUDY SITE

The study site is located at Rivière-Ouelle (47°27'N, 69°58'W), 140 km east of Québec City, on the south shore of the St. Lawrence River (Figure 1). Mean annual temperature is 4.2°C (Environnement Canada, 1993). The coldest month is January, with a mean of −11.3°C and the warmest is July with 18.8°C. The annual precipitation is 930 mm, of which 645 mm falls as rain. Mean annual evapotranspiration is estimated to be 525 mm (Wilson, 1971). Nevertheless, there is a water deficit during the summer months (June, July, and August). Westerly winds are dominant over the year, but easterly winds are more frequent from February to April and during snowfalls. The bog covers an area of 15 km^2; about half of it is used for peat harvesting.

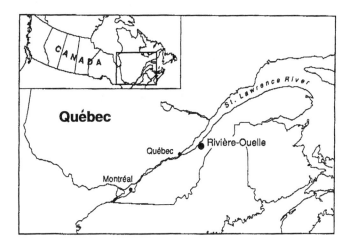

Figure 1 Location of study site.

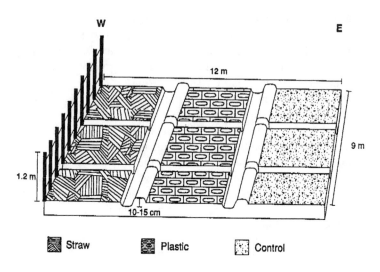

Figure 2 Schematic view of a main plot showing the arrangement of windbreaks, ground
covers, and microtopography. On this plot, the windbreak is effective in summer
while westerly winds are dominant. Depressions and mounds created with the
machinery are parallel to the subplots, while tracks left by the manure spreader
are perpendicular. For the effect of snow trapping, the windbreak would be on
the right side, and for both effects there would be windbreaks at each end.

METHODS

Preparation of the Experimental Site

The experiment was conducted on two bays that had been abandoned for
10 years. Bays are the areas between two drainage ditches; in this case they are
40 m wide. The drainage ditches along the bays were blocked in May 1993.
Microtopography was created using an excavator that pushed the soil across the
bay before spreading any living material. Flat areas were bordered by a depression and a mound, the depressions being about 10 to 15 cm deep (Figure 2).

Collecting and Spreading Living Plant Material

Living plant material was collected in the nearby natural bog with a shredder
used by the producers to prepare the peatland surface before harvesting. The
shredder went over an area of about 500 m² (15 × 35 m), and the top spit layer
was chipped on a thickness of 10 to 15 cm. This provided a material containing
pieces from 1 to a few centimeters long with some bigger chunks. A manure
spreader was used to distribute the living plant material on the experimental site
to a surface eight times larger than the natural site (ratio collected-surface/surface-
covered, 1:8). Because some difficulty occurred when adjusting the manure
spreader, the plant material was unevenly spread on the ground. The spreader
also left tracks, which added some microtopography at the site.

Windbreaks and Ground Covers

The experiment testing the effect of windbreaks and ground covers on the recolonization success of *Sphagnum* species, other mosses, and vascular plants was designed as a split-plot experiment in a completely randomized design. For windbreaks, wooden snow fences were set perpendicular to the prevailing winds (Figure 2). The fences were 1.2 m high, with a porosity of 50 percent. Windbreaks were expected to work as snow traps in winter. Consequently, there were four possibilities regarding the position of the windbreaks: (1) the effect of windbreaks in summer, (2) the action of fences as snow traps in winter, (3) the combination of both, (4) a control with no windbreak. The experimental site was divided into 12 main plot units (15 × 9 m), to which the four levels of the windbreak factors were randomized (three replicates for each windbreak condition).

Each main plot unit was subdivided into three subplots units, to which the three cover types (no cover, plastic cover, and straw) were randomized. The plastic cover consisted of plastic snow fences (porosity 66 percent) that were unrolled on the ground, while straw was spread by hand at a density of approximately 1500 kg/ha. As 3600 to 4000 kg/ha are needed to achieve 100 percent ground covering (Greb, 1966; Unger, 1976), both the straw cover and the plastic cover were estimated to have approximately the same porosity.

Data Collection

The effects of treatments (windbreaks and cover types) toward the establishment of a vegetation cover were assessed by counting the number of plant individuals present at the end of the growing season. For the purpose of this experiment, three groups of plants were considered: (1) *Sphagnum* species, (2) other mosses (only the *Dicranum* and *Polytricum* genera), (3) vascular plants. Plants were counted in several sampling units (generally three) in each replicate. Sampling units consisted of four quadrants of 25 × 25 cm, systematically positioned in a 1-m square frame placed on flat surfaces. Mean numbers obtained for each sampling unit were averaged to give the values used for the analysis of variance for each subplot.

Counts were also made in depressions. Because they account for a small amount in terms of surface, sampling units in depressions consisted of only two quadrats. Those data were not included in the analysis of variance comparing windbreaks and cover types but were used to compare plant reestablishment between depressions and flat surfaces.

As mentioned earlier, the use of a manure spreader resulted in an uneven thickness of material. Thus, when counting plant individuals, each sampling unit was attributed a rating for material thickness as follows: (1) scant (material covering less than 50 percent of the ground), (2) thin layer (material covering over 50 percent of the ground), and (3) thick layer (100 percent of the ground covered by a layer more than 2 cm thick). These data were used to evaluate the effect of spread material thickness on plant reestablishment.

Finally, a vegetation survey was also conducted at the nursery borrow site at the end of the growing season, in order to estimate the damage caused by the machinery when chipping and collecting the living plant material.

Statistical Analysis

The effects of the two factors tested (windbreaks and cover types) on the number of *Sphagnum* individuals, other mosses, and vascular plants were compared using the GLM procedure of SAS (SAS Institute, 1988). Once the overall effect of a factor was shown significant with no interaction with the other factor, differences between the levels of a factor were located using a Tukey test (Sokal and Rohlf, 1981). Data were square-root transformed to meet the assumptions of normality and to reduce heterogeneity of variance [$y = \sqrt{(x + 0.5)}$]. The significance level for the ANOVA and the Tukey test was set at 0.05.

RESULTS AND DISCUSSION

Windbreaks

As no direct measurements of physical parameters (near ground temperature, evaporation, wind reduction) were done, we cannot assess the effect of windbreaks on the microclimate at the ground surface. However, our results show that windbreaks did not affect significantly the number of plants on the bare peat surface after the first growing season (Table 1). On the other hand, the presence of windbreaks substantially increased snow depth as expected. The site was visited at the end of March 1994, when there was little snow left on the harvested area. On the lee side of snow fences snow depth was still up to 1 m.

Ground Covers

Ground covers promoted plant reestablishment for the three groups of plants measured (Table 1; Figure 3). Straw tended to give a higher number of individuals per m² than the plastic cover. In all situations, vascular plants were far less numerous than mosses and *Sphagna*, but they were more abundant under the straw than in the control plots (p <0.05). Our observations, however, force us to qualify those results. At the end of the growing season in October 1993, we noted a clear difference in the shape or health of *Sphagnum* capitula between both types of cover, despite comparable numbers. Under the plastic cover, *Sphagna* capitula were small and barely visible. Since summer growth was minimal, the plastic cover only allowed *Sphagna*, and to some extent other plant diaspores, to survive but not flourish. However, under the straw we found big and healthy *Sphagna* capitula and mosses. Their growth even led to the formation of small clusters of

Table 1   Anova Tables for the Experiment on the Effect of
Windbreaks and Ground Covers on Plant Reestablishment

Source	df	Type III SS	Mean square	F value	p > F
		Number of *Sphagnum* individuals			
Windbreaks	3	68	22.8	0.66	0.6012
Error a	8	278	34.7	—	—
Cover	2	114	56.8	9.47	0.0019
Interaction	6	22	3.7	0.62	0.7137
Error b	16	96	6.0	—	—
		Number of moss individuals			
Windbreaks	3	107	35.8	1.07	0.4139
Error a	8	267	33.4	—	—
Cover	2	60	30.0	9.21	0.0022
Interaction	6	26	4.3	1.31	0.3101
Error b	16	52	3.3	—	—
		Number of vascular plant individuals			
Windbreaks	3	9	3.0	1.46	0.2969
Error a	8	17	2.1	—	—
Cover	2	38	18.8	7.91	0.0041
Interaction	6	14	2.3	0.95	0.4885
Error b	16	38	2.4	—	—

Note: The experiment was designed as a split-plot experiment in a completely randomized design.

capitula. Straw not only increases the chance of survival but also promotes the growth of plant fragments, which is an important point to consider regarding the reestablishment of a moss carpet. It is not known yet what is the real effect of straw. It probably decreases evaporation by lowering ground surface temperature and wind effect, but it may also help plant reestablishment by introducing some nutrients by leaching. Other studies already pointed out such beneficial effect of straw on plant reintroduction (Bradshaw, 1987).

Microtopography

Microtopography had a positive effect on diaspore establishment. Whatever the ground cover, the number of plant individuals for the three groups of plants more than doubled in depressions, compared to flat areas (Figure 4). The presence of depressions reduced the positive effect of ground covers, as we obtained a comparable success with and without any cover. However, the best overall results were found in depressions with the application of straw. Wetter conditions found in the depressions and the fact that in those microsites wind and solar radiation are less important help explain the positive effect of microtopography. Depres-

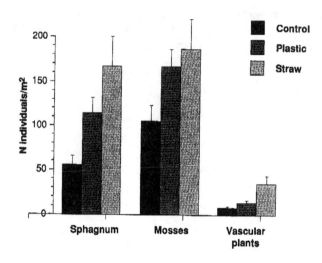

Figure 3   Effect of ground cover types on the establishment of the three groups of plants. Mean number of individuals per m² (± standard deviation). Both ground covers, plastic and straw, have a significant positive effect (p <0.05) on the number of *Sphagnum* and mosses when compared to the number of plants observed in the control plots according to the Tukey test. More vascular plants (p <0.05) were found under the straw ground cover than in the control plots.

sions represent sheltered sites where plant fragments can survive and grow and from where a moss carpet could establish and expand to adjacent flat areas and mounds. On the other hand, we did not expect plant fragments to reestablish readily on raised parts, and no data were collected on mounds. Indeed, visual observations showed that very few diaspores survived on those sites. Mounds are made of loose, lower density peat, and, though no measurements were made, its bigger pore size favors rapid drainage and impedes the presence of capillary water (Armson, 1977). Such dry conditions make the mounds unsuitable sites for diaspores survival. The efficiency of plant reestablishment in depressions, when compared with flat areas, compensates for the lack of regeneration on the mounds. Depressions can act as starting points for the establishment of a moss carpet that will spread to mounds, provided they are not too high.

Thickness of Living Material

The uneven thickness of living plant material spread on the experimental site appeared as an unexpected factor that may have influenced our results. However, for each ground cover type (control, plastic, and straw), about the same proportion of quadrats were measured under the different thicknesses of living plant material. Therefore, the sampling method does not interfere with the results on ground covers (Table 2). The numbers of living individuals after one growing season varied with the thickness of the spread material. With no ground cover (control), the number of individuals for the three groups of plants rose with the greater thickness of spread material (Figure 5). In the presence of ground covers, plastic

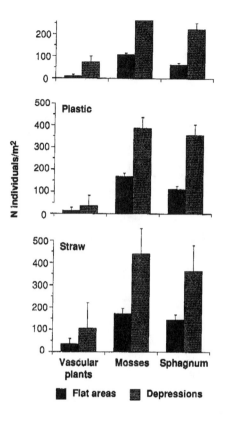

Figure 4 Effect of microtopography on the establishment of the three groups of plants. Mean number of individuals per m² (± standard deviation). N varies from 4 to 46 sampling units. The highest numbers of individuals per m² are found in depressions with straw where they reach over 105, 364, and 439 for vascular plants, *Sphagnum* species, and other mosses, respectively.

Table 2 Distribution of Quadrats, in Percentage, According to the Thickness of Living Plant Material Spread

	Scant %	Thin layer %	Thick layer %
Control	15	72	13
Plastic	17.5	72.5	10
Straw	14	67	19

Note: The almost equal distribution of quadrats under the different thicknesses of plant material does not influence the results for the ground cover types, except for a small underestimation of the positive effect of straw, as the thin layer had a better success (Figure 3).

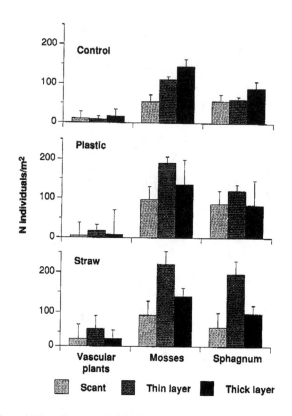

Figure 5   Effect of living plant material thickness on the establishment of the three groups of plants. Mean number of individuals per m² (± standard deviation). N varies from 4 to 33 sampling units.

or straw, the reintroduction had a better success when a thin layer (material covering more than 50 percent of the ground) of plant material was applied. When it was scanty (less than 50 percent of ground covering) or too abundant (more than 2 cm thick), ground covers had no positive influence and even had a negative effect in some cases. This suggests that when the material is too abundant, diaspores located at the bottom suffer from being buried, while those at the top lose contact with the ground and hence soil moisture. Otherwise, it is normal to have fewer plants when there is less plant material.

Plant Recovery at the Collection Site

Overall success is limited if part of a natural bog has to be destroyed to restore a harvested site. It is important that natural areas not be damaged irreversibly by collecting the top spit layer. For the purpose of this experiment, the ratio collected surface/surface covered by the living plant material was 1:8. Experiments started in 1994 have shown that this ratio could be up to 1:20, provided that diaspores are smaller and the living plant material devoid of chunks. On the other hand, results of a vegetation survey conducted after the growing season, three months after the top spit layer was harvested, indicate that the removal of the vegetation did no irreversible damage to the bog. In fact, vascular plants recovered well. A mean of 261 shoots of vascular plants per m^2 was found at the borrow site, compared with 105 under the best conditions on sites under restoration (in depressions with straw). The situation is similar for *Sphagnum* species with 385 capitula and for the mosses with 468 individuals, compared to 364 and 439, respectively, in depressions with straw at the experimental site. The shoots of vascular plants came from the root system and/or rhizomes left in place. For technical reasons, we cannot pick up all the chipped material; hence, many living *Sphagna* and moss fragments are left on the ground and many of them survived. Another interesting point is that the composition of vascular plants reflects that found in natural conditions. *Chamaedaphne calyculata* (L.) D. Don and *Kalmia angustifolia* L. were well represented, while the other species were less abundant. *Ledum groenlandicum* Retzius was the only species that was under-represented.

CONCLUSION

Little is known about peat bog restoration in North America, and this experiment represents one of the first trials aiming to reintroduce plant diaspores on a large scale (1300 m^2). Colonization by *Sphagnum* and other moss species is very slow, and many years of observation are needed to assess the success of this method in terms of reestablishing a moss carpet. However, the following conclusions can be drawn from this experiment after one growing season, which can guide further experiments:

1. Windbreaks showed no effect on the success of plant reestablishment at this site. Direct measurements of physical parameters should be done in order to assess the effect of windbreaks on ground surface microclimate.
2. The use of a ground cover allowed more individuals per m^2 to reestablish when a thin layer of plant material was spread on bare peat sufaces. Straw gave better results than plastic, as suggested by the better health of individuals and the formation of small clusters after one growing season. Straw has the double advantage of being very cheap and easily available.
3. The top spit layer of a natural bog has good potential for regeneration of the three groups of plant considered. The reestablishment of the vegetation, though, was more successful when ground covers were provided to increase water availability for plants.
4. The thickness of living plant material spread on the bare surface was an important variable. Too little or too much material inhibited the positive effect of ground covers.
5. Microtopography proved to be efficient in terms of number of individuals per m^2 for each group of plant by providing wetter and sheltered sites.
6. The plant recovery at the natural nursery site suggests that collecting living plant material does no irreversible damage to the environment.
7. Commonly available machinery accomplished the work involved in the restoration process.

ACKNOWLEDGMENTS

This research project was made possible with the financial support of several public and private organizations: the Ministère de l'Environnement et de la Faune du Québec, the Centre québécois de valorisation de la biomasse, the Canadian Wildlife Service, the Université Laval, the Centre de recherche et de développement de la tourbe (New-Brunswick), the Canadian Sphagnum Peat Moss Association, the Association québécoise des producteurs de tourbe, Fafard & Frères Ltée, Johnson & Johnson, Lambert Peat Moss and Premier Peat Moss. We wish to thank especially the staff of Lambert Peat Moss at the Rivière-Ouelle plant for their collaboration in providing technical support and Suzanne Campeau for the statistical analyses.

REFERENCES

Armson, K. A., *Forest Soils: Properties and Processes*, University of Toronto Press, 1977, 390 pp.

Bradshaw, A. D., The reclamation of derelict land and the ecology of ecosystems, in *Restoration Ecology. A Synthetic Approach to Ecological Research*, Jordan, W. R., III, Gilpin, M. E., and Aber, J. D., Eds., Cambridge University Press, 1987, p. 53.

Campeau, S. and Rochefort, L., *Sphagnum* regeneration on bare peat surfaces: Field and greenhouse experiments, *J. Appl. Ecol.,* in press.

Clymo, R. S. *Sphagnum*-dominated peat bog: a naturally acid ecosystem, *Phil. Trans. Roy. Soc. Lond. Ser. B*, 305, 487, 1984.

Dickey, G. L., Crop water use and water conservation benefits from windbreaks, *Agric. Ecosystems Environ.,* 22–23, 381, 1988.

Elling, A. E. and M. D. Knighton, *Sphagnum* moss recovery after harvest in a Minnesota bog, *J. Soil Water Conserv.*, 39, 209, 1984.

Environnement Canada, Normales climatiques au Canada 1961–1990, Vol. 5, Québec, Service de l'environnement atmosphérique, 1993, 157 pp.

Enz, J. W., Brun, L. J., and Larsen, J. K., Evaporation and energy balance for bare and stubble covered soil, *Agric. For. Meteorol.,* 43, 59, 1988.

Greb, B. W., Effect of surface-applied wheat straw on soil water losses by solar distillation, *Soil Sci. Soc. Am. Proc.,* 30, 786, 1966.

Heisler, G. J. and DeWalle, D. R., Effects of windbreaks structure on wind flow, *Agric. Ecosystems Environ.,* 22–23, 41, 1988.

Ingram, H. A. P., Hydrology, in *Mires: Swamp, Bog, Fen and Moor*, Gore, A. J. P., Ed., Elsevier, New York, 1983, p. 67.

Rosenberg, N. J., Effects of windbreaks on the microclimate, energy balance and water use efficiency of crops growing on the Great Plains, in Shelterbelts on the Great Plains. Proceedings of a Symposium, Great Plain Agriculture Council Publication, Denver, CO, 1976, p. 49.

Rosenberg, N. J., Blad, B. L., and Verma, S. B., *Microclimate: The Biological Environment*, 2nd ed., John Wiley & Sons, 1983, 495 pp.

Salonen, V. and Laaksonen, M., Effects of fertilization, liming, watering and tillage on plant colonization of bare peat surfaces, *Ann. Bot. Fenn.,* 31, 29, 1994.

SAS Institute Inc., SAS/STAT User's Guide: Release 6.03 Edition, SAS Institute Inc., Cary, NC, 1988, 1028.

Schouwenaars, J. M., Hydrological research in disturbed bogs and its role in decisions on water management in the Netherlands. *Proc. of The International Symposium on the Hydrology of Wetlands in Temperate and Cold Regions*, Publication of the Academy of Finland 4, 1988a, p. 170.

Schouwenaars, J. M., The impact of water management upon groundwater fluctuations in a disturbed bog relict, *Agric. Water Manage.*, 14, 439, 1988b.

Sokal, R. R. and Rohlf, F. J., *Biometry*, 2nd ed., W. H. Freeman, New York, 1981, 859 pp.

Spencer, M., Famous, N. C., and Anderson, D. S., Natural regeneration patterns of Bryophytes on abandoned vacuum harvested peatlands, Communication presented at the ASLO & SWS Meeting, Edmonton, Alberta, Canada, 1993.

Unger, P. W., Surface residue, water application, and soil texture effects on water accumulation, *Soil Sci. Soc. Am. Proc.*, 40, 298, 1976.

Wilson, C., *Le climat du Québec*. Atlas climatique, Ottawa, Service météorologique canadien, 1971.

CHAPTER **12**

A Ground-Penetrating Radar Study of Peat Landforms in the Discontinuous Permafrost Zone Near Fort Simpson, Northwest Territories, Canada

Inez M. Kettles and Stephen D. Robinson

CONTENTS

INTRODUCTION

Canada has vast peatland areas, many of which are permanently, but discontinuously, frozen (Zoltai et. al., 1988). For environmental research purposes, including monitoring the effects of climate change on the landscape, and for land use planning, it is frequently necessary to assess the thermal character of peatlands. In this paper, we evaluate ground-penetrating radar (GPR) as a means for accurate mapping of frozen and unfrozen peatlands. Although GPR has been used for peatland surveys in Canada (Warner et al., 1990; Pelletier et al., 1991; Theimer et al., 1994), in the United States (Baraniak, 1983), and in Scandinavia (Ulriksen, 1980; Hanninen, 1992), few were conducted in the discontinuous permafrost zone (Hayley, 1988; Doolittle et al., 1992).

The Fort Simpson area was chosen for study for a number of reasons. First, approximately 35 percent of the region is covered by peatlands, of which 44 percent is permanently frozen (Aylsworth et al., 1993). Second, these discontinuously frozen peatlands have numerous morphologically distinct forms (Zoltai and Tarnocai, 1975). Finally, the Mackenzie Valley has been designated as an integrated research monitoring area for Geological Survey of Canada's Global Change Program. Peatland mapping and characterization are important aspects of this research effort.

STUDY LOCATION AND SETTING

The Fort Simpson area, which lies at the confluence of the Liard and Mackenzie Rivers (61°52′N, 121°21′W), is underlain by flat-lying Devonian shales and siltstones (Douglas, 1959) (Figure 1). The bedrock is overlain by till and, in low lying areas, glaciolacustrine fine-grained sediments (Vincent, 1989; Smith, 1992). Glaciolacustrine sediments are, in places, overlain by postglacial deltaic sands and silts, upon which aeolian dunes have commonly formed (Minning et al., 1980; Smith, 1992). Modern alluvial sands and gravels occur along the Liard and Mackenzie Rivers.

Peat deposits have developed on flat lying deposits of glaciolacustrine sediments and till, in low-lying areas between sand dunes, and in depressions in alluvial deposits. The 110 boreholes available for the region show an average peat thickness of 2.4 m (range 1.5 to 5.0 m) in bogs and an average of 1.55 m (0.6 to 3.2 m) in fens (Aylsworth et al., 1993).

Mean annual air temperature at Fort Simpson is −4.0°C (Atmospheric Environment Service, 1982) and ground temperatures are generally near 0°C but vary depending on surface cover (Judge, 1973).

PEATLAND MORPHOLOGY

The bog-fen complex of the Fort Simpson area comprises numerous peat landforms typical of the discontinuous permafrost zone (Tarnocai, 1973). Most common is flat low-lying fen that is saturated and without permafrost. Rutter et al. (1973) reported that terrain underlying fens remained unfrozen to depths of at least 6 m (the borehole bottoms). Peat plateaus, elevated 1 or 2 m above the adjacent fens, are generally better-drained, flat-topped, treed expanses of bog peat

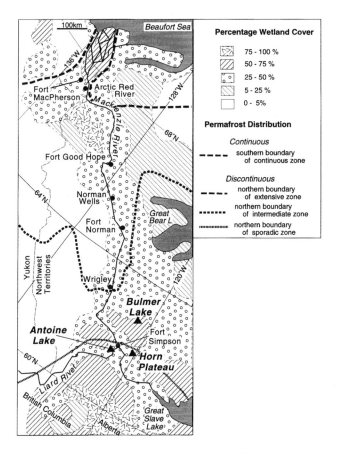

Figure 1 Peatland cover and permafrost distribution in the Mackenzie Valley. Study sites in the Fort Simpson area are marked by black triangles.

containing permafrost, often with segregated ice lenses. Sediment underlying bogs with thick peat (>1.8 m) is generally frozen to at least 6 m in this region (Rutter et al., 1973). At higher elevations, and hence, in colder environments, peat plateaus may contain ice-wedge polygons and have permafrost that extends into the underlying sediments; these are termed polygonal peat plateaus.

Also present are palsas, fen ridges, and collapse scars. Palsas are mounds of frozen peat, 5 to 30 m in diameter, typically raised up to 4 m above the surrounding unfrozen fen. Lenses of pure ice are common with the peat and underlying frozen sediment, especially at the peat–sediment interface (Zoltai and Tarnocai, 1975). Fens may also contain a series of sinuous ridges typically 1 to 3 m wide and up to 1 m high. These fen ridges commonly contain permafrost lenses that may extend in the underlying sediments where peat cover is thin. A collapse scar fen results from the thaw and collapse of a whole or part of a peat plateau or palsa to the level of the surrounding fen.

USE OF GROUND-PENETRATING RADAR AS A MAPPING TOOL

Ground-penetrating radar surveying is a fast, reliable, and inexpensive method for non-destructive mapping of shallow subsurface materials (Davis and Annan, 1989). The technique involves the transmission of short pulses of high frequency electromagnetic (EM) energy into the ground, and measurement of the energy reflected from interfaces between subsurface materials with contrasting electrical properties. Reflections may be caused by changes in sediment type, water content, density, and temperature. Evenly spaced measurements taken along a transect allow the construction of an electromagnetic cross-section or profile, with the vertical axis representing two-way signal travel time and the horizontal axis indicating survey position. Signal propagation velocities are determined in the field by conducting common midpoint (CMP) surveys, allowing depth to replace two-way travel time on the vertical axis. This radar profile, in conjunction with borehole information, permits the interpretation of subsurface characteristics.

Surveying in the discontinuous permafrost zone presents unique problems to the interpretation of GPR profiles. Where frozen pore water is present in subsurface materials, signal absorption is often decreased, thereby increasing signal penetration depth (Arcone and Delaney, 1985). Also, propagation velocities through frozen materials are often two to three times greater than through similar unfrozen materials, especially if the ice content is high. As a result, it is necessary to present two depth scales with radar profiles that cross both frozen and unfrozen materials.

The pulse EKKO IV GPR system with interchangeable antennas, from Sensors and Software Inc., of Mississauga, Ontario, was used for these surveys. The higher frequency antennas (100 MHz) provided high resolution in the near-surface, while the lower frequency antennas (50 MHz) yielded greater depth penetration, with an accompanying loss in resolution. Instrument readings were taken at 1-m intervals along transect lines set out using survey chains. Simple point-averaging noise reduction filters, automatic gain control, and topographic corrections were the only processing enhancements applied to the profiles. To confirm peat depth and some physical characteristics of peat, cores were obtained from each site using a Macaulay peat corer (Jowsey, 1966) in unfrozen fens, and a CRREL coring kit (Veillette and Nixon, 1980) in frozen bogs.

SURVEY SITES

Three study sites within a 100 km radius of Fort Simpson were selected to represent the diversity of local peat landforms. The Antoine Lake site (61°43′N, 121°45′W, elevation 190 m) (Figure 2) is part of a large bog-fen complex overlying clayey sand, with the treed bog raised about 2 m above the level of the fen. Several low-lying, treed palsas are found within the fen. Five radar profiles were conducted across peat plateau, fen, and palsa.

Figure 2 Photograph of the Antoine Lake bog-fen complex. A black star represents bore-hole site and 100 m position on profile (see Figure 4). Survey line progresses eastward (right to left) over peat plateau and fen.

Near Bulmer Lake (62°43'N, 121°06'W, elevation 244 m) (Figure 3), an isolated, round peat plateau, 225 m in diameter and overlying clay till, abuts a large lake. A collapse scar, approximately 30 m wide and 60 m long, has formed within the peat plateau. The radar survey included one profile over peat plateau, the surrounding fen and fen ribs, and another across the collapse scar fen within the peat plateau.

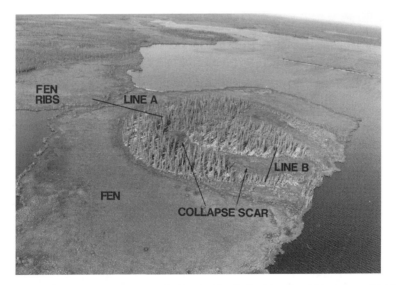

Figure 3 Photograph of the Bulmer Lake site. Profile for Line A, which crosses (right to left) peat plateau, fen, and rib fens, is shown in Figure 5. Profile for Line B, crossing (bottom to top) peat plateau, collapse scar fen, and peat plateau, is shown in Figure 6.

The top of the Horn Plateau (61°56'57.6"N, 120°02'25"W, elevation 685 m) is elevated nearly 500 m above the surrounding lowlands and likely experiences a slightly cooler climate. As a result, some peat deposits developed on clay till take the form of polygonal peat plateau, more typical of continuous permafrost areas at least 200 km further north. This site is one of the most southerly occurrences of polygonal peat plateau in the Mackenzie Valley. The site contains flat-lying, hummocky, treed, and untreed bog; the latter containing ice wedge polygons. The treeless area has been burned and is covered by fossil and lichen-covered, charred tree stumps about 15 cm high.

Basal peat from the three sites — Antoine Lake, Bulmer Lake, and Horn Plateau — were dated at 8680 ± 100 (GSC-5744), 9190 ± 110 (GSC-5750), and 7800 ± 80 (GSC-5727), respectively.

RESULTS AND DISCUSSION OF GPR SURVEYS

GPR profiles and core logs considered representative of peat landforms from the three study sites are shown in Figures 4 to 8. All profiles crossed at least two landform types in order to best examine their interfaces. The left- and right-hand depth scales are for frozen and unfrozen material, respectively. Sections of each profile interpreted to be frozen have been shaded.

Peat Plateau

At all sites, radar returns from frozen peat plateau are characterized by high frequency (narrow pulse in the vertical axis), semicontinuous reflectors (see Figure 4 (Antoine Lake survey positions 100 to 125 and 274 to 286 m) and Figure 5 (Bulmer Lake, 10 to 42 m)). The chaotic nature of these returns can likely be attributed to the presence of ice lenses. Returns from within the upper few meters of peat plateau are more laterally continuous than deeper returns, but a reflector representing the peat–sediment contact can be difficult to distinguish under frozen conditions. The deepest, laterally continuous reflector at Bulmer Lake (Figure 5, P-b, between 10 and 40 m, at a depth of 5 m) may represent the base of peat. Where peat plateau is surveyed at Antoine Lake site there is no laterally contin-uous reflector at or below 6 m (Figure 4) where the base of peat is expected. (The estimate of peat thickness is based on the known peat thickness in the nearby fen (4.3 to 4.9 m) and the peat plateau height (1.5 to 2.0 m)). At both sites, the presence of radar returns well below the expected base of peat indicates that there is little contrast in electrical properties at the frozen peat/sediment interface. This also suggests that the underlying sediment is frozen to depths of at least 10 to 15 m, possibly with ice lenses giving rise to several of the chaotic reflectors.

Fen

Radar profiles crossing fen are shown for the Antoine Lake (Figure 4, survey positions 126 to 192 and 210 to 274 m) and Bulmer Lake (Figure 5, 42 to 92 m)

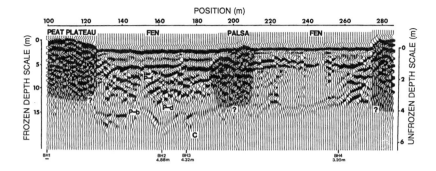

Figure 4   A 50 MHz GPR profile from the Antoine Lake bog-fen complex. Due to different propagation velocities (0.036 and 0.102 m/ns for unfrozen and frozen, respectively), two different depth scales must be used for frozen (left vertical axis) and unfrozen (right vertical axis) terrain. Areas interpreted as being underlain by permafrost are shaded. Positions 100 to 125 m and 274 to 286 represent peat plateau, positions 192 to 212 m a low-lying palsa, and the remainder fen. Interpreted reflectors are base of peat (P-b), clay (C), and zones of decomposed peat (P-d) and fibrous peat (P-f). Locations where peat depths were probed for reflector verification are shown as BH2-BH4. Core log for BH1 is presented in Figure 8.

sites. Unlike peat plateau, the base of peat is easily distinguished in fen. This likely reflects greater electrical contrasts and the higher radar resolution associated with slower propagation velocities, both of which are characteristic of unfrozen materials. The contact between peat and clayey sand (C) across the fen is marked at Antoine Lake by a continuous reflector (P-b) between a depth of 3 m and 4.9 m (Figure 4). This reflector shows the sediment surface rising gently across the fen, with a channel-like depression noted between 153 and 172 m.

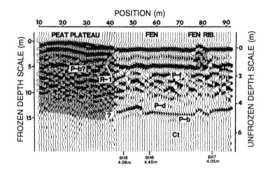

Figure 5   A 50 MHz GPR profile from the Bulmer Lake site (Line A on Figure 3). Positions 10 to 42 m represent peat plateau, 77 to 83 m a fen rib with a frozen core, and the remainder fen. Radar propagation velocities were calculated to be 0.038 and 0.097 m/ns for unfrozen and frozen material, respectively. Interpreted reflectors are base of peat in unfrozen fen (P-b), possible base of peat in frozen bog (P-b?), clay till (Ct), and zones of decomposed peat (P-d) and fibrous peat (P-f). Refer to Figure 9 for a description of reflector R-1. Peat depths probed in the unfrozen fen (BH5-BH7) are also shown. Areas interpreted as being underlain by permafrost are shaded.

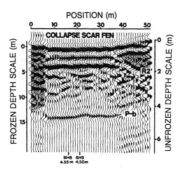

Figure 6   A 50 MHz GPR profile from the Bulmer Lake site (Line B on Figure 3) Positions
0 to 7 m represent peat plateau, 7 to 39 m collapse scar fen, and 39 to 50 m
peat plateau. Radar propagation velocities were calculated to be 0.038 and 0.097
m/ns for unfrozen and frozen material, respectively. Areas interpreted as being
underlain by permafrost are shaded. Note the wedge of permanently frozen peat
which begins at 32 m and thins towards the collapse scar (refer to text for a
description of the radar delineation of this frozen wedge). Probed peat depths
(BH8-BH9) are also shown. Refer to Figure 8 for BH9 log.

Depths of peat measured in fen cores (BH2, 3, and 4) correspond closely to the
depths of the reflector P-b on the radar profile at the borehole sites.

At the Bulmer Lake site (Figure 5), the contact (P-b) between peat and clay
till (Ct) appears to be flat lying, with both measured (BH5, 6, and 7) and interpreted
peat thicknesses being at least 4 m. The P-b reflector is stronger at Bulmer Lake
than at Antoine Lake, even though peat thicknesses were similar. This discrepancy
may reflect a stronger electrical contrast at the interface between peat and clay
till than between peat and clayey sand. Rapid signal attenuation below the P-b
reflector at both sites indicates that the sediment is unfrozen beneath the fen.

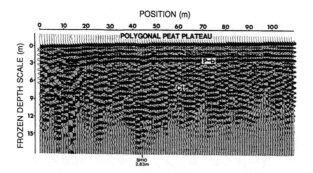

Figure 7   A 100 MHz GPR profile across a polygonal peat plateau at the Horn Plateau
site. The high propagation velocity over the frozen bog (0.113 m/ns) results in a
compressed depth scale. Note the marked difference in reflector patterns for
peat (horizontal reflectors in upper 3 m, including the base of peat (P-b)), and
the underlying frozen clay till (Ct)(sub-horizontal, less continuous reflectors).
Borehole location is shown, with the base of peat at 2.83 m. Refer to Figure 8
for borehole log.

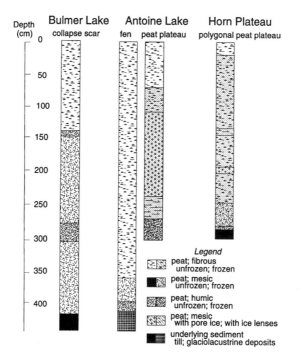

Figure 8   Selected detailed cores from the three survey sites.

At both sites, immediately above the P-b reflector lies a "quiet zone" (P-d) generally less than 1 m thick, which appears relatively homogeneous and transparent to GPR. In the cores collected, some of which are shown in Figure 8, peat was generally more decomposed near the base, as would be expected. Fibrous peat near the fen surface is characterized by the upper, subhorizontal reflectors (P-f). A much thicker "quiet zone" occurs between 223 and 250 m in Figure 4. As no detailed peat core is available at this location, the cause of the quiet zone is not known.

Collapse Scar Fen

A collapse scar, where over 4 m of peat was measured overlying clay till (BH8-BH9), was surveyed near Bulmer Lake (Figure 6, positions 7 to 37 m). Radar returns for the collapse scar fen indicate patterns which are very similar to the nearby fen (Figure 5). Interpretations pertaining to peat depth in Figure 6 are based on observations of peat in the BH9 core (Figure 8) and the configuration of reflectors on the radar profile. The underlying clay till (Ct) is interpreted to be unfrozen as the signal is quickly attenuated. This rapid attenuation commonly characterizes unfrozen, fine-grained sediment (Davis and Annan, 1989).

Fen Rib

The fen at Bulmer Lake contains numerous fen ribs (Figure 3), one of which lies between stations 77 and 83 m on the GPR profile (Figure 5). Although the surrounding fen was unfrozen at the time of the survey, probing below the fen rib showed the upper meter of peat to be frozen. On the profile, the basal reflector (P-b) experiences only slight "pull-up" over the rib, indicating that the permafrost lens causes only a minor increase in near-surface propagation velocity. Radar return patterns for the rib appear to be more closely related to those of unfrozen fen than frozen peat plateau.

Palsa

The Antoine Lake survey also crossed a low-lying, treed palsa (Figure 4, survey positions 192 to 212 m). The high-frequency and chaotic radar returns are similar to those typical of peat plateau and indicate a frozen core, possibly containing segregated ice lenses. The similar nature of radar returns above and below the presumed base of peat in the palsa again suggests that permafrost may extend into the underlying sediments.

Polygonal Peat Plateau

At the Horn Plateau polygonal peat plateau site (Figure 7), the base of the peat can be traced on the profile as the lowest continuous horizontal reflector (P-b), at approximately 3 m. A borehole (Figure 8, BH10) showed 2.83 m of frozen ice-rich peat overlying frozen clay till. Although a high propagation velocity within the frozen material prevents good resolution of detailed peat stratigraphy, borehole information shows that well-decomposed peat is restricted to the basal 40 cm.

Signal propagation extending below 3 m depth and well into the underlying clay till (Ct) suggests that it is frozen to at least the depth of radar penetration (>15 m). With the exception of the thin active layer, no unfrozen material was encountered at the site. Reflectors from within the frozen clay till are subhorizontal and less continuous than those from within the peat. For unknown reasons, the radar returns from the treed bog are more chaotic and of slightly lower frequency at depth. Several profiles conducted at this site crossed small ice wedge depressions (<1.5 m wide), yet the radar returns from these wedges were inconsistent. In a few instances (e.g., at 97 m on Figure 7), the ice wedges resulted in a quiet zone at depth on the radar profile, while at other locations (not shown), their presence failed to have a noticeable influence on the radar signal. In most previous studies (A. S. Judge, personal communication, 1994), larger ice wedges were displayed as deep reverberations or wave trains.

Frozen–Unfrozen Interfaces

Some insight into the thermal stability of frozen peatlands may be obtained from study of permafrost distribution and the configuration of frozen–unfrozen

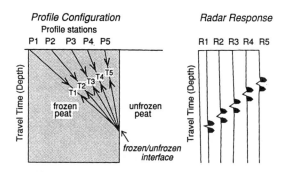

Figure 9 The configuration of radar returns as a vertical interface between frozen and unfrozen peat is approached during a GPR survey. This is the interpreted cause for reflector R-1 in Figure 5. See text for further discussion.

interfaces in peat landforms, as shown through GPR surveying. At Bulmer Lake (Figure 5), there is a steeply dipping reflector (R-1), between the 33 and 42 m stations, where the survey progressed toward the edge of the peat plateau. When corrected for vertical exaggeration, the slope of "R-1" is approximately 45°. This reflector characterizes the radar returns obtained from the frozen–unfrozen boundary as the survey approaches a vertical interface (see Figure 9). Manual probing confirmed that the near-surface frozen–unfrozen interface at this location is near-vertical.

The pattern of radar returns near the contact between frozen peat plateau and unfrozen collapse scar (Figure 6), however, differs markedly from those obtained at the peat plateau–fen interface shown in Figure 5. Deep reflectors between 39 and 50 m exhibit characteristics (lower frequency and fairly continuous) more typical of unfrozen fen than frozen peat plateau (Figures 4 and 5). Additionally, the gently rising reflector between 3 and 4 m depth (unfrozen scale) appears similar to the reflector that marks the base of peat in the adjacent, unfrozen, collapse scar fen. These observations suggest a thin, near-surface wedge of perennially frozen peat near the plateau edge. The wedge pinches out toward the fen, with the wedge base being represented by reflector R2. Also, the lowest continuous reflector, representing the flat-lying base of peat (P-b), is gradually "pulled-up" between positions 39 and 50 m as the wedge of frozen peat thickens and increases the signal velocity. Vertical thaw interfaces between landforms were also noted by Tarnocai (1973) and Zoltai and Tarnocai (1975). A similar pattern of peat plateau thermal degradation was noted along the Hudson Bay railway in the discontinuous permafrost zone in northern Manitoba (Hayley, 1988).

SUMMARY AND CONCLUSIONS

GPR is useful for obtaining a more continuous perspective of the distribution of permafrost and subsurface materials in peatlands in the discontinuous permafrost. The delineation of frozen vs. unfrozen material on the GPR profiles was facilitated by the distinctive character of their respective radar returns. In frozen terrain, radar responses were generally of higher frequency (narrower pulse), more chaotic in nature (less lateral continuity), and were observed to have deeper penetration (less signal attenuation) than returns from unfrozen peat. In unfrozen peat, it is generally possible to delineate the base of peat and also some information pertaining to its stratigraphy. In near-surface unfrozen peat, where organic materials are poorly decomposed, patterns of radar returns allowed the better resolution of peat stratigraphy. In contrast, more decomposed peat near the base of the deposit was characterized by a zone transparent to GPR, without traces of internal stratigraphy. Along most surveyed lines, the peat was over 4 m thick within the fens. More detailed analysis of physical and chemical properties of peat in the cores is needed before the radar records can be further interpreted.

It was generally not possible to interpret the exact base of peat or accurate stratigraphic information from GPR records for frozen peat, except at the Horn Plateau site. In terrain with frozen pore water, the electrical contrast at the peat–sediment contact is not as strong as in unfrozen material. It is possible, however, to infer that a different material has been encountered based on changes in reflector patterns. Although the propagation velocity and the depth of penetration is greater in frozen ground, resolution is poorer due to the high penetration depth to return time ratio, making delineation of individual stratigraphic reflectors difficult. The distinct lack of signal attenuation can be used to suggest underlying materials are frozen at least to depths where the signal weakens. Radar surveys showed peat plateaus, with thicknesses often over 6 m, and polygonal peat plateaus, to be permanently frozen into the underlying sediments to depths of at least 10 to 15 m (e.g., Figure 4, 100 to 125 m). Peat within palsas is also likely frozen, in many places, into the underlying sediments. Fen ribs contain lenses of permafrost about 1 m thick, while fens, collapse scar fens, and sediment underlying them remain unfrozen.

Results of GPR surveys suggest that near-surface frozen–unfrozen interfaces at most study sites were generally vertical. These results are in close agreement with configurations mapped previously. At a collapse scar–peat plateau interface, thawing at the base of peat at the plateau edge was inferred.

ACKNOWLEDGMENTS

The authors extend special thanks to Jean Bisson for his able assistance in the field. Also thanks to Jean Pilon for his loan of the GPR equipment; Jan Aylsworth, Brian Moorman, Larry Dyke, and Mark Nixon for helpful discussions; Manfred Hebel for assistance in the office; Alan Judge and an anonymous reviewer for helpful comments on an earlier version of this report; T. Malterer and J. Jeglum for reviewing this manuscript; Roger McNeely in the Radiocarbon Laboratory, Geological Survey of Canada, for dating peat samples; and the personnel of Canadian Helicopters and the Water Survey of Canada, Fort Simpson, N.W.T. for their cooperation.

REFERENCES

Arcone, S. A. and Delaney, A. J., Model Studies of Surface Noise Interference in Ground Probing Radar, U.S. Cold Regions Research and Engineering Laboratory, CRREL Report 85-19, Hanover, NH, 1985.

Atmospheric Environment Service, Canadian Climate Normals, Temperature and Precipitation 1951–1980 — The North, Y.T. and N.W.T., Canadian Climate Centre, Environment Canada, Toronto, 1982.

Aylsworth, J. M., Kettles, I. M., and Todd, B. J., Peatland Distribution in the Fort Simpson Area, Northwest Territories with a Geophysical Study of Peatland-Permafrost Relationships at Antoine Lake, Geological Survey of Canada Paper 93-1E, 1993, 141 pp.

Baraniak, D. W., Exploration for surface peat deposits using ground penetrating radar, in Proc. Int. Symp. on Peat Utilization, Bemidji, MN, 1983, p. 105.

Davis, J. L. and Annan, A. P., Ground penetrating radar for high-resolution mapping of soil and rock stratigraphy, *Geophys. Prospecting,* 37, 531, 1989.

Doolittle, J. A., Hardisky, M. A., and Black, S., A ground-penetrating radar study of Goodstream Palsas, Newfoundland, Canada, *Arctic Alpine Res.,* 24, 173, 1992.

Douglas, R. J. W., Trout River, District of Mackenzie, N.W.T., Geological Survey of Canada Map, 28-1958, 1959.

Hanninen, P., Application of ground penetrating radar techniques to peatland investigations, in Proc. Fourth Int. Conf. on Ground Penetrating Radar, Geological Survey of Finland Special Paper 16, 1992, p. 217.

Hayley, D. W., Maintenance of railway grade over permafrost in Canada, in Proc. 5th Int. Permafrost Conference, Trondheim, Norway, 1988, p. 43.

Jowsey, P. C., An improved peat sampler, *New Phytol.,* 65, 245, 1966.

Judge, A. S., Thermal Regime of the Mackenzie Valley: Observations of the Natural State, Environmental-Social Committee, Northern Pipelines, Task Force on Northern Development, Report 73-38, 1973.

Minning, G. V., Netterville, J. A., and Rutter, N. W., Surficial geology and geomorphology of Fort Simpson, District of Mackenzie, Northwest Territories, Geological Survey of Canada Map, 3-1978, 1980.

Pelletier, R. E., Davis, J. L., and Rossiter, J. R., Peat analyses in the Hudson Bay Lowlands using ground penetrating radar, Digest of the IEEE, 1991 International Geoscience and Remote Sensing Symposium, Helsinki, Finland, Volume 4, 1991, p. 2141.

Rutter, N. W., Boydell, A. N., and Savigny, K. W., Terrain Evaluation, Mackenzie Transportation Corridor, Southern Part. Environmental Social Committee, Northern Pipelines, Task Force on Northern Oil Development, Report 73-36, 135, 1973.

Smith, D. G., Glacial Lake Mackenzie, Mackenzie Valley, Northwest Territories, Canada, *Can. J. Earth Sci.,* 29, 1756, 1992 .

Tarnocai, C., Soils of the Mackenzie River Area, Environmental Social Committee, Northern Pipelines, Task Force on Northern Oil Development, Report 73-26, 136, 1973.

Theimer, B. D., Nobes, D. C., and Warner, B. G., A study of geoelectric properties of peatlands on ground-penetrating radar surveying, *Geophys. Prospecting,* 42, 179, 1994.

Ulriksen, P., Investigation of peat thickness with radar, in Proc. 6th Int. Peat Congress, Duluth, MN, 1980, pp. 126–129.

Veillette, J. J. and Nixon, F. M., Portable Drilling Equipment for Shallow Permafrost Sampling, Geological Survey of Canada Paper 79-21, 1980.

Vincent, J.-S., Quaternary geology of the northern Canadian Interior Plains; in Chapter 2 of Quaternary Geology of Canada and Greenland, Fulton, R. J., Ed., Geological Survey of Canada, No. 1 (also Geological Society of America, The Geology of North America. v. K-1), 1989.

Warner, B. G., Nobes, D. C., and Theimer, B. D., An application of ground penetrating radar to peat stratigraphy of Ellice Swamp, southwestern Ontario, *Can. J. Earth Sci.,* 27, 932, 1990.

Zoltai, S. C. and Tarnocai, C., Perennially frozen peatlands in the western Arctic and Subarctic of Canada, *Can. J. Earth Sci.,* 12, 28, 1975.

Zoltai, S. C., Taylor, S., Jeglum, J. K., Mills, G. F., and Johnson, J. D., Wetlands of boreal Canada, in *Wetlands of Canada*, National Wetlands Working Group Canada Committee on Ecological Land Classification, Ecological Land Classification Series No. 24. Sustainable Development Branch, Environment Canada, Ottawa, and Polyscience Publications Inc., Montreal, 1988.

Hydrology and Biogeochemistry

CHAPTER 13

Hydrological Processes of Natural, Northern Forested Wetlands

E. S. Verry

CONTENTS

INTRODUCTION

Ecosystem function refers to the collection of biological, chemical, and physical processes that govern the flow of energy and materials through ecosystems

1-56670-177-5/97/$0.00+$.50

(Murphy and Meehan, 1991). Thus, hydrological processes in forested wetlands include the movement and behavior of water through the wetland. We use our understanding of the way water moves in forested wetlands, in combination with its chemical properties to develop systems of wetland classification, to predict and improve vegetation growth, to assess the processes of nutrient and element cycling, to evaluate the role of wetlands in regional hydrology, and to form our beliefs about wetland value and best use.

The source of water delivered to wetlands and water behavior in wetland ecosystems are our best foundations for interpreting biological and chemical processes in wetlands. Wetland functions are molded by landscape controls on water chemistry and water delivery to and from the wetland, water table position and regime in the wetland, water movement rates, water state (solid, liquid, and vapor), and the energy of sun and gravity. This paper examines the influence of these functions in forested wetlands, and by comparison in nonforested wetlands. It will not address in great detail the geology, biology, and hydrology disciplines, but will offer some key concepts to help us understand ecosystem process in forested wetlands and wetland function in landscapes.

WETLAND ESTABLISHMENT

Two conditions are necessary for wetland formation, (1) a physiography or climate that is conducive to slow water movement and (2) water available at the earth's surface. Suitable physiography includes flat areas such as glacial lake beds, sandy outwash plains, and river flood plains. Depressions with no or limited outflow, such as ice block depressions, are also suitable. But peatlands can develop on significant slopes where precipitation (P) is much greater than evapotranspiration (ET), which occurs as a frequent mist giving rise to wet conditions with only slow runoff.

Sufficient water is available at the earth's surface where annual P exceeds ET. Even in continental areas where summer rains are less than ET, the hot days and cool nights produce frequent dew to maintain the viability of *Ericacea*, *Sphagnum*, *Eriophorum*, and *Carex*. Water is also available at the surface where the basin area to wetland area ratio is large (30:1 or 20:1) so that the wetland P + surrounding runoff (RO) is greater than ET.

Wherever groundwater rises near the surface or discharges at the surface, there will be sufficient water for wetlands to form if flow rates are slow and erosion is minor. In fact, the accumulation of peat in groundwater-fed wetlands represents the filling of a hydrostatic pressure field existing in the air prior to peat accumulation. This groundwater pressure field represents the limit of fen peat development, but ombrogenous peat can develop secondary pressure fields on top of the fen peat (Ivanov, 1981; Glaser et al., 1990).

Water source determines wetland type. The quality, quantity, and periodicity of water at the wetland site drives the ecological development of wetlands. While clouds are the origin of precipitation and all wetlands receive direct precipitation, contributions from the land or seascape surrounding the wetland strongly influence the chemistry and periodicity of water delivered. Dilute rain water, calcium-rich groundwater, calcium-poor groundwater, mineral-soil runoff low in calcium (or high in phosphorous and nitrogen), and sodium chloride rich ocean water determine the trophic state of wetlands.

Climate determines wetland form. Temperature sums, annual minimums, and maximums, P and ET interactions, freezing, and freeze-thaw fluctuations mold wetland types into specific expressions of wetland form. Thus, tundra, palsa, aapa, raised, and blanket mires form on the large land areas in the mid to north latitudes. Groundwater input can, of course, substitute for large and direct precipitation and also lead to pool and flark water tracks (see Moore and Bellamy, 1974 and National Wetlands Working Group, 1988 for an explanation of peatland terms).

WETLAND HYDROLOGY

Water Source

Water enters wetlands as precipitation, runoff from surrounding mineral soils, diffuse groundwater inflow, defined flow from springs, the ebb and flood of tides, and river floodwater. Water sources are driven by the weather, including the frequency and magnitude of precipitation, the force of winds (as in water driven shoreward from ocean sounds or inland lakes), the force of tides (frequency and magnitude), and the force of hydraulic potentials in groundwater aquifers.

Two geological cross-sections in Figure 1 illustrate the influence of rock composition on peatland type. Cross-section (a) illustrates how slightly calcareous clay loam till can yield ice-block depression peatlands of bog (nutrients derived primarily from precipitation) or poor fen (additional calcium and magnesium derived from the till layer) type, as the size of contributing watershed, and thus the amount of calcium, increases. Also shown in cross-section (a) is groundwater whose flow lines arch deep to limestone bedrock and then return calcium-rich groundwater to the strong fen. Numbers under the peatland type names are theoretic values for specific conductance in microsiemens cm^{-1}. Cross-section (b) shows a series of three bogs with slightly increasing specific conductance values, but a lack of calcium either in the underlying sands or the granite bedrock restricts increases in specific conductance to evapotranspiration concentration of precipitation.

Bog locations in the sands (Figure 1b) may have initiated peat development when water tables were higher, and well-decomposed peat in the depression now slows water loss by seepage. Banding within the sand may also retard seepage. Thus, bogs can occur with a groundwater source; however, the ionic strength of the groundwater is only slightly stronger than precipitation. Examples of groundwater-fed bogs occur on Wisconsin's central sand dome, where acid cranberry bogs are common, and in the eastern half of Michigan's Upper Peninsula, where base-poor sands such as the Baraga sand plain support a groundwater system

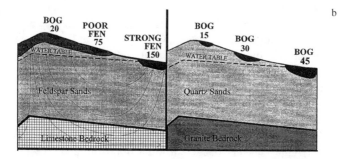

Figure 1 Peatlands in ice-block depressions (solid black) become a different wetland type as the amount of calcium and other bases dissolved in the incoming water increases (specific conductance values shown beneath type name in micro-siemens cm^{-1}). Geological cross-section (a) shows a progression of peatland types as the contributing watershed area to the peatland increases in the slightly calcareous clay loam till. Similarly, groundwater that has dissolved the underlying limestone enters the strong fen (see arrowed flow lines). In contrast, a geological cross-section without calcium sources yields a series of bogs, even though one is fed by very calcium-poor groundwater (b).

having few bases dissolved in the water. In both cases the bogs occur near groundwater divides, high in the watershed system where groundwater has not traveled very far and has not had the opportunity to dissolve bases.

Evapotranspiration

Evapotranspiration from all wetlands is near potential rates (within 15 percent) as calculated from climatic data (Thornthwaite, 1955; Eggelsmann, 1963; Bay, 1966). Ingram (1983) concludes that bogs dominated by nonvascular mosses may evapotranspire somewhat less than potential rates, while fens with many more vascular plants evapotranspire somewhat more water than potential rates. However, Lafleur and Rouse (1988) measured equal summer time evapotranspiration rates (determined by the Bowen energy ratio) from a moss-dominated and a sedge-dominated peatland near James Bay (50°N, 80°W). In growth chambers at 19°C there was no difference in lysimeter-measured evapotranspiration in large peat pots with moss only or moss and significant amounts of sedges (Nichols and Brown, 1980). In a black spruce, *sphagnum* bog in Minnesota (47^1/$_2$°N, 93^1/$_2$°W), complete water balance residuals for evapotranspiration were essentially equal to Thornthwaite potential evapotranspiration (Verry and Timmons, 1982).

A "crop coefficient" can also be applied in peatland studies of evapotranspiration during the fall period of plant senescence. In northern Minnesota, latent heat flux (LE) consumed 60 to 80 percent of the net radiation (Rn) during May to August (daily mean air temperature varied from 15 to 25°C). However, the value of LE/Rn ranged from 0.4 to 0.6 during the senescence period (September to October) when air temperatures fell below 10°C. During the entire measurement period (late May to early October), the daily ET ranged from 0.9 to 6.0 mm d^{-1}, with a seasonal average of 3.6 mm d^{-1} using eddy correlation techniques

(Verma et al., 1993). Previous Thornthwaite estimates of growing season potential evapotranspiration ranged from 3.5 to 3.9 mm d^{-1} (Verry and Timmons, 1982).

The sum of evaporation and transpiration from wetlands has been compared to the evaporation from open water (lysimeter tubs, ponds, lakes). In studies of lakes where emergent vegetation zones are compared to open water zones, water loss by evapotranspiration is greater than loss by evaporation (Penfound and Earle, 1948; Timmer and Weldon, 1967; Brezny et al., 1973; Van der Weert and Kamerling, 1974; Benton, 1978). Evapotranspiration from cattails (*Typha* sp.) is either more, equal to, or less than evaporation from open water, depending on stage of growth and stem density (Braslavskii and Vikulina, 1963; Eisenlohr, 1966; Shjeflo, 1968; Linacre et al., 1970). In large growth-chamber peat pots, Nichols and Brown (1980) found that *Sphagnum* covered peat or *Sphagnum* and sedge covered peat evapotranspired more than open water pots, using lysimeter measurements of water loss.

In contrast, Rouse et al. (1977) found evapotranspiration from a wet sedge meadow to be 8 percent less than that from a shallow lake near the Hudson Bay coast (57°45'N, 88°45'W), using Bowen ratio methods. Farther north and west (64°27'N, 97°47'W), Roulet and Woo (1986) found that a compacted-peat fen and a small shallow lake had approximately the same amount of water loss using Bowen ratio and lysimeter measurement, respectively. Both a moss-dominated and sedge-dominated wetland near James Bay (50°N, 80°W) had less evapotranspiration (Bowen ratio method) than an estimate of open water evaporation based on the Pennman formula (Lafleur and Rouse, 1988). Cold summer soils (< 10°C) and permafrost at high latitudes may provide a considerable heat sink that robs net radiation energy in northern wetlands compared to shallow lake water so that less energy is available for evapotranspiration from the wetland than from the lake. In contrast, net radiation energy was augmented by nearly 50 percent with heat exchange from the air to the evaporating surface in peatland-placed *Sphagnum* moss tanks near Ottawa, Canada (45°N, 75°W) where there was little exchange of heat between the underlying peat and the evaporating surface (Williams, 1968).

Actual evaporation is primarily a function of net radiation (although ground heat and exchange of heat with the air are locally very important). For instance, water balance studies from peatlands in northern Minnesota (47°N latitude), and eastern Finland (61°N latitude) show equal precipitation, 775 mm, but evapotranspiration of 500 and 250 mm, respectively. Peatland runoff from ombrogenous mires in eastern Finland (525 mm) is more than double peatland runoff in northern Minnesota (225 mm). The result is relatively higher growing-season water tables, reduced soil aeration, and slower tree growth.

Large-scale, empirical watershed data cannot be acquired with any greater precision than ± 12 to 15 percent. However, detailed plot studies report smaller significant differences caused by water table depth and vegetation type (see for instance Juusela et al., 1969; Romanov, 1968). When water rises above the peat surface (or hollow bottoms), there is a slight reduction in actual ET. When the water table drops lower than 30 to 40 cm below the bottom of hollows (or a level

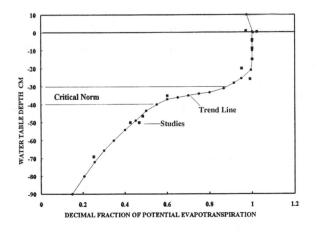

Figure 2 Actual evapotranspiration as a decimal fraction of potential evapotranspiration in relation to water table depth. Data are from Romanov (1968) as interpreted by Ahti (1987), Nichols and Brown (1980), and Mannerkoski (1985). Error about the relationship line is estimated at ± 30 percent. (From Romanov, V. V., *Evaporation from Bogs in the European Territory of the USSR*, Transl. for Israel Program for Scientific Translations Ltd., 1968, 186 pp.; from Ahti, E., *Commun. Inst. For. Fenn.*, 141, 64 pp., 1987; from Nichols, D. S. and Brown, J. M., *J. Hydrol.*, 48, 289, 1980; and Mannerkoski, H., *Effect of Water Table Fluctuation on the Ecology of Peat Soil*, Publication of the Department of Peatland Forestry, University of Helsinki, No. 7, 1985, 190 pp. With permission.)

peat surface), there is a greater reduction in ET (reaching 75 percent at 70 cm). Well-documented data from controlled field and growth chamber studies are summarized in Figure 2 (see also similar relationships in Vompersky and Sirin, this volume.)

Streamflow Hydrology

Streamflow Duration

The excess of water entering the wetland over evapotranspiration from the wetland occurs as either streamflow or groundwater recharge (water yield in total). Water source and landscape setting determine the magnitude and consistency of water yield from wetlands. Two basins at the Marcell Experimental Forest in north central Minnesota (47°32′N, 93°28′W) illustrate the impact of surface and groundwater sources on streamflow regime. These are basins number 3 and 5, each has a surface watershed of 53 ha, and each contains 16 ha of peatland (30 percent of the watershed). One is a groundwater-fed forested fen, and one a surface water-fed forested bog. Both basins lie in ice block depressions, the first in outwash sands with the peat deposit nestled within the regional aquifer, and the second in outwash sand capped with 3 m of clay loam till holding the peat deposit above it. Annual flow-duration curves for these basins of equal surface area are quite different (Figure 3).

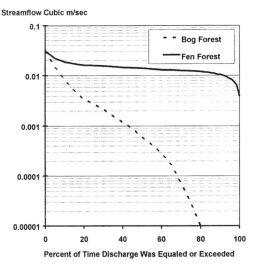

Figure 3   Streamflow duration curves for a groundwater fen and perched bog basin. Both
basins have 53 ha of surface area and both have 30 percent peatland. (From
Verry, E. S. and Boelter, D. H., in *Proc. Hydrology of Marsh-Ridden Areas,* Minsk,
The Unesco Press, IAHS, Paris, 1975, p. 469. With permission.)

The bog basin receives water from surrounding mineral soil subsurface flow,
as well as from rain and snow, but storage in the peatland and in the mineral soil
(sandy loam to clay loam texture in the till) is small. Thus, streamflow leaving the
basin is high during heavy rain or snow melt and declines rapidly during periods
of high ET demand. In many years the stream stops flowing during midsummer
or late winter for as much as 25 percent of the year (Figure 3). Its flow duration
curve is relatively steep, reflecting a lack of long-term storage in the basin.

The groundwater-fed fen has a flow duration curve that is nearly flat, but has
the same high flow rates and relatively low streamflow rates only 5 percent of
the time (Figure 3). The high flow duration from the fen is caused by large inputs
of groundwater from a groundwater basin that is ten times the area of its surface
watershed area. It is not the wetland (in this case a forested peatland) that sustains
streamflow, but rather the large sand aquifer that stores and releases groundwater
that originated as surface water percolating through either deep sands or the poor
fen laggs (wet, shrub edges) of perched bog basins. Sustained flow duration curves
have also been documented by Roulet (1989, 1990) for a southern Ontario
groundwater-discharge wetland, and for spring-fed swamps in the northeastern
United States (Carter and Novitzki, 1988).

Both surface basins (bog and fen) have the same peak flow because peak
flows are dominated by surface processes (snowmelt or large rains) and because
both basins have the same amount of peatland. Low flows and annual water yield
are determined by either the low detention storage available in the organic surface
soils or by the high detention storage available in groundwater aquifers.

Flow duration curves for wetland streams can be intermediate when wetlands are connected ephemerally to adjacent hydrologic units. For instance, Taylor and Pierson (1985) showed that the spring recession in flow extended over a month in a southern Ontario cedar swamp seasonally connected to a perched aquifer. A similar spring recession from a low Arctic fen lasted only two weeks when overflow water from adjacent lakes and streams entered the fen (Roulet and Woo, 1988). When these exchanges cease, streamflow from the peatland ceases (Roulet, 1990).

Water Balance

Monthly Water Budgets

The near potential evapotranspiration from wetlands can subtract substantially from precipitation inputs in the midlatitudes. Basin number 2 on the Marcell Experimental Forest in northern Minnesota is 10 ha in total size. The upland is two-thirds of the basin, and the black spruce peatland (bog with a poor fen lag) is one-third of the basin. The monthly water balance in this basin illustrates the dramatic consumption of precipitation by evapotranspiration demand (Figure 4). In January 1969, there was a high amount of snowfall, but there were normal amounts of rain in May through September. Even so, the high potential ET (Thornthwaite and Mather, 1955) reduced streamflow to 10 or 20 percent of precipitation.

Trees or No Trees

Although precise water budget data comparing peatlands with or without trees have not been developed, paired basin studies (Basins 1 and 2 at the Marcell

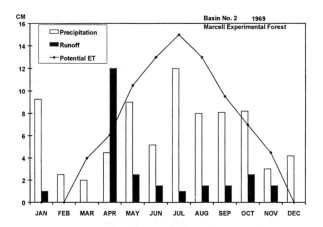

Figure 4   Streamflow from a perched peatland basin is severely limited by evapotranspiration even though summer rainfall is large. (From Boelter, D. H. and Verry, E. S., Peatland and Water in the Northern Lake States, USDA Forest Service, North Central Forest Experiment Station, General Technical Report, NC-31, 1971, 22 pp. With permission.)

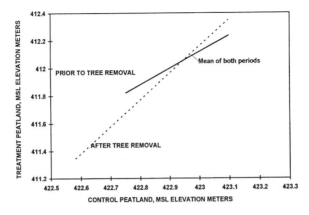

Figure 5   Peatland water tables with an initial growing season norm within 15 cm of the surface will fluctuate more after trees are harvested, but the average water table position is virtually the same. (From Verry, E. S., Water table and streamflow changes after stripcutting and clearcutting an undisturbed black spruce bog, in Proc. Sixth Intl. Peat Congress, Duluth, MN, Intl. Peat Society, Helsinki, 1981. With permission.).

Experimental Forest) have been conducted that evaluated the impact of tree removal from natural, undrained, forested peatland on water table position and, presumably, evapotranspiration.

Cutting trees from a natural peatland, with a water table normally well within 30 cm of hollow bottoms, will increase the range of week-to-week fluctuation in water table elevation. Water tables will be higher during and just following rains (perhaps 10 cm higher) and lower during drying periods (perhaps 20 cm lower), but the average water table elevation and annual water yield will remain the same when annual precipitation is within ± 30 percent of the mean (Verry, 1981) (Figure 5). Fluctuations between periods of rain and drying normally occur on a weekly basis or two week basis; thus, the additional fluctuation imposed by sedge-moss surfaces as compared with trees (assuming similar streamflow outlet characteristics) is averaged out within the monthly bars shown in Figure 4.

Cutting trees on fine-grained mineral soil wetlands (fine sands and silts), where the water table is normally lower than 30 cm from the surface under a mature tree canopy, will cause an increase in average water table elevation and annual water yield (Trousdell and Hoover, 1955). This response is similar to that following tree harvest on well-drained, upland mineral soils. Twenty to thirty percent increases in water yield are common and are caused by a reduction in canopy interception and tree transpiration (Bosch and Hewlett, 1982; Brooks et al., 1991).

The Critical Norm is the average summer time water table position below the surface (in drained peatlands, it is midway between ditches). The cause of the dichotomy in water table and streamflow response to tree cutting is the existence of a nearly saturated soil profile in natural, undrained peatlands, and the existence of a deep, mostly unsaturated soil profile (prior to tree removal) in fine-textured mineral soils. The location of this critical norm is shown in Figure 2. Here the nearly horizontal part of the response curve occurs between a depth-to-water table of 30 to 40 cm. If the water table during the growing season is normally above this level, the water table is supplying water to the surface continually by the process of capillary rise, and ET rates are near potential. Cutting trees will change the range of water table fluctuation, but ET rate will still be near potential. Thus, significant changes in streamflow volume cannot occur. If the water table during the growing season is below this critical norm (30 to 40 cm), the presence of a deep unsaturated profile will allow the increased input from canopy interception and reduced transpiration following tree cutting to find shelter from the sun and wind and will cause the water table to rise and water yield to increase.

The ability of maturing trees to keep the water table below the critical norm during the growing season (except for storm peaks), has been termed *biologic drainage* by Heikurainen and Päivänen (1970), and is a confirmation in organic soils of the streamflow response processes occurring in unsaturated mineral soils (Bosch and Hewlett, 1982). Large-scale documentation of this process is provided by Mustonen and Seuna (1971). One should be cautious about all of the interpretations in this section when areas north of 60°N or south of 35°N are considered because of the lesser or greater impact of solar radiation on the water balance.

Variable/Static Source Areas

Hewlett and Hibbert (1967) have demonstrated the importance of variable source areas to the generation of streamflow. In their description of a North Carolina stream system, they described the expanding wetted-channel length as repeated rains brought the local water table into the bottom of dry stream channels. Thus, more and more channel became ready to immediately respond with streamflow at the next rain. Wetlands can behave similarly. That is, the dry peatland (Figure 4) may need to fill with water before it contributes to streamflow in perched peatland basins with relatively flat peat surfaces. Thus, repeated rains may bring more and more channel-connected peatlands into the active streamflow system (Brooks et al., 1991). On the other hand, fens that are nearly always wet or domed bogs that have developed above their mineral soil base are usually wet and represent a static landscape feature that will respond immediately to rain-generated streamflow. Both specific seepage points that sustain surface streamlets and diffuse groundwater inflow to a southern Ontario cedar swamp kept the swamp near saturation, and maintained stream baseflow with little seasonal variation (Roulet, 1990).

However, the magnitude of storm peak response depends on the nearness of the water table to the surface prior to rainfall. Peak stormflow response from a perched peatland in Minnesota was three times larger when the water table was near the surface compared to peak stormflow response when the initial water table was 15 cm below the surface (Bay, 1970).

The 1971 hydrograph from Basin number 2 on the Marcell Experimental Forest in north central Minnesota (see Water Balance) illustrates the relative mportance of the basin components to streamflow from the entire basin. Because the wetland (black spruce, perched bog) is the source of the channel leaving the basin, it responds immediately to snowmelt or large rains, while flow from the surrounding mineral soil uplands lags behind and usually produces much less streamflow.

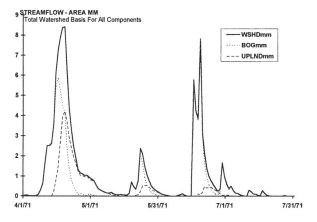

Figure 6a Hydrograph separation from Mire No. 2 (9.7 ha total watershed, 6.5 ha of mineral soil upland, and 3.2 ha of ombrotrophic bog). Spring snowmelt and large summer storm response.

Figure 6a illustrates the relative contribution of the mineral soil upland (two-thirds of the basin) and organic bog (one-third of the basin) to the total basin hydrograph. The April peak is snowmelt that clearly shows the initial and strong contribution of the bog, followed several days later by flow from the upland soils that sustain the total-basin, snowmelt hydrograph well into the month of May. The importance of the wetland to produce streamflow in the growing season is well illustrated by the large storm peak in June (Figure 6a) and again in late October following a dry summer period (Figure 6b). Although it is not illustrated, the relative contribution of bog and upland is probably nearly equal during 20-year plus recurrence events. They are approaching equality during the April-May snowmelt event.

The hydrograph responses for fens and bogs are nearly identical, except that the fen hydrograph rides on a usually much higher base flow. Thus, wetland response to storms is immediate if the basin is full enough to contribute to streamflow. How then do we compare basins with significant wetlands to those without significant wetlands? First, we must understand that stormflows generated from wetlands behave virtually identically to stormflows passing through reservoirs. Both reservoirs and wetlands are flat, and it is their flat morphology that spreads water out, slows flow rate, and reduces peak flow rate in the receiving channel (Chow, 1964; Conger, 1971; Moore and Larson, 1979; Verry et al., 1988; Johnston et al., 1990; Krug et al., 1992).

Any flat physiographic feature — wetland, lake, or reservoir — will spread flood waters and slow them down. This effect is most important when wetlands or lakes make up from 5 to 20 percent of a basin, and when these features are

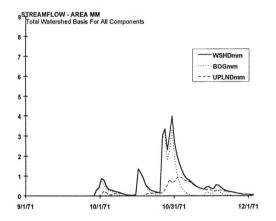

Figure 6b Hydrograph separation from Mire No. 2 (9.7 ha total watershed, 6.5 ha of mineral soil upland, and 3.2 ha of ombrotrophic bog). Fall storm response following a dry late summer.

part of the regional drainage system where lakes are in channels and wetlands are in the flood plain. At this density, they can reduce flood peak rates by up to 75 percent (Figure 7).

Moore and Larson (1979) have developed similar equations for basins with wetlands and lakes in southern and western Minnesota and in eastern North and South Dakota (North American Great Plains). Their equation predicts the mean annual flood peak as a function of an intercept, the ⅔ power of basin area and the −½ power of the area of wetland and lake (a value of 1 was added to wetland and lake percentage to avoid 0 percent). They acknowledged that the last exponent reflects an indirect topographic index as well as the percent of undrained wetlands and lakes. Thus, it considers general differences between flat and rolling areas within a basin.

WATER TABLE POSITION AND VEGETATION GROWTH

A basic tenet of peatland drainage is that tree height and wood production are increased as water tables are lowered. The term Growing Season Norm is used to identify the average water table during the summer midway between ditches. However, the relation between water table position and maximum vegetation height is universal and equally applicable to plant ecology and wetland classification endeavors.

Defining water table position is hampered by seasonal and storm-generated rises and falls. However, water table position can be presented more precisely in a way similar to flow duration curves for streamflow. Two long-term recording wells at the Marcell Experimental Forest in north central Minnesota and drainage experience from the literature and personal experience give rise to a family of water table duration curves whose fiftieth percentile value is correlated with maximum vegetation height on peatlands (Figure 8).

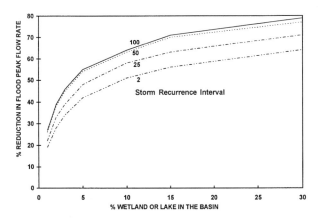

Figure 7 Storm peak flows are significantly reduced if basins contain more than 15 percent wetland or lake connected to the major drainage system. (From Conger, D. H., *Estimating Magnitude and Frequency of Floods in Wisconsin,* U.S. Geological Survey, Madison, WI, Open File Report, 1971, 200 pp. With permission.)

The curves in Figure 8 represent growing season measurements taken between May 15 and September 30. Typically, values above the 95 percent duration value represent storm days. Values below the 20 percent duration represent extended drying periods of 2 weeks or longer. The curves cover a range of 80 cm, 10 cm above the hollow surface and 70 cm below the hollow surface (an additional 10 cm should be added to the range above the surface in order to include water levels during large snowmelt periods in March and April). The low values are typical of northern Minnesota droughts from 1961 through 1992, but would not include droughts typical of the 1930s; thus, values below −70 cm should be expected and can persist for a year or two.

The surface line in Figure 8 was drawn for peatlands where the surface is essentially static; it does not float. However, the relationships in Figure 8 should also hold for floating peatlands. That is, the water table regime curves and the peat surface would move up and down together. Thus, a floating peatland could well experience a greater range of absolute well water elevations.

The heavy lines associated with maximum vegetation heights of 2 and 15 m are taken from Mire Number 2 and from the Bog Lake peatland at and near the Marcell Experimental Forest. They represent 3 and 25 years of continuous record, respectively. The Bog Lake Peatland is dominated by *Sphagnum* moss, Schuchzerzia, Rhynchospera, and scattered stunted tamarack trees to 2 m in height. It is classified as a treed poor fen. The Mire Number 2 is a *Sphagnum* moss, low to medium site black spruce peatland where the trees reach 15 m in height. Other curves are hand drawn to approximate similar-shaped distributions.

The fiftieth percentile can be used to characterize this family of water table duration curves. Thus, a peatland with a fiftieth percentile value of about 6 cm below the peat (hollow) surface will support vegetation up to a meter tall (sedges, grasses, *Ericacea*, etc.). A peatland with a fiftieth percentile of 10 cm below the surface will support vegetation to 2 m, and so on: 17 cm/15 m, 30 cm/20 m, 43

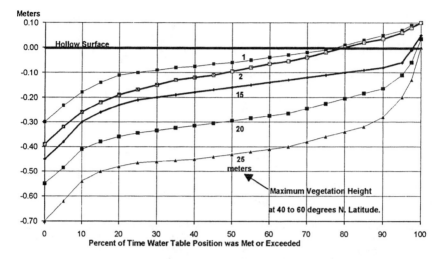

Figure 8 Relation of water table regime to maximum vegetation height in northern peatlands (Verry, unpublished data, 1993). Bold lines are based on long-term data, and thin lines on literature-guided speculation.

cm/25 m. This rapid response of vegetation height to relatively small changes in water table position is no surprise to peatland foresters. However, the general trend (Figure 9) is widely applicable in wetland science; for instance, wetland classification systems sometimes use the term *swamp* to describe wetlands supporting commercial size trees. While use of commercial tree harvest criteria seems incongruent in a wetland classification system, the ecological basis of simple maximum vegetation height as a function of water table position is a very reasonable criterion for separating sites with or without large trees. Water table position (a fiftieth percentile of 15 cm, with 15-m tall vegetation) alone can be the criterion. Names, such as *swamp*, have different meanings depending on colloquial usage; nevertheless, a commonly accepted word or words (such as *treed* for peatlands with small scattered trees, and *forest* for peatlands with large well-spaced trees) seem achievable.

Further ecology-based support for this approach is presented by Vompersky (1991). In his evaluation of the distinction between peatland forests and treed peatlands, he suggests that when the annual net production of trees (g m^{-2} yr^{-1}) is greater than the annual production of smaller plants, the wetland should be classed as a peatland forest. When the annual production of the smaller plants is more than the trees, the wetland should be classified as treed peatland if small trees are present.

Drainage can, of course, change the maximum tree height by lowering water table position. Other factors, such as nutrients and temperature sums, will also impact maximum vegetation height. However, the critical water table range is in the 0- to 15-cm zone, after which tree height can be promoted with several factors. Other uses of the water table position/vegetation height relation can be made. For instance, the occurrence of sedge/grass, shrub, or tree zones in riparian areas

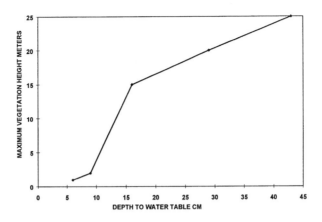

Figure 9   Maximum vegetation response to changes in fiftieth percentile water table position is rapid in the 0 to 15 cm depth below the peat surface.

suggests that vegetation management must consider water table position first before desired future conditions in riparian areas are prescribed.

The plant response curve of Figure 9 does not continue upward. Feddes, Kowalik, and Zaradny (1978) have shown that crop yields on sand, sandy loam, and clay soils produce maximum biomass as the length of time soil water matrix potential remains at −10 Kpa (field capacity) increases. Thus, rigorously controlled peatland drainage seeks to maintain a well-drained soil profile near field capacity. As doughtiness increases and less favorable soil water potentials prevail, plant biomass production decreases.

Olsta et al. (1990) have shown that the lowest level that water tables can drop in a peat–muck soil and still maintain maximum evapotranspiration is 60 cm. This is 20 cm lower than that shown in Figure 8, but near the ± 30 percent error band associated with the relation in Figure 8. In general, organic soils with a greater degree of decomposition (von Post H-values in the 7 to 10 range) will have a capillary fringe that reaches the surface from a water table at 60 cm, while less decomposed peats will have a capillary fringe that reaches the surface with the water table at 30 to 40 cm (see also Vompersky and Sirin, this volume.)

SNOW AND ICE

Snow and ice are major factors in the movement of materials in wetlands. Snow loads of only 30 to 45 cm depth compress senescent emergents and living moss in peatlands, and the development of ice (frost) in frigid peatlands helps define some aspects of the acrotelm/catotelm concept (active zone/dead storage zone concept). In north central Minnesota, snow packs in late December accumulate to 30 cm 2 out of 3 years, and continue to accumulate to at least 45 cm in most years; this is sufficient to compress annual net vegetation production. The depth of soil frost penetration at the end of the winter season (February,

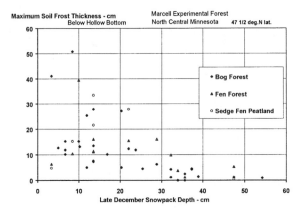

Figure 10   Peatland frost depth near the end of the winter is a function of the depth of snow in late December.

March) is a function of the snow depth occurring in late December, whether the peatland is a bog or fen forest or a sedge fen peatland (Figure 10).

Frost typically penetrates 15 cm, and in 2 out of 30 years can penetrate 40 to 50 cm. Frost penetration is the primary mechanism for material movement from the acrotelm to the catotelm. Because pure water freezes first, colloidal peat material and plant fragments migrate to the bottom of the soil frost layer. This is easily seen in a section of the soil frost cut from the peatland. These sections visually show clear ice at the surface, usually extending 10 to 15 cm before darker colored ice forms at the bottom of the frost layer. Thus, dating of peat cores may be more accurate if the usual bottom of frost formation is used as the beginning of peat accumulation. This may be particularly important when peat cores from frigid climates are compared to peat cores from climates where frost is not a major factor in plant material accumulation.

In frigid climates, frost disappearance in the spring is another hydrological process that impacts water table position, nutrient cycling, and tree harvest operability. Water tables in frigid peatlands may be 5 to 10 cm higher during early snow melt, when meltwater rides above the soil frost. A noticeable drop in water table occurs when soil frost leaves the peatland. Snow melt from upland mineral soils is laden with phosphorous from ice-ruptured leaves, but rapid uptake in new *Sphagnum* growth above the frost layer in peatlands quickly utilizes the available phosphorous and diminishes phosphorous export from the wetland (Verry and Timmons, 1982).

The disappearance of frost from year to year may span over 6 weeks, depending on weather conditions (Figure 11). However, frost disappearance is easily predictable. In each of 30 years at the Marcell Experimental Forest in north central Minnesota, frost disappeared in each case after three consecutive nights with above freezing air temperatures (Figure 11). Thus, harvesting and log transportation equipment can be removed from peatlands before soil is severely rutted or compacted.

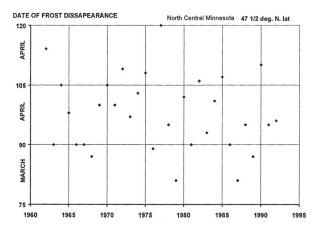

Figure 11 The disappearance of peatland concrete frost varies over a period of 6 weeks across a 30 year record. However, each frost disappearance is predictable, because each date above occurred after three consecutive nights of above-freezing air temperatures.

NUTRIENT CYCLING IN WETLANDS

Nutrient cycling in wetlands is strongly influenced by the chemistry and amount of flow in the water source and by the aerobic/anaerobic conditions in wetland waters. The pH regime and oxidation-reduction regime in wetlands determine which elements or ions precipitate or stay in solution. The lack of the circum-neutral pH buffer (calcium bicarbonate) and the presence of organic acids establish low pH values in bogs, while the presence of increasingly higher amounts of calcium (and magnesium) bicarbonate largely defines the broad class of peatlands known as fens (poor, transitional, modal, and strong). The balance between water input and evapotranspiration — water yield — then determines the overall concentration level of ions and particulates. Wetlands with strong groundwater flow physically export organic materials at high rates. The relative size of the wetland in a basin and the amount of water source flow through a wetland largely control the wetland's ability to sequester nutrients. Examples of proton (H^+), carbon, and nutrient budgets illustrate the impact that water source quality and amount have on nutrient cycling in wetlands.

Proton Cycling

The lagg and bog of Mire No. 2 at the Marcell Experimental Forest provide one view of how a wetland ecosystem handles sources and sinks of acidity. The proton balance is an integration of the many processes involving acid–base and oxidation–reduction reactions. Feedback loops are controlled by the concentration of H as it influences the capacity of the system to generate or consume acidity by such processes as nutrient uptake, decomposition, weathering, ion exchange,

Table 1 Acidity Balance for Marcell Mire S-2
(3.2 ha)

Component	Acidity meq m^{-2} yr^{-1}
Sources	
Wet deposition	$-<1$
Dry deposition	-1.5
Upland run-off	-44
Nutrient uptake	827
Organic acid production	263
Total	1044
Sinks	
Denitrification	12
Decompostion	784
Weathering	76
Outflow	142
Total	1014

From Urban, N. R., et al., in *Geographic Variation in Northeastern North America,* Hutchinson, T. C. and Mleema, K. M., Eds., Springer-Verlag, Berlin, 1987, 577. With permission.

and others. Table 1 illustrates the acidity balance for Marcell Mire S-2 (Urban et al., 1987).

In bogs receiving primarily precipitation as their water source, production of organic acids is the dominant source of acidity (263 meq m^{-2} yr^{-1}), and these organic acids serve to buffer the bog water near pH 4. Net biological uptake is the difference between plant nutrient uptake that releases H^+ to the wetland water (827 meq m^{-2} yr^{-1}), and plant decomposition that consumes protons (H^+) from the water (784 meq m^{-2} yr^{-1}), leaving a net acidity of 43 meq m^{-2} yr^{-1}. Outflow of water and mineral soil weathering (dust blown into the bog) consume additional protons, but overall a net acidity of 30 meq m^{-2} yr^{-1} remains and keeps bog water near pH 4.

Wetlands receiving large amounts of calcium bicarbonate-rich groundwater may have a groundwater source of negative acidity (that is alkalinity largely as bicarbonate ions HCO_3^-) between 400 and 500 meq m^{-2} yr^{-1}. In these cases, fen water would buffer between pH 5.4 and 9, depending on the strength of bicarbonate and hydroxyl ions in the groundwater source. Thus, water source chemistry and amount can change the fundamental processes of nutrient cycling by controlling the pH of the soil solution.

Carbon Cycling

Wetland soils in the mid to northern latitudes of the northern hemisphere make up 3 percent of terrestrial soils, but these northern wetlands contain approximately 24 percent of the total global carbon pool (Maltby and Immirzi, 1993).

Table 2   Estimated Carbon Budget for Marcell Mire S-2 (24 ha)

Component	g C m⁻² yr⁻¹
Inputs to the acrotelm	
Living moss respiration (pass through)	118
Sphagnum moss net primary productivity	190
Tree and shrub net primary productivity	175
Herbaceous net primary productivity	20
Total	503
Outputs from the acrotelm	
Peat accumulation (C storage)	28
TOC in water yield	37
Methane evolution	6
CO_2 soil surface efflux	471
CO_2 moss respiration	118
Total	660

From Verry, E. S. and Urban, N. R., *Suo*, 43, 147, 1993. With permission.

Regionally, wetlands in northern latitudes (> 45°N) can compose 10 to 30 percent of the landscape. Thus, the amount of carbon in these systems is significant, and the interaction of these systems with climate may also be significant. Water table position and soil temperature largely control the amount of the greenhouse gases, carbon dioxide, and methane released to the atmosphere.

Mire S-2 in north central Minnesota again gives an example of the carbon budget in a peatland ecosystem (Table 2). However, weather and hydrology cause great variation in the individual terms for various wetlands in the northern hemisphere. Methane evolution from Mire 2 is only 6 g C m⁻² yr⁻¹. Mire 2 is relatively dry compared to nonforested wetlands, even though its soil temperature is normally above 15°C at 10 cm during the height of the growing season. Crill et al. (1988, 1992) show that methane evolution from Canada and Minnesota can range between 0.1 and 66 g C m⁻² yr⁻¹ as a function of water table position, vegetation height and shading, and thus soil temperature.

Dissolved organic carbon (DOC) makes up most of the export of waterborne carbon from wetlands (DOC is 90 percent or more of Total Organic Carbon TOC; Sallantaus, 1988). The interplay of water supply (water source) and water balance (relative dryness and thus relatively higher concentrations) also gives a wide range in DOC (or TOC) export from peatlands. Sallantaus (1988) gives DOC export values from peatlands ranging from 6 to 78 g C m⁻² yr⁻¹ in a western Finland ombrotrophic mire and a New Zealand fen, respectively (Moore, 1987).

The apparent accumulation rate of C in the catotelm at Mire S-2 is 28 g C m⁻² yr⁻¹; however, Tolonen et al. (1992) report a range from 8 to 41 g C m⁻² yr⁻¹ for many peatlands in Europe and North America; it is similar to a northern hemisphere average estimate by Gorham and Janssens (1992) of 29 g C m⁻² yr⁻¹.

While carbon accumulation rate is not directly correlated with peatland type (water source chemistry or amount), it is associated with regional temperatures and precipitation (see Klimanov and Sirin, this volume).

Major Waterborne Nutrients

The amount of nutrient (N or P) retained in a wetland is dependent on the hydraulic loading to the wetland and on the relative size of the wetland with respect to the entire basin. An example of a low hydraulic load bog (Minnesota) and a high hydraulic load fen (Ontario) illustrates how variation in water amount impacts the ability of wetlands to retain nutrients. On an area basis, the entire basin of the Minnesota bog yields 250 mm of water yield, while the entire basin of an Ontario fen yields 540 mm of water yield. An additional factor is the upland:wetland area ratios for each site. The Minnesota bog has a ratio of 2:1, while the Ontario fen has a ratio of about 7:1. Overall the Ontario fen has a hydraulic loading of about seven times that of the Minnesota bog.

On an annual basis, the bog retains 50 percent of the total nitrogen and 61 percent of the total phosphorus, while the fen retains only 10 percent of the total nitrogen and 4 percent of the total phosphorus. Much of the difference is caused by the much larger export of organic forms of nitrogen and phosphorus in the particulate form from the fen that passes seven times the amount of water. Organic nitrogen retention in the bog is 36 percent and −76 percent in the fen, while organic phosphorus retention in the bog is 61 percent and a −15 percent in the fen (Table 3).

While only small basin bog data are available for the metals, little more than half of the calcium is retained, while a little less than half of K, Na, and Mg are retained (Table 3). In large basins with large amounts of calcium- and magnesium-rich groundwater, the basins will show a net export of calcium and magnesium as limestone and dolomite in basin aquifers are leached.

These patterns of nutrient retention hold in general for wetlands (Johnston et al., 1990; Detenbeck et al., 1993), unless stream channels through wetlands do not allow channel water to spread throughout the wetland.

CONCLUSIONS

Ecosystem function refers to the chemical and physical processes that govern the flow of materials through ecosystems. In wetlands, hydrological processes play a major role in the movement of materials in the liquid, gas, and solid phases. Wetlands form where physiography or climate are conducive to slow water movement, and that water is available at the earth's surface. It is the quality, quantity, and periodicity of water at the site that drive the development of wetlands and determine wetland type.

In peatlands, bogs form primarily where precipitation exceeds evapotranspiration, the precipitation-derived water is poor in bases (Ca, Mg, etc.), and organic acid production yields high H^+ levels buffered near pH 4. In some locations,

Table 3   Annual Nitrogen, Phosphorous, and Metal Input, Output, and Retention Values for a Minnesota Black Spruce Bog

Nutrient	In bog (kg/ha⁻¹/year⁻¹)	Out bog (kg/ha⁻¹/year⁻¹)	Ret. bog (kg/ha⁻¹/year⁻¹)	Ret. bog (%)	In fen (kg/ha⁻¹/year⁻¹)	Out fen (kg/ha⁻¹/year⁻¹)	Ret. fen (kg/ha⁻¹/year⁻¹)	Ret. fen (%)
Nitrate — N	2.0	0.3	1.8	86	6.4	3.1	3.3	51
Ammonium — N	2.3	0.7	1.5	69	3.7	0.4	3.3	89
Organic — N	8.4	5.4	3.0	36	6.3	11.1	-4.8	-76
Total — N	12.7	6.4	6.3	50	16.4	14.7	1.7	10
Ortho — P	0.4	0.2	0.2	60	0.9	0.4	0.5	53
Organic — P	0.8	0.3	0.5	61	2.1	2.4	-0.3	-15
Total — P	1.2	0.5	0.7	61	3.0	2.8	0.2	4
Potassium	11.1	6.1	6.1	45	—	—	—	—
Sodium	4.2	2.4	2.4	43	—	—	—	—
Calcium	20.9	9.1	9.1	56	—	—	—	—
Magnesium	6.0	3.8	3.8	36	—	—	—	—

Note: Precip. = 750 mm; ET = 500 mm; water yield = 250 mm; upland:wetland ratio = 2:1; and annual nitrogen and phosphorus values for an Ontario white cedar swamp (fen): precip. = 1000 mm; ET = 460 mm; water yield = 540 mm; and upland:wetland ratio = 7:1.

From Verry, E. S. and Timmons, D. R., *Ecology*, 63, 1456, 1982; and from Devito, K. J. and Dillon, P. J., *Water Resources Res.*, 29, 2675, 1993. With permission.

usually near major watershed divides, where outwash sands (quartz) or bedrock (granite) are base-poor, bogs form in depressions fed by groundwater little different in chemical composition from ET-concentrated rain water. Fens represent a broad range of peatlands where the chemistry of the groundwater source has an increasingly higher concentration of bases (Ca, Mg, etc.) and a corresponding increase in the bicarbonate ion (HCO_3^-) that buffers water between pH 5.4 and 9. The groundwater source also varies in the strength of its total flow. Water table position relative to the peat surface rises as more groundwater enters a wetland, but the net impact on water table position is also controlled by the ease of water to leave the wetland at its outlet under the influence of wetland slope and physical constrictions. Similar outlet controls can be exhibited in bogs that are not elevated in raised domes above their surroundings. Mineral soil wetlands or mineral soil wetlands with thin organic surface layers exhibit similar ranges in water chemistry and water table position.

Water table position relative to the wetland soil surface determines the life form of mature vegetation. The fiftieth percentile of the water table duration curve is directly correlated with mature vegetation height. Small changes in average depth-to-water yield large changes in mature vegetation height: 6 cm to 1 m (grasses and sedges), 10 cm to 2 m (stunted trees and shrubs), 17 cm to 15 m (trees), 30 cm to 20 m, 43 cm to 25 m.

Evapotranspiration from wetlands is within ±15 percent of potential rates as long as the water table is within 30 to 40 cm of the peat surface (the Critical Norm) and the capillary fringe extends to the surface. Evapotranspiration declines below this critical norm to only 10 percent of potential rates when the water table resides a meter below the surface. In late fall when plants senesce, ET from wetlands may decline 40 percent below potential rates for a short period before cool temperatures minimize ET. Dense emergent wetlands generally evapotranspire more water than open water ponds or lakes. However, exceptions occur where emergents are thinly spaced yet still provide a relatively deep boundary layer. Another exception is where wetland soils are colder than pond bottoms because of differential heating of dark pond bottoms or the presence of permafrost in the wetland that robs net radiation potentially available for evapotranspiration and uses it instead to warm the cold soils.

Streamflow from wetlands is strongly diminished (and may cease for 20 percent of the year) if rain and snow (surface water) are the only sources of water input (25 to 75 percent of precipitation). In contrast, wetlands receiving groundwater from sand aquifers beneath the wetland have strong consistent amounts of streamflow. Depending on the ratio of ground watershed to surface watershed, streamflow from groundwater-fed wetlands may be 100 to 1000 percent of the precipitation falling only on the surface watershed of the wetland.

Tree harvesting in wetlands may or may not cause a rise in the absolute water table elevation. If, prior to harvest, the water table is within 30 cm of the surface (hollow bottoms in hummocky peatlands), harvesting trees will not cause a rise in the average water table elevation. The range in week-to-week fluctuations will be greater: a little higher after rains, and a little lower after a week or two of no rain, but the growing season average water table elevation will remain essentially the same. Since the water table before and after harvest is above the critical norm and the capillary fringe reaches the surface, only the transitory impacts of diminished canopy interception, drying winds at the surface, and increases in sedge biomass after harvesting are experienced, but they tend to cancel each other out at the same prior-to-harvest average elevation.

If, prior to harvest, the water table is lower than 30 cm below the surface, then tree harvest will increase total water yield just like many upland harvest experiments have shown, and the water table will rise. Especially on mineral soil wetlands with a relatively smooth surface compared to hummock-hollow peatlands, the higher water table is more obvious. Soil rutting or compaction may also expose more surface water. How much the water table rises depends on the specific yield of the soil and the restrictions on flow from the wetland outlet. Absolute rises in the average water table elevation and increases of 30 percent in water yield are common. New trees have the potential to reverse this process and lower water tables as they grow, particularly if the existing water table is maintained 30 cm or more below the surface by drainage.

Peak stormflow response from wetlands is dependent on antecedent water table position and is higher with higher antecedent water tables. However, peak flows from wetlands are attenuated because of their flat topography and just like lakes or reservoirs can reduce flood peak flows up to 80 percent if wetlands compose 30 percent of the basin and are connected to the basin stream system. Peak flow rate reductions of 40 to 50 percent occur even if wetlands and lakes comprise only 15 percent of the basin. Their impact diminishes rapidly below this portion of the basin.

Snow and ice are significant factors moving materials in northern wetlands. First the weight of the snow pack compresses annual vegetation and mosses to the wetland surface where its decomposition accelerates. The formation of soil frost then moves organic colloids produced during decomposition into the catotelm. Clear ice forms first, with progressively darker colored water freezing at the bottom of the frost layer, which typically extends 10 to 20 cm below hollow bottoms in peatlands. This bottom-of-the-frost layer is the effective peat surface in frigid climates.

The ability of wetlands to retain nutrients is a function of their hydraulic loading, pH, and redox regimes. Nutrient retention for total N and P can be as high as 60 percent on an annual basis in low-water flux bogs or as low as 4 percent in high-water flux fens where organic matter movement in the water is high.

REFERENCES

Ahti, E., Water balance of drained peatlands on the basis of water table simulation during the snowless period, *Commun. Inst. For. Fenn.*, 141, 64 1987.

Bay, R. R., Evaluation of an evapotranspirometer for peat bogs, *Water Resources Res.*, 2, 437, 1966.

Bay, R. R., Water table relationships on experimental basins containing peat bogs, IASH–UNESCO, Wellington, NZ, December 1970, pp. 360–368, 1970.

Benton, A. R., Jr., Evapotranspiration from water hyacinth [*Eichornia crassipes* (Mart.) Solms] in Texas reservoirs, *Water Resources Bull.*, 14, 919, 1978.

Boelter, D. H. and Verry, E. S., Peatland and Water in the Northern Lake States, USDS Forest Service, North Central Forest Experiment Station, General Technical Report NC-31, 1977, 22 pp.

Bosch, J. M. and Hewlett, J. D., A review of catchment experiments to determine the effect of vegetation changes on water yield and evapotranspiration, *J. Hydrol.*, 55, 3, 1982.

Braslavskii, A. P. and Vikulina, Z. A., Evaporation Norms from Water Reservoirs, Israel Prog. for Sci. Transl., Jerusalem (Transl. from Russian by I. Schechtman), 1963.

Brezny, O., Mehta, I., and Sharma, R. K., Studies on evapotranspiration of some aquatic weeds, *Weed Sci.*, 21, 779, 1973.

Brooks, K. N., Folliott, P. F., Gregersen, H. M., and Thames, J. L., *Hydrology and the Management of Watersheds,* Iowa State University Press, Ames, 1991, 392 pp.

Carter, V. and Novitzki, R. P., Some comments on the relation between ground water and wetlands, in *The Ecology and Management of Wetland,* Vol. 1, Hook, D. D., Ed., Croom Helm, London, 1988, pp. 68–96.

Chow, V. T., *The Handbook of Hydrology,* McGraw-Hill, New York, 1964.

Conger, D. H., Estimating Magnitude and Frequency of Floods in Wisconsin, U.S. Geological Survey, Madison, WI, Open File Report, 1971, 200 pp.

Crill, P., Bartlett, K. B., Harriss, R. C., Gorham, E., Verry, E. S., Sebacher, D. I., Madzar, L., and Sanner, W., Methane flux from Minnesota peatlands, *Global Biogeochem. Cycles,* 2, 371, 1988.

Crill, P., Bartlett, K., and Roulet, N., Methane flux from boreal peatlands, *Suo,* 43, 173, 1992.

Detenbeck, N. E., Johnston, C. A., and Niemi, G. J., Wetland effects on lake water quality in the Minneapolis/St. Paul metropolitan area, *Landscape Ecol.,* 8, 39, 1993.

Devito, K. J. and Dillon, P. J., The influence of hydrologic conditions and peat oxia on the phosphorus and nitrogen dynamics of a conifer swamp, *Water Resources Res.,* 29, 2675, 1993.

Eggelsmann, R., Die potentielle und aktuelle evaporton eines Seeklima-Hochmoores, General Assemembly, Berkeley, CA, *Int. Assoc. Sci. Hydrol. Publ.,* 62, 88, 1963.

Eisenlohr, W. S., Water loss from a natural pond through transpiration by hydrophytes, *Water Resources Res.,* 2, 443, 1966.

Feddes, R. A., Kowalik, P. J., and Zaradny, H., *Simulation of Field Work Use and Crop Yield,* John Wiley & Sons, New York, 1978.

Glaser, P. H., Janssens, J. A., and Siegel, D. I., The response of vegetation to chemical and hydrological gradients in the Lost River Peatland, northern Minnesota, *J. Ecol.,* 78, 1021, 1990.

Gorham, E. and Janssens, J., The paleorecord of geochemistry and hydrology in northern peatlands and its relation to global change, *Suo,* 43, 1, 1992.

Heikurainen, L. and Päivänen, J., The effect of thinning, clear cutting, and fertilization of the hydrology of peatland drained for forestry, *Acta For. Fenn.,* 104, 1, 1970.

Hewlett, J. D. and Hibbert, A. R., Factors affecting response of small watersheds to precipitation in humid areas, in *Forest Hydrology,* Pergamon Press, New York, 1967, p. 275.

Ingram, H. A. P., Hydrology, in *Ecosystems of the World,* Vol. 4A, *Swamp, Bog, Fen and Moor,* Gore, A. J. P., Ed., Elsevier, Amsterdam, 1983, p. 67.

Ivanov K. E., *Water Movement in Mirelands,* Academic Press, New York, 1981, 276 pp.

Johnston, C. A., Detenbec, N. E., and Niemi, G. J., The cumulative effect of wetlands on stream water quality and quantity. A landscape approach, *Biogeochemistry,* 10, 105, 1990.

Juusela, T., Kaunisto, S., and Mustonen, S., Turpeesta tapahtuvaan haiduntaaan vaikuttavista tekijoista (On factors affecting evapotranspiration from peat), *Commun. Inst. For. Fenn.,* 67, 1, 1969.

Krug, W. R., Conger, D. H., and Gebert, W. A., Flood Frequency Characteristics of Wisconsin Streams, U.S. Geological Survey, Water-Resources Investigations Report 91-4128, Madison, WI, 1992, 185 pp.

Lafleur, P. M. and Rouse, W. R., The influence of surface cover and climate on energy partitioning and evaporation from a subarctic wetland, *Boundary-Layer Meteorol.,* 44, 327, 1988.

Linacre, E. T., Hicks, B. B., Sainty, G. R., and Grauze, G., The evaporation from a swamp, *Agric. Meteorol.,* 7, 375, 1970.

Maltby, E. and Immirzi, P., Carbon dynamics in peatlands and other wetland soils: regional and global perspective, *Chemosphere,* 27, 999, 1993.

Mannerkoski, H., *Effect of Water Table Fluctuation on the Ecology of Peat Soil,* Publication of the Department of Peatland Forestry, University of Helsinki, No. 7, 1985, 190 pp.

Moore, I. D. and Larson, C. L., Effects of drainage projects on surface runoff from small depressional watersheds in the north central region, Water Resources Research Center, University of Minnesota, WRRC Bulletin 99, 1979, 225 pp.

Moore, P. D. and Bellamy, D. J., *Peatlands,* Springer-Verlag, New York, 1974, 221 pp.

Moore, T. R., Dissolved organic carbon in forested and cutover drainage basins, Westland, New Zealand, in Proc. Forest Hydrology and Watershed Management, Vancouver 1987, IAHS-AISH Publ. no. 167, 1987, p. 481.

Murphy, M. L. and Meehan, W. R., Stream Ecosystems, American Fisheries Society Special Publication 19, 1991, 17 pp.

Mustonen, S. and Seuna, P., Metsaojituksen vaikutuksesta suon hydrologiaan (Influence of forest draining on the hydrology of peatland), Publication of the Water Research Institute, No. 2, National Board of Waters, Helsinki, 1971, 63 pp.

National Wetlands Working Group, *Wetlands of Canada,* Ecological Land Classification Series, No. 24, Sustainable Development Branch, Environment Canada, Ottawa Ontario, and Polyscience Publications Inc., Montreal, Quebec, 1988, 452 pp.

Nichols, D. S. and Brown, J. M., Evaporation from a sphagnum moss surface, *J. Hydrol.,* 48, 289, 1980.

Olszta, W., Szajda, J., and Zawadzki, S., Effect of the sucking power of peat-muck soil on the evapotranspiration value, *Zeszyty Problemowe Postepów Nauk Rolniczych,* 390, 152, 1990.

Penfound, W. T. and Earle, T. T., The biology of the water hyacinth, *Ecol. Monogr.,* 18, 447, 1948.

Romanov, V. V., Isparenie s bolot Evropeiskoi territorii SSSR (Evaporation from Bogs in the European Territory of the USSR), Transl. for Israel Program for Scientific Translations Lt., 1968, 186 pp.

Roulet, N. T., Groundwater flux in a headwater wetland in southern Ontario, in *Ontario Wetlands: Interia or Momentum,* Bardecki, M. J. and Patterson, N., Eds., Federation of Ontario Naturalists, Don Mills, Ontario, 1989, p. 301.

Roulet, N. T., Hydrology of a headwater basin wetland: groundwater discharge and wetland maintenance, *Hydrol. Proc.,* 4, 387, 1990.

Roulet, N. T. and Woo, M. K., Runoff generation in a low Arctic drainage basin, *J. Hydrol.,* 101, 213, 1988.

Rouse, W. R., Microclimate of arctic tree line, 2. Soil microclimate of tundra and forest, *Water Resources Res.*, 20, 67, 1984.

Rouse, W. R., Mills, P. F., and Stewart, R. B., Evaporation in high latitudes, *Water Resources Res.,* 13, 909, 1977.

Sallantaus, T., Water quality of peatlands and man's influence on it, in *Symp. on the Hydrology of Wetlands in Temperate and Cold Regions,* Vol. 2, Joensuu, Finland, Publications of the Academy of Finland, 5/1988, Helsinki, 1988, 80 pp.

Shjeflo, J. B., Evapotranspiration and the Water Budget of Prairie Potholes in North Dakota, U.S. Geological Survey Professional Paper, 585-B, 1968.

Taylor, C. H. and Pierson, D. C., The effect of a small wetland on runoff response during spring snowmelt, *Atmospher-Ocean*, 23, 137, 1985.

Thornthwaite, C. W. and Mather, J. R., *The Water Balance,* Laboratory of Climatology, Publ. 8, Centerton, NJ, 1955.

Timmer, C. E. and Weldon, L. W., Evapotranspiration and pollution of water by water hyacinth, *Hyacinth Contr. J.,* 6, 34, 1967.

Tolonen, K. H., Vasander, H., Damman, A. W. H., and Clymo, R. S., Preliminary estimate of long-term carbon accumulation and loss in 25 boreal peatlands, *Suo*, 43, 277, 1992.

Trousdell, K. G. and Hoover, M. D., A change in groundwater level after clearcutting of loblolly pine in the coastal plain, *J. For.,* 53, 493, 1955

Urban, N. R., Eisenreich, S. J., and Gorham, E., Proton cycling in bogs, in *Geographic Variation in Northeastern North America,* Hutchinson, T. C. and Mleema, K. M., Eds., NATO ASI Series Vol. G16, Effects of Atmospheric Pollutants on Forests, Wetlands, and Agricultural Ecosystems, Spring-Verlag, Berlin, 1987, p. 577.

Van der Weert, R. and Kamerling, G. E., Evapotranspiration of water of water hyacinth (*Eichhornia crassipes*), *J. Hydrol.,* 22, 201, 1974.

Verma, S. B., Ullman, F. G., Shurpali, N. J., Clement, R. J., Kim, J., and Billesbach, D. P., Micrometeorological measurements of methane and energy fluxes in a Minnesota peatland, *Suo,* 43, 285, 1993.

Verry, E. S., The hydrology of wetlands and man's influence on it, in *Symp. on the Hydrology of Wetlands in Temperate and Cold Regions,* Vol. 2, Joensuu, Finland, Publications of the Academy of Finland, 5/1988, Helsinki, 1988, p. 41.

Verry, E. S., Water table and streamflow changes after stripcutting and clearcutting an undisturbed black spruce bog, in Proc. Sixth Intl. Peat Congress, Duluth, MN, Intl. Peat Society, Helsinki, 1981, p. 493.

Verry, E. S. and Boelter, D. H., The influence of bogs on the distribution of streamflow from small bog-upland catchments, in *Proc. Hydrology of Marsh-Ridden Areas,* Minsk, The Unesco Press, IAHS, Paris, 1975, p. 469.

Verry, E. S., Brooks, K. N., and Barten, P. K., Streamflow response from an ombrotrophic mire, in *Symp. on the Hydrology of Wetlands in Temperate and Cold Regions,* Vol. 1, Joensuu, Finland, Publications of the Academy of Finland, 5/1988, Helsinki, 4/1988, 1988, p. 52.

Verry, E. S. and Timmons, D. R., Waterborne nutrient flow through an upland-peatland watershed in Minnesota, *Ecology,* 63, 1456, 1982.

Verry, E. S. and Urban, N. R., Nutrient cycling at Marcell bog, Minnesota, *Suo,* 43, 147, 1993.

Vompersky, S. E., Forest and bog: peculiarities of the cycle of matter and manifestation of a biosphere role, *Lesovedenie,* 6, 54, 1991.

Williams, G. P., Heat balance over saturated sphagnum moss, in Proc. 1st Can. Congr. on Micrometerology, Ottawa, Ontario, Feb. 1968, Part 1, 1968, p. 173.

CHAPTER **14**

Hydrology of Drained Forested Wetlands

Stanislav E. Vompersky and Andrej A. Sirin

CONTENTS189

INTRODUCTION

A rich experience of reclamation of forested wetlands for use in forestry has accumulated in a number of countries. However, some important questions are still not completely resolved. These questions include substantiation of optimum drainage and of the influence of peatland drainage on the hydrological cycle, water resources, and river runoff. In this paper we present a brief review on the present state of knowledge regarding these problems.

Forested wetlands can be divided into two main groups. The first group represents wetlands with temporary, recurrent moisture surplus, for which there may be either no peat formation or only thin peat layers. The second group represents wetlands with constant moisture surplus, which leads to peatland formation. Peatlands are the object of forest drainage more often, which is why they receive more attention in this chapter.

In both cases, the sources of surplus moisture can include atmospheric precipitation, surface or subsurface flow from adjacent mineral lands, and discharge of artesian and groundwaters. Peatland sites may be influenced predominantly by one or several of these sources of water. Riparian forest peatlands will not be considered here. In most of the countries that have significant drainage, they are not drained and are either reserved as protected strips or harvested with partial cutting methods.

Drainage for forestry is mainly done in wetlands already forested, but draining of open peatlands for forestry purposes, which is especially common on the British Isles is also considered. The majority of publications of the problems under review are listed, but the examples given are mostly from Russian experience, which is less known and less accessible to the scientific society outside of Russia.

PROBLEM OF OPTIMUM REGULATION OF WATER REGIME OF FORESTED WETLANDS

Since drainage is a forestry measure that pursues the goal of creating more productive stands, it is important to know (1) what soil water regime should be created by drainage and (2) what is technically possible.

Drainage for forestry is carried out by a system of open ditches. Its regulative role is limited, especially on wet mineral soils where the level of groundwater drops below the bottom of the ditches, except for spring flooding periods. The classical approach of drainage network construction disregards weather fluctuations. Attempts to improve moisture availability during drought by creating dams on ditches were made, for instance, in Ukraine (Suljko et al., 1973) and Finland (Päivänen, 1984). However, this proved to be ineffective (Päivänen, 1984).

Thus, the forest drainage network is designed only to protect against extreme moisture surplus. If this system is designed for very wet years, then it could be moisture deficient during dry years. If the system is designed for years of average precipitation, then there will be no protection to the trees during rainy years. Consequently, the scientific problem is to determine what system will provide the longest period of optimum moisture possible during the growing season.

In comparison to mineral soils, pore space in peatland soils occupies up to 90 to 95 percent of its volume, and macropores predominate there (Romanov, 1961; Päivänen, 1973). Small moisture rise through the capillaries depends on these pores, which have high water conductivity. Large pores, which are free from water, provide peatland soils with good gas exchange, even when they are close to the groundwater level. Below the groundwater level there is usually no free oxygen, as a rule (Vompersky and Busarova, 1966; Jurkevitch et al., 1966; Orlov and Kosheljkov, 1971; Mannerkoski, 1985).

Two important principles of the drainage of forested peatlands will become clear from above. First, comparatively small lowering of the water table is enough for the drainage of such peatlands, and second, it is important not to permit tree roots to be flooded and subjected to anaerobic conditions.

WHAT DEPTH OF WATER TABLE CAN BE CONSIDERED FAVORABLE TO TREE GROWTH ON PEATLANDS?

Drainage creates a different water regime through the area; the water level in the interditch profile has the shape of an "arch" that influences tree stand productivity. If the ditch network is deep and widely-spaced, the depth to water at ditch side will be much deeper, but the depth in the centers of strips between ditches only slightly deeper than for shallow and narrowly-spaced ditches. In the last case, the amplitude of level fluctuations is narrower.

Drainage does not eliminate the causes of paludification. When the drainage network starts to become ineffective, the trees near the ditch, which initially showed greatest growth response, will be most severely retarded in growth, and some will even die (Vompersky, 1968).

Traditionally, foresters have tried to fix the depth of the groundwater table at the minimum average at which the drained forest shows its highest productivity. As in Table 1, depths of groundwater level on peat soils during the growing season, which correspond to the highest productivity in the given site, are predominantly close to 40 to 50 cm, while they usually range from 30 to 70 cm. For paludified mineral soils, these values of water depth should be 25 to 30 percent higher, according to Davydov and Pisarjkov (1970). However, the average depth of the groundwater level is regarded to be the most general estimation. Its main ecological weakness is due to water level fluctuations (Mannerkoski, 1985) and also to varying effects in tree response for soils with different hydrophysical and nutritient properties.

Another approach for the determination of the optimum depth of groundwater is based on soil–water–energy interrelationships. Finnish scientists (Paavilainen and Virrankoski, 1967; Heikurainen, 1967c; Päivänen, 1973; Ahti, 1987 and others) have made valuable contributions to this field, but here this approach is illustrated with our data (Vompersky and Eizerman, 1990).

Many authors consider the upper limit of optimum moisture to be when air occupies at least 10 percent of the volume of peat soil (Paavilainen, 1967; Päivänen, 1973; Sudnitsin et al., 1986). In this case (Figure 1), this limit corresponded to matrix potential of soil moisture, equalled to 1 kPa for oligotrophic *sphagnum* peat with the percent of decomposition ranging from 5 to 8 percent (USSR-60-N method; see Malterer et al., 1992), and 3 kPa for eutrophic tree-grass peat, with the degree of decomposition ranging from 35 to 45 percent. Päivänen (1973) used approximately the same criteria of optimum moisture for the similar peats, where pF equalled 1 and 1.5, respectively. Heikurainen (1973) rated optimal pF for different peats 2.2.

Table 1 Average Depth of Groundwater Level During Growing Season on Drained Peatlands, Which Is Considered to Provide Maximum Productivity of Coniferous Stands

Country	Average depth of groundwater level, cm	Reference	Notes
Belorus	40	Erkin, 1934	Very poor soils
	50	Budyka, 1954	
	70–80		
Estonia	40–50	Hainla, 1959	
Finland	70	Kokkonen, 1931	With fertilizers
	40–60	Huikari and Paarlahti, 1967	
	30–40	Päivänen, 1990	
Lithuania	40–65	Kapustinskaite, 1969	For mineral soils
	60–80		with thin peat layer
Poland	30	Kosturkiewiecz, 1966	
Norway	30	Meshochok, 1969	For young tree stand
	40–60	Braekke, 1974	For grown-up tree stand
	32		For pine
	40		For spruce
Russia (Northwest)	30–40; 50	Dubakh, 1936, 1945	
	40–60	Elpatjevsky, 1964	
	50–60	Vompersky, 1964, 1968	
Sweden	43 (20–56)	Heikurainen, 1973	
U.S. (Southeast)	45	Schlaudt, 1960	Slash pine, loblolly pine

It is more difficult to substantiate the lowest limit of optimum moisture, because it depends not only on physical soil properties and plant species but also on meteorological conditions. Decrease of soil moisture content is accompanied by the replacement of water capillary movement with moving on the surface of soil particles. As we can see in Figure 1, visible decrease of moisture conductivity

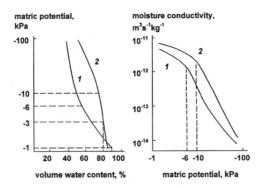

Figure 1 Main hydrophysical characteristics for drained forest oligotrophic (1) and (2) eutrophic peats. (From Vompersky, S. E. and Eizerman, N. I., *Lesovedenie*, 5, 17, 1990 (in Russian with English summary). With permission.)

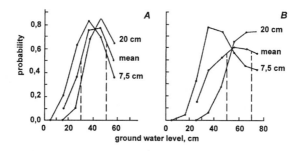

Figure 2   The probability of optimum drained peat moisture, depending upon groundwater level for oligotrophic (A) and eutrophic (B) peats. (From Vompersky, S. E. and Eizerman, N. I., *Lesovedenie,* 5, 17, 1990 (in Russian with English summary). With permission.)

comes approximately at the potential P, which equals 6 kPa for *sphagnum* oligotrophic peat and 10 kPa for eutrophic tree-grass peat. We assumed these values to be the lowest limit of moisture optimum.

On the basis of three years observations in water wells, we revealed those water levels, for which the optimum range of soil moisture was recorded most often. The average water level (among all these water levels) for the depths of 7.5 and 20 cm was assumed to be optimum for the upper 20 cm of soil layer, and it was considered to be the representative water level for the whole plot.

As shown in Figure 2, the optimum moisture on *sphagnum* peat is found to be the most probable at the water level depth of 40 cm, within the possible level ranging from 30 to 50 cm. The optimum moisture on tree-grass peat is found to be the most probable at the water level depth of 60 cm, within the possible level ranging from 50 to 70 cm. The first drainage depth can be recommended for oligotrophic-mesotrophic peatlands with weakly humified peats, and the second for eutrophic peatlands with well decomposed peats.

However, hydrophysical substantiation of optimum drainage depth has only theoretical significance. The search for optimum water levels, which were differentiated according to the types of peatland forests, peat types, and its degree of decomposition, did not need to be justified, as had been suggested (Heikurainen, 1964). The main arguments include the following: the estimations of range of optimum soil moisture are debatable; it is impossible to provide simultaneously the optimum moisture in the whole zone of aeration over interditch space; and forest productivity depends only indirectly on drainage depth, which these researchers are trying to substantiate with too much precision.

It is known that tree roots on fertile soils penetrate annually into soil layers with heightened risk of flooding, where they do not reach the age of 2 or 3 years before they die (Vompersky, 1967). Relatively high productivity is maintained in such sites, along with shallow water level and low soil aeration (Lahde, 1971). Contrary to this fact, pine on well-drained oligotrophic peatlands will not develop root layers deeper than to the depth of 20 to 30 cm, where a good aeration is created (Paavilainen, 1963). This has been explained by low biological soil activity and low potassium and phosphorus content on such sites (Kaila, 1956; Binns, 1962; Smolyak and Pobedov, 1962).

Therefore, at present, the general approach of maintaining an average level within certain limits is considered as quite satisfactory for a drainage norm. This would include some criteria of probability for different levels. The problem for regions with steady excessive moisture is to prevent water rise above the drainage standard. Moisture deficit is possible here only in the topmost 5 cm surface layer, and it is known that capillary properties of peat become better after drainage (Vompersky, 1968; Päivänen, 1982b; Zalitis, 1983).

In Latvia, Buss and Zalitis (1968) determined that drained forest reached its maximum productivity when the upper 40 cm of soil layer was not flooded during 60 to 70 percent of the growing season in pine forest, and during 80 to 90 percent of the growing season in spruce forest. In the northwest part of Russia, the highest productivity of pine stand was noted when the upper 30 cm of soil layer was not flooded during 85 percent of the growing season (Vompersky et al., 1975). Zalitis (1983) summarized data collected in Latvia (Figure 3) and found out that the correlation between pine stands productivity and probability that the water level does not rise above 40 cm is higher for sites with a more sizable gradient of groundwater levels. A higher gradient leads to quicker horizontal movement of water and better aeration of soil.

The question about the value of the drainage norm and the criteria of its probability does not have a simple answer for direct conditions. For example, Toth and Gilard (1988) from Canada defined a drainage norm to be 40 cm and the duration of any separate soil flooding episode to be no more than 14 d. Total duration of the calculated season varies among different authors. In particular,

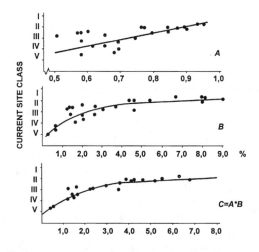

Figure 3   Relationship between forest productivity (current site class) and probability that water level does not rise >40 cm (A), the slope of water table (B), the composition of these two indexes (C) for pine stand in Latvia. (From Zalitis, P., *Principles of Rational Forest Drainage in Latvijskaya SSR,* Zinatne Publishers, Riga, 1983, 230 pp. (in Russian). With permission.)

Vompersky (1968) suggested determining it by the time during which 80 to 90 percent of tree annual increment would have been formed, not considering the time of formation of the first and the last 5 to 10 percent of annual increment. The ecological significance of these periods is not high, but maintenance of a drainage norm within them will lead to notable narrowing of ditch spacing and raising of drainage costs. In any case, the assumed standards and the probability criteria should be based on local data.

GROUNDWATER REGIME AND FOREST DRAINAGE NETWORK PARAMETERS

Perhaps, the most difficult task is to achieve in practice the given water regime of soils through drainage. What network construction can better provide the required drainage depth in particular conditions? How do the depth factors and the distance between ditches interact?

The silvicultural method (Figure 4) for determining forest drainage intensity does not depend directly on the question of what water regime will be created. Generally, when a fixed ditch depth is taken for granted, the distance between them is calculated, pursuing the goal of achieving a certain silvicultural effect or maximum economic profit. Under such criteria, the drainage depth is deliberately shallower than the ecologically optimum depth.

Theoretical hydrological methods for calculating network parameters (e.g., Pisarjkov and Timofeev, 1964) have not been applied because of the rough estimates of parameters used in calculation and the resulting insufficient precision. It is difficult to improve these estimations because of the necessity to estimate a great quantity of physical and biological factors of water balance. That is why, during the latest decades, modeling methods of groundwater level dynamics were developed on the basis of experimental peatland drainage data.

Meshechok's (1969) investigation for the southern part of Norway, which was later broadened by Braekke (1983) for its northern part, was an example of such an approach. These authors suggested empirical dependences of water level in the middle of the interditch space on the ditch depth and the distance from one ditch to another for treeless peatlands and young forest cultures. Some meteorological and other factors were taken into account. Nomograms of possible changes of network parameters, which provide the given drainage depth (Figure 5) were the result of their work.

Another method of modeling was developed by the use of the analysis of water level observations for the different ditch depths (0.5 to 1.8 m) and spacings (30 to 125 m), which lasted about 10 years during the large-scale experimental forest drainage in the center of the European part of Russia (Vompersky et al., 1986). The function of distribution in time of soil-groundwater level for an arbitrary point of interditch space, depending on the ditch depth and spacings, and soil surface gradient was worked out in an analytical form (Vompersky et al., 1986).

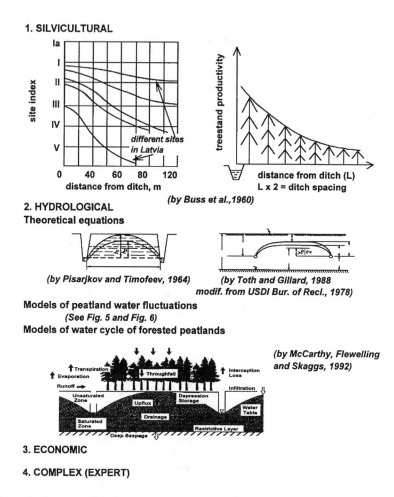

1. SILVICULTURAL

different sites in Latvia

distance from ditch, m

site index

treestand productivity

distance from ditch (L)
L x 2 = ditch spacing

(by Buss et al.,1960)

2. HYDROLOGICAL
Theoretical equations

(by Pisarjkov and Timofeev, 1964)   *(by Toth and Gillard, 1988*
modif. from USDI Bur. of Recl., 1978)

Models of peatland water fluctuations
(See Fig. 5 and Fig. 6)
Models of water cycle of forested peatlands

(by McCarthy, Flewelling
and Skaggs, 1992)

3. ECONOMIC

4. COMPLEX (EXPERT)

Figure 4   Methods of ditch spacing and ditch depth determining.

The model permits the selection of optimum drainage network parameters, which provide, according to probability criteria (Py, PΔy), the required regime of soil-groundwater level fluctuations in the interditch space during growing season. Nomograms for the ditch depth selection and for the distance between ditches, according to various required conditions, were worked out for practical use (Figure 6). For the regions with different (from K = 1.92 — studied area) hydrothermal coefficient (the ratio of heat expenditure for precipitation–evaporation to the available radiation energy during May to September), a special curve for correction is given.

The method of iteration approximation applied by Ahti (1987) and the drawing of a synthetic hydrograph carried out by Toth and Gilard (1987) are noted among other approaches for the dynamics of modeling of groundwater level. With the help of these methods, Canadian authors calculated drainage network parameters, according to the given drainage standard and its fluctuation.

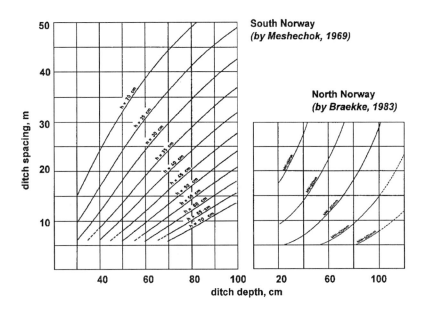

Figure 5   Diagrams to choose ditch spacing and depth on ombrogeneous peatlands. h —
certain drainage norms, MW — mean water tables midway between ditches.

The principal drawback of the above methods of modeling of water level
dynamics and of optimum network parameters recommended from these models
is the problem of applying them directly to other areas. There is a need to verify
them experimentally. That is why the complex (expert) method that is still widely
used in the design for specific forest drainage purposes (e.g., Wally Creak Area
Forest Drainage Project — Jeglum, 1991) is also the economical one. The economical method determined the ditch network parameters based on the aim of
receiving the lowest cost of additional tree increment by forest drainage (Sabo,
1961; Keltikangas, 1971).

However an outgrowth of hydrological modeling for forest drainage purposes
could be proposed, stimulated by additional hydrological problems connected
with forest drainage. These problems are raising the water table after cutting of
old drained forest lands, which poses difficult hydrological questions in both
drained and undrained peatland conditions (Heikurainen, 1967a,b; Päivänen,
1981, 1982a; Verry, 1981; Dube et al. 1995a,b, and others) and reconstruction of
old ditch network.

The transition from statistical and conceptual models to the description of all
physical processes (from water movement in the saturated zone to transpiration
and interception) for the watershed of a single ditch as an element of a drainage
system is quite possible for modern modeling of the hydrological cycle. It is
believed that vast experience with hydrological modeling for agricultural purposes
and river watersheds will be used; this was the case with the model, DRAINMOD,
developed for agricultural lands (Skaggs, 1980) and modified in its last versions
(McCarthy, 1990; McCarthy and Skaggs, 1991) and with the model DRAINLOB

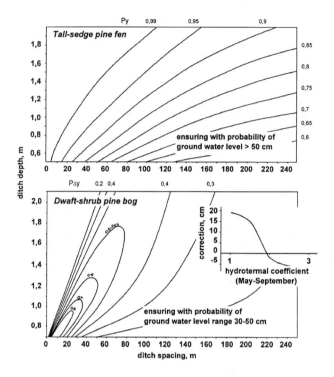

Figure 6   Diagram for determining ditch spacing and depth combination. (From Vompersky, S. E. et al., Socio-Economic Impacts of the Utilization of Peatlands in Industry and Forestry, Proc. IPS Symp., Oulu, Finland, 1986, 119 pp. With permission.)

(McCarthy et al., 1992; Amatya et al., 1994) for drained forested wetlands. Measured data for different peatland types, drainage parameters, and climatic conditions is quite sufficient. Numerical experiments could help not only forecast the hydrological consequences of forest practices but also elucidate the relationship between drainage parameters and soil moisture and help to predict offside hydrological impacts of forest drainage.

INFLUENCE OF FOREST DRAINAGE ON HYDROLOGICAL CYCLE AND RIVER RUNOFF

The hydrological influences of forest drainage are a topic of much concern both for proponents and opponents of drainage of wetlands for forestry purposes. Forest drainage results in two main hydrological changes: there is an increase of the density of the hydrographic network, and there is an increase in the water storage capacity of peat after lowering the groundwater level. Changes in the hydrological regime occur during the first 1 or 2 years against the background of additional income of peat waters after lowering of their levels, and later against the background of tree stand changes and progressive deterioration of drainage ditches (Heikurainen, 1976; Sirin, 1983; Vompersky and Sirin, 1986).

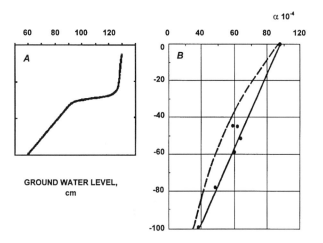

Figure 7   Dependence of specific evaporation α (evaporation/radiation) on groundwater level for oligotrophic (A) and eutrophic (B) peatlands. (From Romanov, V. V., *Evaporation from Mires in the European Territory of USSR*, Gidrometeoizdat Publishers, Leningrad, 1962, 228 pp. (in Russian) Transl. from Russian, Jerusalem, 1968, 183 pp. With permission.)

In most cases, forest drainage increases both summer and winter low runoff (base flow). The main reason for this is the increase in the volume of peat waters, which take part in runoff formation, with prolonged lowering of water levels after snow melt and precipitation events. The creation of artificial slopes for water levels prolongs the runoff period even during dry periods. Besides this, on virgin peatlands slopes quickly become critical, and the runoff stops. Pisarjkov (1981) proved this theoretically for peatlands. Low runoff does not increase after the drainage of forests on hydromorphic mineral soils (Aire, 1977; Zalitis, 1983) and is rather small from drained forests with thin peat layers (Chesnokov, 1981).

The reduction of the total evapotranspiration after forest drainage also may cause the increase of summer low runoff, especially on oligotrophic peatlands with weakly decomposed *sphagnum* peat. The curves of dependence of specific evaporation (i.e., the value of evaporation per unit of absorbed energy from the level of peat water (Romanov, 1968)) are shown in Figure 7. A sharp decline of physical evaporation from the soil surface and transpiration from aboveground cover during the lowering of capillary fringe in the zone of *sphagnum* fibrous and weakly decomposed peat are typical for raised bogs. The decline of total evaporation during the lowering of the water level is smoother on swamps and fens owing to better capillary properties of their peats.

Meso- and eutrophic peatlands usually have a more developed natural tree stand, which also is conducive to relatively less change in evaporation during their drainage in comparison with oligotrophic peatlands. Drainage influences evaporation from soil and transpiration of aboveground vegetation. But this is small in comparison to the total evapotranspiration of well-treed peatlands with the high levels of the treestand's interception and transpiration.

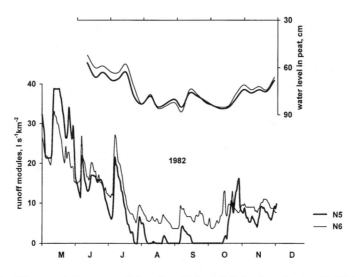

Figure 8   Difference between runoff from two ditches of drained peatlands with (N 5) and without (N 6) direct discharge of piezometric water from underlying peat deposits. (From Vompersky, S. E. et al., *Runoff Formation and Regime in a Process of Forest Drainage,* Nauka Publishers, Moscow, 1988, 168 pp. (in Russian). With permission.)

For groundwater discharge sites, drainage may increase the inflow of groundwaters (Zalitis, 1983). This leads quite often to their direct discharge in the drainage network. Figure 8shows the difference in two hydrographs of summerautumn runoff, taken from two ditches situated next to each other. Their interditch space is identifiable by tree stand, peat, and regime of groundwaters. But a small (several dozens of centimeters) difference in peat thickness (why the ditch of the watershed N 6 penetrates into underlying watered sands) explains the upward piezometric discharge into one (N 6) of them (Vompersky et al., 1988).

All of the above result in increases in low runoff levels, which are observable in separate peatland plots, whole peatland complexes, and whole river catchments. Close relationships between drainage intensity and summer low flow from concrete ditches for different forest peatland sites were shown by observations in Finland (Huikari et al., 1966; Ahti, 1987), Russia (Vompersky, 1974; Babikov, 1978), Czech Republic (Ferda and Novak, 1976; Ferda and Cermak, 1981), Latvia (Aire, 1977) and other countries.

Difference in hydrographs of summer runoff from drained and undrained parts of two oligotrophic bogs is illustrated by Figure 9a. Figure 9b demonstrates that drainage of the whole peatland complex in the European part of Russia increases daily (June 19, 1985) modules during dry weather summer period by 1.5 times, in comparison to the undrained complex.

Forest drainage caused the increase of low flows in summer and winter from studied river basins in Finland (Seuna, 1982; and others), Great Britain (Robinson, 1986), Sweden (Berquist et al., 1984, Lundin, 1984, 1988, 1994), Canada (Berry, 1991), United States (Trettin, 1994), Russia. In Figure 10 are shown several

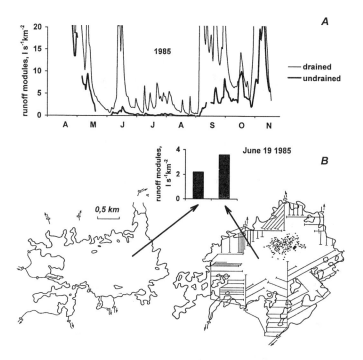

Figure 9   Difference between summer low runoff from drained and undrained raised bogs. (From Vompersky, S. E. et al., *Runoff Formation and Regime in a Process of Forest Drainage,* Nauka Publishers, Moscow, 1988, 168 pp. (in Russian). With permission.)

situations including low flow from river basins with small (7 percent) — Velesa, large (30 percent) — G9 part covered by forest drainage, and also a completely drained watershed (V5) almost entirely consisting of forested peatlands. It is notable that the difference in modules of low flow (both summer and winter) from two river basins (G9 and Velesa) equals approximately the modules from the drainage system V5 fed primarily by drained peatlands.

Mustonen (1971) has revealed the direct dependence of summer low runoff values from the area covered by forest drainage on the base of mass comparison of runoff of small rivers in Finland. Thus, forest drainage, as a rule, increases low runoff, especially in summer. This increase can rise up to several times, but on the average it makes the first dozens of per cents. However, it is difficult to suggest an effective method to forecast the level of these changes.

The influence of peatland drainage on peak flow is more complicated and complex. This problem attracted attention in 1930 to 1940 after a series of catastrophic spring floodings in Scandinavia and the British Isles (McDonald, 1973; Ahti, 1977).

Impact of forest drainage on peak runoff is connected to the combined hydrologic influences of the increase in peatland water storage capacity and increasing of the density of hydrographic network. These factors interact differently on

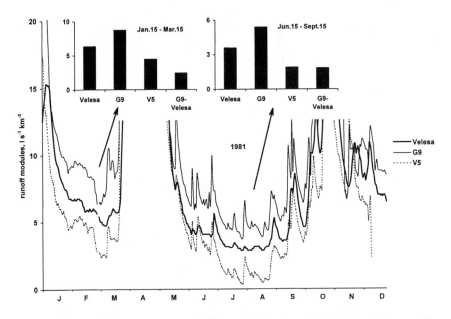

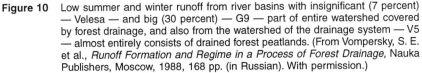

Figure 10   Low summer and winter runoff from river basins with insignificant (7 percent) — Velesa — and big (30 percent) — G9 — part of entire watershed covered by forest drainage, and also from the watershed of the drainage system — V5 — almost entirely consists of drained forest peatlands. (From Vompersky, S. E. et al., *Runoff Formation and Regime in a Process of Forest Drainage,* Nauka Publishers, Moscow, 1988, 168 pp. (in Russian). With permission.)

various catchments and in different hydrometeorological situations, partly or completely compensating each other.

However, for very high peak flow events of low probability, increased water storage capacity of drained peatlands is of little importance, because storage is filled at an early stage of the event (Mustonen and Seuna, 1971). On the other hand the ditch network reduces the lag time of runoff from snowmelt and precipitation events on the river basin as a whole; also peatland margins stop their regulating role (Mustinen and Seuna, 1971; Vompersky et al., 1988).

In comparison with rainfall runoff, formation of snowmelt peak flow is still further complicated. The influence of drainage on snow accumulation, peat freezing and snow melting is very ambiguous (Heikurainen, 1976; Kaliuzny et al., 1988; Makovsky and Chindjaev, 1988; Swanson and Rothwell, 1989). Thus, in the literature are found several explanations of changes of peak runoff after drainage, supported by investigation results (McDonald, 1973; Starr and Päivänen, 1981). Data on the increase of maximum values of peak flow from peatland plots (Vompersky et al., 1988; Jackson, 1987) and river catchments, as a whole (Mustonen, 1975; Chesnokov, 1981; Berquist et al., 1984; Robinson, 1986; Lundin, 1988 and others) predominate among investigations. However, there are opposite conclusions (Multamaki, 1962; Seuna, 1982) and contradictory data within the frames of separate investigations (Lundin, 1984; Berry, 1991). There is no definite relationship between changes in peak flow and extent of drainage (Verry, 1988). Besides this, location of drained areas within a river basin could have an influence on peak flow change (Mustonen and Seuna, 1971; Acreman, 1985).

In these circumstances it is promising to use mathematical models of water flow, imitating the main processes of runoff formation in relation to temporal distribution of meteorological elements and parameters for a particular watershed (Nazarov and Sirin, 1988). The experience of modeling in peatland hydrology ranges from the creation of models of "black box" (Dooge and Kin, 1972), models of transformation, when the algorithm of simulation in natural conditions and under drainage varies (Pakutin, 1985) to conceptual (Predmore and Brooks, 1980) and physical-mathematical models (Guertin et al., 1987). However, the models of river runoff, which are adapted particularly for our problem, are of greatest interest to us.

The Swedish conceptual model HBV (Bergstrom, and Forsman, 1973; Bergstrom, 1975) with modification PULSE (Bergstrom et al., 1985) are among such models. It was used to estimate the influence of forest drainage on river runoff from small and middle catchments of Sweden and Finland (Brandt, 1987, 1990; Johansson, 1993; Johansson and Seuna, 1994). The two dimensional, physical-mathematical, Danish-French-British model SHE was approved for small watersheds in Sweden (Iritz, 1993). Russian conceptual-physical-mathematical model FWH (Nazarov and Sirin, 1988; Sirin et al., 1991) was applied for river runoff simulation from the Velesa River basin of 870 km^2. The results of numerical experiments, which are presented in Figure 11, correspond well with the experimental data referred to above. The influence of natural characteristics of catchments and location of drained areas were shown, and the changes of water balance structure after different forestry treatments were calculated (Nazarov and Sirin, 1988). Also, using modeling, the curves of probability of runoff (needed for hydrological design) after drainage could be obtained.

In addition to this, modeling could reveal differences of forest drainage impacts under various weather conditions, possible climate change (Kutchment et al., 1990; McCarthy et al., 1992; Amatya et al., 1994), and hypothetical conditions of extreme spring flooding (Kutchment et al., 1990; Johansson, 1993; Iritz, 1993). It must be mentioned that most hydrological models aimed at peak flow simulation provide less opportunity for low flow analysis.

The results of numerical experiments with the change of plant cover after drainage (Johansson, 1993) supplemented limited experimental information on long-term drainage consequences. The values of total summer evapotranspiration (ET) from peatland pine stands of various productivities are shown in Figure 12. Increases of total ET and the decrease in the additional amount of runoff, which took place after drainage with the increase in tree stand productivity, have been noted by a number of authors (Heikurainen, 1976; Braekke, 1983; Ledger and Harper, 1987). This has been confirmed by field (Grip et al., 1984; Lindroth and Grip, 1984; Persson, 1989) and laboratory (Kaunisto, 1983; Saarsalmi, 1984) experiments with "energy" forest cultures.

These conclusions are connected only with peatland plots. Possible lowering of summer and winter minimum river runoff during the second and following

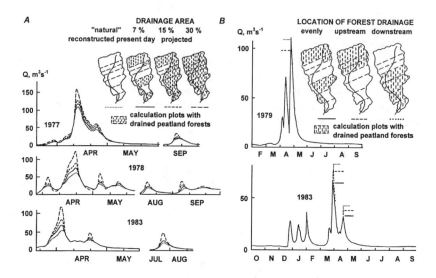

Figure 11   Impact of forest drainage extent (A) and location (B) on the hydrograph of the Velesa River (F = 870 km²) in Middle European Russia for years with different hydrometeorological conditions on the basis of numerical experiments by FWH model. (From Sirin, A. et al., *Ambio,* 20, 334, 1991. With permission.)

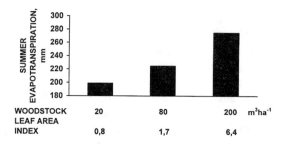

Figure 12   Relation between summer evapotranspiration and biomass characteristics of drained peatland pine stands. (From Sirin, A. A., Investigation of runoff formation and regime from drained forest peatlands, Synopsis of Candidate thesis, Institute of Geography, USSR Academy of Science, Irkutsk, 1989, 16 pp. (in Russian). With permission.)

decades after drainage is connected mainly with the deterioration of the drainage network (Seuna, 1981; Ledger and Harper, 1987). The large-scale reductions in river runoff as a result of increased productivity of drained forests are assumed by a number of Russian authors to be unjustified (Vompersky et al., 1988). On a real river catchment, the influence of productive drained forests could be smoothed by clearcutting, as plots are entering rotation periods at different times.

Hydrological consequences of cutting of peatland forests (Seuna, 1988) do not differ in essence from the same consequences for forests on uplands. However, the sharp reduction of moisture losses due to evaporation after clearcutting stimulates paludification. This impairs conditions for the following reforestation and could have some special influences on regional hydrological cycle and river runoff.

CONCLUSION

In conclusion, peoples' fear about the negative influence of forest drainage on water resources are not substantiated. Peatland drainage increases the low runoff levels (base flow), thereby increasing the contribution of runoff to rivers during periods of low precipitation. There is a danger of spring flood increase; however, it can be forecast to some extent on the basis of river runoff models.

The most urgent questions for further investigation, in our opinion, are (1) the substantiation of optimum parameters of drainage network and (2) the quantitative estimation of dynamics of water balance components of forest peatlands *in situ* in relation to their different uses and to possible climate change. The fact that forest drainage has long-term objectives must be taken into consideration. The most fruitful approach is that of physical-mathematical modeling, trying to use the data on water regime of drained forested peatlands that have accumulated for many years in different countries.

ACKNOWLEDGMENTS

The authors would like to acknowledge the contribution of John Jeglum (Forestry Canada, Ontario Region), who helped to make the manucript more understandable to North American readers, as well as the advice and assistance of Erkki Ahti (Finnish Forest Reserach Institute), Bjorn Hånell (Dept. of Forest Ecology, Swedish University of Agricultural Sciences), Juhani Päivänen (University of Helsinki), and Carl Trettin (Center for Forested Wetlands Research, USDA Forest Service).

REFERENCES

Acreman, M. C., The effects of afforestation on the flood hydrology of the upper Ettrick Valley, *Scott. For.* 39, 89, 1985.

Ahti, E., Runoff from open peatlands as influenced by ditching. I. Theoretical analysis, *Commun. Inst. For. Fenn.*, 92, 1, 1977.

Ahti, E., Water balance of drained peatlands on the basis of water table simulation during the snowless period, *Commun. Inst. For. Fenn.*, 141, 1, 1987.

Aire, A., About the runoff regulation properties of drained forests in Latvijskaya SSR, *Lesovedenije*, N, 67, 1977 (in Russian).

Amatya, D. M., Skaggs, R. W., and Gregory, J. D., Hydrologic modeling of drained forested watersheds, in Water Management in Forested Wetlands, Proc. of a Workshop Apr. 26–28, 1994, Atlanta, Georgia, USDA Forest Service, Technical Publication RS-TP 20, 1994, p. 27.

Babikov, B. V., Forest drainage and water feeding of rivers, *Lesnoye khozajstvo*, 10, 24, 1978 (in Russian).

Bergstrom, S., The development of a snow routine for the HBV-2 model, *Nordic Hydrol.,* 6, 73, 1975.

Bergstrom, S. and Forsman, A., The development of a conceptual deterministic rainfall-runoff model, *Nordic Hydrol.,* 4, 147, 1973.

Bergstrom, S., Carlsson, B., Sandberg, G., and Maxe, L., Integrated modeling of runoff, alkalinity and pH on a daily basis, *Nordic Hydrol.,* 16, 89, 1985.

Berquist, B., Lundin, L., and Andersson, A., Effects of Peatland Drainage on Hydrology and Limnology in the Basin Siksjobacken. Institute of Limnology, Uppsala Univ. LIU:B4, 1984, 116 (in Swedish with English summary).

Berry, G. L., Hydrology of drained and undrained black spruce peatlands: streamflow and hydrological balance, Forestry Canada, Ontario Region, Sault Ste. Marie, Ontario, COFRDA Report 3317, 1991, 25 pp.

Binns, W. O., Some aspects of peat as a substrate for tree growth, *Irish For.,* 19, 32, 1962.

Braekke, F. H., The effect of fertilization and drainage intensity on height growth of Scots pine and Norway spruce in North Norway, in Proc. Int. Symp. Forest Drainage 1974 Jyvaskyla-Oulu, Finland, Vol. 1, IPS-IUFRO Working Party, Helsinki, 1974, p. 207.

Braekke, F. H., Water table levels at different drainage intensities on deep peat in northern Norway, *For. Ecol. Manage.,* 5, 169, 1983.

Brandt, M., Peatland Drainage and Floods, Swed. Meteor. and Hydr. Inst., FoU-Notiser No. 55, Norrkoping, 1987, 17 pp. (in Swedish).

Brandt, M., Human Impacts and Weather-Dependent Effects on Water Balance and Water Quality in Some Swedish River Basins, Dissertation, Stockholm: Royal Institute of Technology, Department of Land and Water Resources, TRITA-KUT 90, 1059, 1990.

Budyka, S. H., *Forest Hydrotechnical Reclamations,* Belorussia SSR Academy of Science, 1954, 100 pp. (in Russian).

Buss, K. K., Klavins, J. J., Majke, P. M., and Sabo, E. D., *Drainage of Forest Lands,* Lesnaja Promyshlennost Publishers, Moscow-Leningrad, 1960, 160 pp. (in Russian).

Buss, K. K. and Zalitis, P. P., On groundwater regime of drained forests, in *Questions of Hydroforestreclamation,* Zinatne Publishers, Riga, 1968, p. 51 (in Russian).

Chesnokov, V. A., Runoff formation features in different periods under forest drainage, in *Water and Thermal Regime of Drained Soils in Karelia,* Karelian Branch USSR Academy of Science, Petrozavodsk, 1981, p. 75 (in Russian).

Davydov, P. I. and Pisarjkov Kh. A., Drainage standards, in Influence of Drainage Network on Regime and Forest Growth, Forestry Technical Academy, Leningrad, Scientific Papers, No. 142, Leningrad, 65, 1970 (in Russian).

Dooge, J. and Kin, R., Mathematical modeling of runoff from small territories of undrained and drained peatlands in Glenamoy, in Hydrology of marsh-ridden areas, Proc. Minsk Symp. IASH-AISH-UNESCO, Publ. 105, 1975.

Dubakh, A. D., Materials for studying groundwater dynamics in forest and their influence on forest growth, in *Increasing of Forest Productivity by Means of Draining Reclamation,* Vol. 1, Goslestechizdat Publishers, Moscow, 1936, p. 40 (in Russian).

Dubakh, A. D., *Hydrotechnical Reclamations of Forest Lands,* Goslestechizdat Publishers, Moscow, 1945, 375 pp. (in Russian).

Dube, S., Plamondon, A. P., and Rothwell, R. L., The effects of cutting on the soil water and aerobic status of forested wetlands, *Can. J. For. Res.,* in press.

Dube, S., Plamondon, A. P., and Rothwell, R. L., Watering-up after clearcutting on forested wetlands of the St-Lawrence lowland, *Water Resources Res.,* 1995, in press.

Elpatjevsky, M. P., Reclamation of paludified pine stands, in *Works of Leningrad Foresty Research Institute,* Vol. 8, Lesnaya Promyshlennost Publishers, Moscow, 1964, p. 150 (in Russian).

Erkin, G. D., *Influence of Drainage on Forest Productivity*, Moscow, 1934, 200 pp. (in Russian).

Ferda, J. and Cermak, R., Vliv odvodneni lesnich pud na zmeny prostredi, *Lesnictvi,* 27, 271, 1981 (in Czech with English summary).

Ferda, J. and Novak, M., The effect of ameliorative measures on the changes of the quality of surface and groundwaters in peat soils, in Proc. 5 Int. Peat. Congr., Vol. 1, Poznan, Poland, 1976, p. 118.

Grip, H., Halldin, S., Lindroth, A., and Persson, G., Evapotranspiration from a willow stand on wetland, in Ecology and Management of Forest Biomass Production Systems, Perttu, K., Ed., Dept. Ecol. and Environ. Res., *Swed. Univ. Agric. Sci. Rep.,* 15, 47, 1984.

Guertin, D. P., Barten, P. K., and Brooks, K. N., The peatland hydrologic impact model: development and testing, *Nordic Hydrol.,* 18, 79, 1987.

Hainla, V. E., Increasing of peatland forests productivity with the help of drainage, in *Publ. For. Int. USSR Acad. Sci. Moscow,* 49, 94, 1959 (in Russian).

Heikurainen, L., Improvement of forest growth on poorly drained peat soils, in *International Review of Forest Research,* Vol. 1, Romberger, J. A. and Mikola, P., Eds., Academic Press, 1964, p. 39.

Heikurainen, L., Effect of cutting on the ground-water level on drained peatlands, in *International Symposium on Forest Hydrology,* Sopper, W. E. and Corbett, E. S., Eds., Pergamon Press, Oxford, England, 1967a, p. 345.

Heikurainen, L., On the influence of cutting on the water economy of drained peatland, *Acta For. Fenn.,* 82, 1, 1967b (in Finnish with English summary).

Heikurainen, L., On the possibilities of optimum drainage in peatlands, in Proc. XIV Congr. IUFRO, IV, Sect. 23, Munhen, 1967c, p. 264.

Heikurainen, L., *Forest Drainage*, P.A. Norstedt & Soners Forlag, Stockholm, 1973, 444 pp. (in Swedish).

Heikurainen, L., Comparison between runoff conditions on a virgin peatland and a forest drainage area, in Proc. 5 Int. Peat Congr. Poznan Poland, Vol. 1, 1976, p. 76.

Huikari, O. and Paarlahti, K. Results of field experiments on the ecology of pine, spruce and birch, *Commun. Inst. For. Fenn.,* 64, 1, 1967.

Huikari, O., Paarlahti, K., Paavilainen, E., and Ravela, H., On the effect of strip-width and ditch-depth on water economy and runoff on a peat soil, *Metsantutkimuslaitoksen Julk.,* 61, 1, 1966 (in Finnish with English summary).

Iritz, L., Studying the Effects of Forest Drainage on Flood Conditions by a Physically Based Distributed Model (SHE), Institute of Earth Sciences, Uppsala University, 1993, 60.

Jackson, R. J., Hydrology of an acid wetland before and after draining for afforestation, western New Zealand, in Proc. Forest Hydrology and Watershed Management, Aug. 1987, Vancouver, British Columbia, IAHS-AISH Publ. No. 167, 1987, p. 465.

Jeglum, J. K., The Wally Creek area forest drainage project in Ontario's clay belt: progress report, in Proc. Symp. Peat and Peatlands. Diversification and Innovation, Vol. 1, Peatland Forestry Quebec City, Canada, Aug. 1989, Publ. Canad. Soc. for Peat and Peatlands, 1991, p. 47.

Johansson B., Modeling the effects of wetland drainage on high flows, Swed. Meteor. and Hydr. Inst. Reports Hydrol., No. 8, Norrkoping, 1993, 34 pp.

Johansson, B. and Seuna, P., Modeling the effects of wetland drainage on high flows, *Aqua Fenn.,* 24, 1, 1994.

Jurkevich, I. D., Smoljak, L. P., and Girin, B. E., About oxygen content in soil water and carbon dioxide in soil air in forested peatlands, *Pochvovedenie,* 2, 41, 1966 (in Russian).

Kaila, A., Phosphorus in various depths of some virgin peat lands, *J. Sci. Agric. Finl.,* 28, 90, 1956.

Kaliuzny, I. L., Pavlova, K. K., and Lavrov, S. A., *Hydrophysical Investigations During Melioration of Lands with Water Surplus,* Gidrometeoizdat Publishers, Leningrad, 1988, 260 pp. (in Russian with English summary).

Kapustinskaite, T. K., *Drainage of Forest Lands in Lithuanian SSR,* Kaunas, 1969, 28 pp. (in Russian).

Kaunisto, S., Biomass production of Salix viminalis and its nutrient and water consumption on differently fertilized peats in greenhouse, *Folia For.,* 551, 1, 1983 (in Finnish with English summary).

Keltikangas, M., Effects of drain spacing on the economic results of forest drainage investments, *Acta For. Fenn.,* 123, 1, 1971 (in Finnish with English summary).

Kokkonen, P., Tutkimuksia kuivatuksen aiheuttamasta turvekerrosten painumisesta, *Valt.Maatalousk. Julk.,* 40, 1, 1931 (in Finnish with German summary).

Kosturkiewiecz, A., Wody gruntowe a bonitacja siedliska drzewastanov sosnowych w malej zlewni lesnej puszyczy Zgorzelekiej, Mater. Konf. nauk-techn. Ruszow, 1966 (in Polish).

Kutchment, L. S., Motovilov, J. G., and Nazarov, N. A., *Sensitivity of Hydrological Systems (Influence of Antrophogenic Changes of River Basins and Climate on Hydrologic Cycle),* Nauka Publishers, Moscow, 1990, 144 pp. (in Russian).

Lahde E., On anaerobic conditions in various virgin peat soils and the significance of the aerobic limit as an indicator of site quality, *Silva Fenn.,* 5, 36, 1971 (in Finnish with English summary).

Ledger, D. C. and Harper, S. E., The hydrology of a drained, afforested peat bog in southern Scotland, 1977–1986, *Trans. Royal Soc. of Edinburgh: Earth Sciences,* 78, 297, 1987.

Lindroth, A. and Grip, H., Design of water balance studies in energy forest on wetland, *Vannet i Norden,* 1, 3, 1984.

Lundin, L., Peatland drainage. Effects on the Hydrology of the Mire Docksmyren, Division of Hydrology, Uppsala University, Report Series A, 3, 1, 1984 (in Swedish with English summary).

Lundin, L., Impacts of drainage for forestry on runoff and water chemistry, in Proc. Int. Symp. on Hydrology of Wetlands in Temperate and Cold Regions, Vol. 1, *Suomen Akatemian Julk.,* 4, 97, 1988.

Lundin, L., Impacts of forest drainage on flow regime, *Stud. For. Suenica,* 1994, in press.

Makovsky, V. I. and Chindjaev A. S., *Forestry-Ecological Basis of Forest Drainage in the Middle Urals,* Publ. Ural's branch of the USSR Acad. Sci., Sverdlovsk, 1988, 96 pp. (in Russian).

Malterer, T. J., Verry, E. S., and Erjavec, J., Fiber content and degree of decomposition in peats: review of national methods, *Soil Soc. Am. J.,* 56, 1200, 1992.

Mannerkoski, H., Effect of Water Table Fluctuation on the Ecology of Peat Soil, Department of Peatland Forestry, University of Helsinki, No. 7, 1985, 190.

McCarthy, E. J., Modification, testing and application of a hydrologic model for a drained forest watershed, Ph.D. Thesis, Department of Biological and Agricultural Engineering, N.C. State University, Raleigh, NC, 1990.

McCarthy, E. J., Flewelling, J. W., and Skaggs, R. W., Hydrologic model for drained forest watershed, *J. Irrigation Drainage Engin.,* 118 (2), 242, 1992.

McCarthy, E. J. and Skaggs, R. W., A simplified model for predicting drainage rates for changing boundary conditions, *Trans. Am. Soc. Agric. Eng. (ASAE)*, 34 (2), 443, 1991.

McDonald, A., Some views of the effect of the peat drainage, *Scott. For.*, 27, 315, 1973.

Meshechok, B., 1969, Drainage of swamps at different ditch distances and ditch depths, *Medd. Nor. Skogforsoekves*, 27, 227, 1969 (in Norwegian with English summary).

Multamaki, S. E., The effects of forest drainage on the outflow ratio of peatland, *Commun. Inst. For. Fenn.*, 55, 1, 1962 (in German).

Mustonen, S., Variations in the minimum runoff from small basins, *Vesientutkimuslaitoksen Julk.*, 1, 1, 1971 (in Finnish with English summary).

Mustonen, S., The influence of drainage for forestry on the hydrology of peatland, *Nordic IHD Rep.*, 8, 68, 1975.

Mustonen, S. and Seuna, P., Influence of forest draining on the hydrology of peatlands, *Vesientutkimuslaitoksen Julk.*, 2, 1, 1971 (in Finnish with English summary).

Nazarov, N. A. and Sirin, A. A., *Model and Computation Algorithms of the River Flow Formation in a Forest Drainage Area*, VINITI Publishers, Moscow, 1988, 108 pp. (in Russian with English summary).

Orlov, A. J. and Kosheljkov, S. P., *Soil Ecology of Pine*, Nauka Publishers, Moscow, 1971, 323 pp. (in Russian).

Paavilainen, E., On water content of peat and groundwater level, *Suo*, 14, 8, 1963 (in Finnish with English summary).

Paavilainen, E., Relationships between the root system of Scots pine and the air content of peat, *Commun. Inst. For. Fenn.*, 63, 1, 1967 (in Finnish with English summary).

Paavilainen E. and Virrankoski, K., Studies on the capillary rise of water in peat, *Folia For.*, 36, 1, 1967 (in Finnish with English summary).

Päivänen, J., Hydraulic conductivity and water retention in peat soils, *Acta For. Fenn.*, 129, 1, 1973.

Päivänen, J., The effect of silvicultural treatments on the groundwater table in Norway spruce and Scots pine stands on peat, in Proc. 6 Int. Peat Congress, 1980, Duluth, MN, Int. Peat Soc., Helsinki, 1981, p. 433.

Päivänen, J., The effect of cutting and fertilization on the hydrology of an old forest drainage area, *Folia For.*, 516, 1, 1982a.

Päivänen, J., Physical properties of peat samples in relation to shrinkage upon drying, *Silva Fenn.*, 16, 247, 1982b.

Päivänen, J., The effect of runoff regulation on tree growth on a forest drainage area, in Proc. 7th Int. Peat Congr., Vol. 3, Dublin, June 18–23, 1984, p. 476.

Pakutin, A. V., Experience on estimation of transformation of spring snowmelt flow from mire massifs after drainage, *Vestnik of Leningrad University, Serie on Geogr. and Geology*, 7, 102, 1985 (in Russian).

Persson, G., Energy forest water balance on a raised bog, *Scand. J. For. Res.*, 4, 29, 1989.

Pisarjkov, Kh. A., About water protection and water regulation importance of drained forested peatlands, *Lesnoj J.*, 2, 13, 1981 (in Russian).

Pisarjkov, Kh. A. and Timofeev, A. F., *Hydrotechnical Reclamations of Forest Lands*, Lesnaya Promyshlennost Publishers, Moscow, 1964, 275 pp. (in Russian).

Predmore, S. R. and Brooks, K. N., Predicting peat mining impacts on water resources — a modeling approach, in Proc. 6 Int. Peat Congress, 1980, Duluth, MN, Int. Peat Soc., Helsinki, 1981, p. 655.

Robinson, M., Changes in catchment runoff following drainage and afforestation, *J. Hydrol.*, 86, 71, 1986.

Romanov, V. V., *Hydrophysics of Mires*, Hydrometeoizdat Publishers, Leningrad, 1961, 350 (in Russian), Transl. from Russian, Jerusalem, 1968, 299.

Romanov, V. V., *Evaporation from Mires in the European Territory of the USSR*, Gidro-meteoizdat Publishers, Leningrad, 1962, 228 pp. (in Russian), Transl. from Russian, Jerusalem, 1968, 183 pp.

Saarsalmi, A., Biomass production and nutrient and water consumption in Salix "Aquatica Gigantea" plantation, *Folia For.*, 602, 1, 1984 (in Finnish with English summary).

Sabo, E. D., Economical efficiency of drainage of forest lands, *Lesnoye Khozajstuo*, 7, 45, 1961 (in Russian).

Schlaudt, E. A., Water regulation for increased timber production, in Proc. 5th World Forestry Congr., Vol. 3, University of Washington, Seattle, Aug.–Sept. 1960, 1960, p. 1723.

Seuna, P., Long-term influence of forestry drainage on the hydrology of a open bog in Finland, *Vesientutkimuslaitoksen Julk.*, 43, 3, 1981.

Seuna, P., Influence of forestry draining on runoff and sediment discharge in the Ylijoki Basin, North Finland, *Aqua Fenn.*, 12, 3, 1982.

Seuna P., Effects of clear-cutting and forestry drainage on runoff in the Nurmes-study, in Proc. Symp. on the Hydrology of Wetlands in Temperate and Cold Regions, Vol. 1, *Suomen Akatemian Julk.*, 4, 122, 1988.

Sirin, A. A., The impact of forest drainage on runoff and methods for its investigation, in Proc. Int. Symp. on Forest Drainage. IPC Commis. III. Tallin, 1983, p. 164.

Sirin, A. A., Investigation of runoff formation and regime from drained forest peatlands, Synopsis of Canadian thesis, Institute of Geography, USSR Academy of Science, Irkutsk, 1989, 16 pp. (in Russian).

Sirin, A., Vompersky, S., and Nazarov, N., Influence of forest drainage on runoff: main concepts and examples from central part of the USSR European territory, *Ambio*, 20, 334, 1991.

Skaggs, R. W., A water management model for artificially drained soils, Technical Bulletin No. 276, North Carolina Agricultural Research Service, North Carolina State University, Raleigh, NC, 1980.

Smoljak, L. P. and Pobedov, B. S., Phosphorus nutrition of tree species on Sphagnum peatlands, in *Collection of Experimental Botany*, Minsk, 1962, p. 86 (in Russian).

Starr, M. R. and Päivänen, J., The influence of peatland forest drainage on runoff peak flows, *Suo*, 32, 79, 1981.

Sudnitsin, I. I., Sidorova, M. A., and Vinokurova, V. M., Ecological-hydrophysical basis of soilwater regime optimization, *Pochvovedenie*, 7, 88, 1986 (in Russian).

Suljko, V. V., Mikhovich, A. I., and Davidjuk, K. S., Two-way water regulation for forest drainage, *Lesnoye Khozjajstvo*, 5, 30, 1973 (in Russian).

Swanson, L. E. and Rothwell, R. L., Substrate freeze-thaw in drained Alberta fen, *Can. J. For. Res.*, 19, 1024, 1989.

Toth J. and Gilard, D., Experimental design and evaluation of peatland drainage system for forestry by optimization of synthetic hydrograph, *Can. J. For. Res.*, 18, 353, 1988.

Trettin, C. C., Hydrologic response of northern wetlands to silvicultural water management systems, in Water Management in Forested Wetlands, Proc. of a Workshop Apr. 26–28, 1994, Atlanta, Georgia, USDA Forest Service, Technical Publication RS-TP 20, 1994, p. 56.

Verry, E. S., Water table and streamflow changes after stripcutting and clearcutting an undrained Black spruce bog, in Proc. 6 Int. Peat Congress, 1980, Duluth, MN, Int. Peat Soc., Helsinki, 1981, p. 493.

Verry, E. S., The hydrology of wetlands and man's influence on it, in Proc. Symp. on the Hydrology of Wetlands in Temperate and Cold Regions, Vol. 1, *Suomen Akatemian Julk.*, 4, 41, 1988.

Vompersky, S. E., Silvicultural efficiency of forest drainage, *Lesnoe Khozajstvo*, 9, 19, 1964 (in Russian).

Vompersky, S. E., On tree stands roots flooding on drained peatlands, *Lesovedenie*, 1, 75, 1967 (in Russian).

Vompersky, S. E., *Biological Basis of Forest Drainage,* Nauka Publishers, Moscow, 1968, 312 pp. (in Russian).

Vompersky, S. E., Investigation of the water balance of drained forests and swamps, in Proc. Int. Symp. Forest Drainage 1974 Jyvaskyla-Oulu, Finland, Vol. 1, IPS-IUFRO Working Party, Helsinki, 1974, p. 405.

Vompersky, S. E. and Busarova, E. I., Physical-chemical conditions of draining lands and their value for tree growth, in *Influence of Moisture Surplus on Forest Productivity,* Nauka Publishers, Moscow, 1966, p. 57 (in Russian).

Vompersky, S. E. and Eizerman, N. I., The use of a thermodynamic approach for optimal drainage intensity approvement, *Lesovedenie*, 5, 17, 1990 (in Russian with English summary).

Vompersky, S. E., Rubtsov, V. V., and Dudorov, A. V., The method for determining forest drainage network parameters by modeling the groundwater regime, in Socio-Economic Impacts of the Utilization of Peatlands in Industry and Forestry, Proc. IPS Symp., Oulu, Finland, 1986, p. 119.

Vompersky, S. E., Sabo, E. D., and Formin, A. S., *Forest Drainage Amelioration*, Lesnaja Promyshlennost Publishers, Moscow, 1975, 295 pp. (in Russian).

Vompersky, S. E. and Sirin, A. A., Impact of forest drainage on runoff, *Vodnuye Resursy*, 4, 47, 1986 (in Russian).

Vompersky, S. E., Sirin, A. A., and Glukhov, A. I., *Runoff Formation and Regime in a Process of Forest Drainage*, Nauka Publishers, Moscow, 1988, 168 pp. (in Russian).

Yurkevich, S. D., Smoljak, L. P., and Girin, B. E., On oxygen content in soil water and carbon dioxide content in soil air of forest peatlands, *Pochvovedenie*, 2, 41, 1966 (in Russian).

Zalitis, P., *Principles of Rational Forest Drainage in Latvijskaya SSR*, Zinatne Publishers, 1983, 230 pp. (in Russian).

CHAPTER **15**

Hydrological–Chemical Interactions in Headwater Forest Wetlands

Alan R. Hill and Kevin J. Devito

CONTENTS

INTRODUCTION

Small valley bottom wetlands are a common element of headwater catchments in humid glaciated landscapes (Hill, 1990a; Jacks et al., 1994). Although these wetlands may only occupy a few percent of the catchment area, their location makes them potentially important in the retention, transformation, or export of mineral elements that may have considerable chemical impacts on downstream reaches and lakes.

The influence of small headwater wetlands on water chemistry in northern landscapes has received increased attention in recent years.Although relatively few detailed budget studies have been done, the available data suggest that these small wetlands have variable retention efficiencies. Devito et al. (1989) and Devito and Dillon (1993) reported low annual retention of total N (TN) and total P (TP) in two central Ontario conifer swamps. Low retention of TN was also found in a headwater swamp in southern Ontario (Hill, 1991). In contrast, a conifer bog in Minnesota retained 65 percent of annual TN inputs (Urban and Eisenreich, 1988) and 61 percent of TP inputs (Verry and Timmons, 1982). The same central Ontario and Minnesota wetlands effectively retained NH_4 (88 to 95 percent) and NO_3 (43 to 70 percent) (Urban and Eisenreich, 1988; Devito et al., 1989). Hill (1991) reported annual retention of 68 percent of the NH_4 input in groundwater but no difference between NO_3 input and output for the southern Ontario head-water swamp. Two of five forest wetlands in southern Sweden were sinks for NO_3–N (39 to 79 percent), whereas the others were nearly in balance (Jacks et al., 1994). Sulphate (SO_4) input-output budgets, or analysis of stream flow chemistry, also indicate that individual wetland sites may have high annual retention (Ogden, 1982; Urban and Bayley, 1986), low retention (Steele and Buttle, 1994), or may alternate between low retention and net release (LaZerte, 1993; Devito, 1995).

Despite considerable research on the biogeochemistry of wetlands, our ability to understand why wetlands differ in element cycling is limited. Much of the available data is based on mass balance studies of individual wetlands, which reveal the overall function but do not provide insight into the hydrological and chemical processes that produce different patterns of chemical behavior. A process-based approach is needed to extrapolate the results from individual wetlands to the landscape scale and to model the biogeochemical response of wetlands to changes in input or wetland structure.

Recently, we have focused on linkages between hydrological flow paths and chemistry as a conceptual framework for explaining differences in the roles of wetlands as sources, sinks, and transformers of elements. The importance of hydrology as a basis for understanding wetland chemistry has been emphasized previously (Hemond, 1980). However, most wetland studies have focused only on the quantification of hydrological inputs and outputs for element mass balances rather than on the relationship between hydrology and chemical transformation processes.

Upland–wetland linkages can influence element retention in wetlands. Retention efficiency may be determined by the transport rate and concentration of inputs (Brinson, 1993; Devito and Dillon, 1993). The magnitude and seasonality of the groundwater connection to wetlands control water table fluctuations and surface saturation (Roulet, 1990). These hydrological factors affect wetland soil anoxia and chemistry. Water table position, hydraulic gradient, and water retention capacity of the substrate also largely control the magnitude, rate, and type of flow path (Hammer and Kadlec, 1986). Differences in residence time and the environment associated with various surface and subsurface hydrological pathways within wetlands can influence element retention.

Both external and internal hydrological pathways determine whether a wetland is an element sink or source. These wetland hydrological functions are influenced by the position of the wetland in the landscape and by basin hydrology and vary spatially within and among catchments and temporally with differences in physiography, climate, and antecedent soil moisture (Carter, 1986; Winter and Woo, 1990). Thus, wetland biogeochemistry must be examined in the context of the watershed hydrogeological setting.

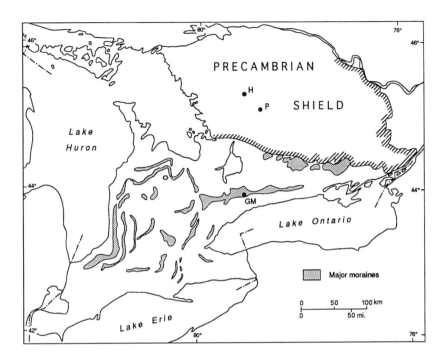

Figure 1   Location of the three wetland sites in Ontario, Canada. GM — Glen Major swamp; H — Harp swamp; P — Plastic swamp.

In this paper we integrate previous research on individual wetlands to show how a hydrological–chemical interaction perspective can explain different patterns of element retention and outflow chemistry in three valley bottom forest wetlands in contrasting hydrogeological settings. Two of the study wetlands are located in headwater catchments of Harp Lake and Plastic Lake near the southern limit of the Canadian Precambrian Shield in central Ontario (Figure 1). The third wetland is located in a headwater catchment on the Oak Ridges moraine at Glen Major in southern Ontario.

Plastic and Harp catchments are underlain by impermeable metamorphic silicate bedrock covered by shallow till. The upland areas of Plastic catchment are forested with white pine, eastern hemlock, and balsam fir, whereas the vegetation in Harp is a deciduous forest of maple and birch. Small wetlands occupy valley bottom locations in these catchments (Figure 2). The two study sites are minerotrophic conifer swamps dominated by white cedar and black spruce, which is the major wetland type in the low boreal region of the Canadian Precambrian Shield (Riley, 1988). A ground layer of *Sphagnum* is well developed in Plastic swamp and is less continuous in Harp swamp.

Catchments in the Glen Major area are underlain by up to 70 m of sands and gravels with till layers. Uplands are covered by mature maple-beech forest and young stands of white pine and birch interspersed with old-fields of grasses and forbs. A mature forest of hemlock and white cedar occurs on wetlands which

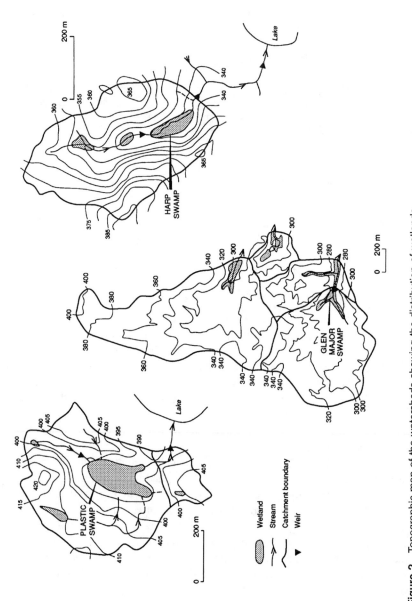

Figure 2 Topographic map of the watersheds showing the distribution of wetlands.

form a zone 20 to 100 m wide along the perennial streams (Figure 2). The Glen Major wetland is typical of forested spring-fed swamps that form an important component of the landscape where aquifers occur in deep glacial and fluvioglacial deposits in eastern Canada and the northeast United States (Carter and Novitzki, 1988; Roulet, 1990).

UPLAND-WETLAND HYDROLOGICAL–CHEMICAL LINKAGES

The hydrological links between Plastic and Harp wetlands and their catchments have been previously considered within a landscape framework that emphasizes a physiographic continuum of overburden depth (Devito et al., 1995). Plastic and Harp swamps are located in catchments with mean till depths of <1 m and 2 to 3 m, respectively, which represent a range characteristic of the southern Canadian Shield. These shallow tills over impermeable bedrock restrict catchment runoff to local flow regimes. Groundwater sources are considered insignificant in the water balance of wetlands in this region (Winter and Woo, 1990). In contrast to the glacial erosional landscape of the Canadian Shield, southern Ontario is a region of glacial deposition. The Glen Major catchment, therefore, represents an extension of the physiographic continuum previously developed for shallow till to areas of deep overburden. An extensive aquifer approximately 218 km^2 in area and 15 m thick occurs in this area of the moraine (Howard and Beck, 1986).

Annual water inputs to the three wetlands reveal differences in catchment-wetland hydrological connections that are important for biogeochemistry. Stream and shallow hillslope groundwater contributed 27 and 56 percent, respectively, of the total annual input to Plastic swamp (Devito, 1995). The inflow stream contributed 50 percent, and shallow groundwater from adjacent hillslopes and deeper groundwater upwelling beneath the swamp contributed 30 and 14 percent, respectively, of the annual input in Harp swamp. At Glen Major, groundwater discharge from the large moraine aquifer represented 95 percent of the annual input (Roulet, 1990). Precipitation declined in importance from 17 percent of annual input at Plastic to 7 percent at Harp and 5 percent at Glen Major.

Large seasonal differences in the importance of hydrological links were observed at Plastic swamp (Figure 3). Although shallow subsurface water from hillslopes is a major component of annual inputs, this input occurred mainly in spring and fall as a result of snowmelt and rain storms. In spring, stream and shallow groundwater represented 31 and 61 percent, respectively, of the input to Plastic, whereas during the summer these contributions were low, and precipitation contributed 85 percent of the water input. For several months during the summer, Plastic swamp was disconnected hydrologically from the entire upland catchment.

Stream and groundwater inputs to Harp swamp were continuous, although they showed seasonal variation (Figure 3). The inflow stream contributed 61 percent of the spring input and 31 percent of the summer input. Hillslope groundwater and groundwater upwelling beneath the swamp represented 42 and 17 percent, respectively, of summer inputs. Small quantities of slowly moving

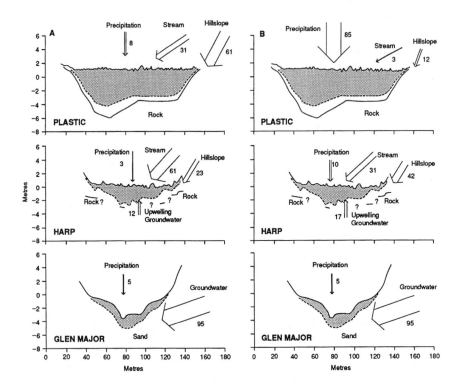

Figure 3   Seasonal variations in hydrologic inputs (percent) to the three swamps. A: Spring period (March to May). B: Summer period (June to August).

groundwater in the deeper strata of moderate depths of till (2 to 3 m) maintain an upland-wetland connection in Harp swamp, which is critical to the surface hydrology of the swamp during low flow periods (Devito et al., 1995).

The large groundwater input to Glen Major was seasonally constant (Figure 3). This constant flow has resulted in the development of organic soils on valley side-slopes of up to 7 percent and has produced well-defined perennial stream channels, which confine storm flows so that flooding of the swamp surface does not occur. The sources of these perennial streams are located within the wetland, and there is therefore no external stream input to the swamp. Storm runoff in this catchment is generated by direct precipitation onto saturated areas of the swamp, and the adjacent moraine slopes do not contribute subsurface storm flow (Roulet, 1991; Waddington et al., 1993).

Differences in the chemistry of hydrological pathways that connect uplands to wetlands influence the chemistry of the three headwater swamps. Significant variations in oxygen-18 and chloride indicate that the large groundwater input to the Glen Major swamp involves two distinct scales of groundwater flow which differ in chemistry. Groundwater discharge from a local flow system at the upland perimeter of the swamp had higher Ca and lower Mg, K, and Na concentrations than regional groundwater, which flowed upward through sands beneath the

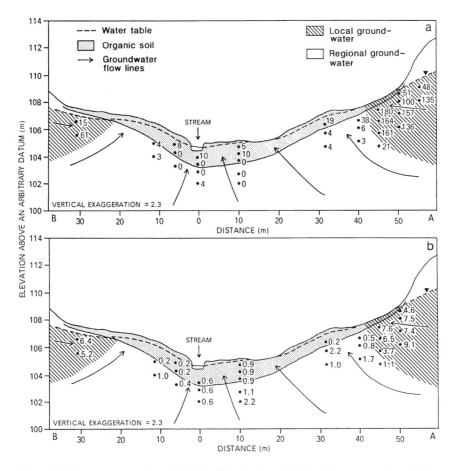

Figure 4 Mean concentration of (a) NO_3–N (μg L^{-1}) and (b) dissolved oxygen (mg L^{-1}) in local and regional groundwater inputs to the Glen Major swamp (June to November 1988).

swamp (Hill, 1990b). The local groundwater had high DO concentrations (6.0 to 10.0 mg L^{-1}) and NO_3–N concentrations of 100 to 180 μg L^{-1}. In contrast, the regional groundwater had low DO levels (0.5 to 2.0 mg L^{-1}) and only trace NO_3–N concentrations (<10 μg L^{-1}) (Figure 4). Variations in cation and NO_3–N concentrations in the Glen Major swamp outlet stream are strongly influenced by these differences in the chemistry of the two groundwater components (Hill, 1990a, b). Variations in chemistry also occurred within local flow systems in the Canadian Shield watersheds. Mean annual SO_4 concentrations in shallow hillslope groundwater inputs to Harp swamp were 9.1 to 11.3 mg L^{-1} and 10.1 to 10.3 mg L^{-1} on the north and south slopes, respectively, whereas deeper upwelling groundwater beneath the swamp was 13.9 mg L^{-1} (Devito, 1995). The high SO_4 beneath the peat may reflect a longer residence of this water.

HYDROLOGY AND CHEMISTRY WITHIN WETLANDS

Peat anoxia and chemistry are controlled mainly by the position of the water table and the rate and magnitude of water flow (Devito and Dillon, 1993; Sparling, 1966). These hydrological characteristics are in turn influenced by hydrological linkages between uplands and wetlands. Differences in catchment hydrogeology between the three Ontario headwater swamps produce a discontinuous upland-wetland link in Plastic swamp and a permanent groundwater connection, which is constant in Glen Major swamp but shows considerable seasonal variation in Harp swamp.

The three swamps have different patterns of water table variation (Figure 5). There were large annual changes in water table position of 98 cm in 1990 to 1991 and 62 cm in 1991 to 1992 in Plastic swamp (Devito, 1995). The water table was frequently above the peat surface in the fall and spring and below the surface in summer and winter. During the summer and early fall when the swamp was hydrologically disconnected from the catchment, the water table varied considerably in response to rain events. In the dry summer of 1990, the water table declined to a minimum elevation of 62.5 cm, whereas in the wetter summer of 1991, the water table dropped to only 23 cm below the surface. Water table elevations in Harp swamp were more stable than in Plastic, with annual changes of 49 and 36 cm in 1990 to 1991 and 1991 to 1992, respectively. Minimum water table heights were 14.1, 6.1, and 1.9 cm below the surface in the three summers of 1990 to 1992, respectively. In the Glen Major swamp the water table elevation beside a saturated area where groundwater emerges continuously varied by only 5 cm in 1988 to 1990 (Figure 5). Annual water table changes of 5 to 15 cm were also recorded in drier areas of the swamp, where the water table was usually 20 to 40 cm below the surface (Roulet, 1990).

Patterns of anoxia and aeration of surface peat associated with water table fluctuations influence element cycling processes and thus element retention and export in these wetlands. Sites in permanently saturated areas of the Glen Major swamp, where the water table was within 10 cm of the surface, had small negative annual rates of net N mineralization (net immobilization) of 4.1 kg ha^{-1} and negligible nitrification in 1985 to 1987 (Hill and Shackleton, 1989). This low rate of N cycling was associated with small seasonal variations in the inorganic N (NH_4–N + NO_3–N) pool of the surface peat at these sites (Figure 6). This pool was measured by analyzing NH_4–N and NO_3–N in fresh peat samples extracted with 2N KCl. Inorganic N was converted from a per gram basis to an areal basis using peat bulk density and depth. The seasonal variability of the surface N pool in Harp swamp was similar to Glen Major. In contrast, the inorganic N pool in Plastic swamp had a much greater range of variation which was related to water table fluctuations. In particular large increases in September 1990 and July 1991 occurred during a period of maximum water table drawdown (Figures 5 and 6). Devito and Dillon (1993) showed that N mineralization may occur at depths >10 cm during periods of water table drawdown in Plastic swamp. High NH_4–N concentrations (>1 mg L^{-1}) and NO_3–N concentrations of >0.5 mg L^{-1} were

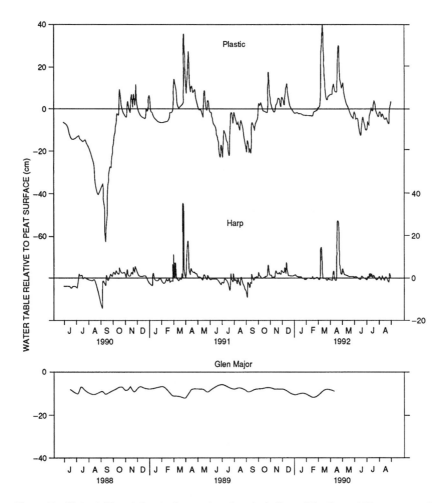

Figure 5   Water table relative to the peat surface in hollows (Plastic and Harp swamps) and in permanently saturated areas (Glen Major).

recorded in porewater at 20 to 30 cm depth as the water table declined to 45 cm below the surface in the late summer of 1987.

Differences in hydrological flow paths within the three headwater swamps influence biogeochemistry. In the Glen Major swamp, local groundwater discharges to the surface as seeps and springs near the upland boundary producing small surface streamlets, which cross the swamp to the streams (Figure 7a). A second pathway involves regional groundwater, which flows upward through the peat to emerge in permanently saturated areas in the swamp. This regional groundwater mixes with the local groundwater in the surface streamlets, which link many of the saturated areas to the stream (Figure7b). Elsewhere in the swamp the streamlets are not connected to the water table (Roulet, 1990). The third flow path consists of regional groundwater, which flows upward through the peat and

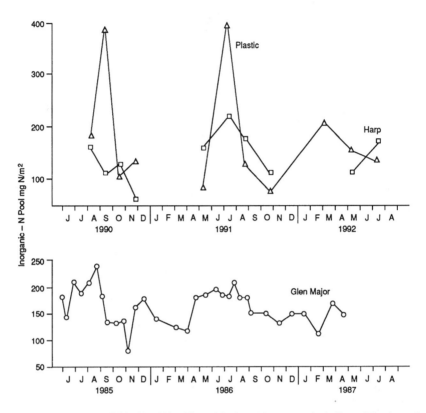

Figure 6   Inorganic N (NH₄-N + NO₃-N) pool in 0 to 10 cm peat in hollows (Plastic and Harp swamps) and in permanently saturated areas (Glen Major swamp).

reaches the stream either as bed seepage or through subsurface pipes that emerge at the stream banks (Figure 7b).

Groundwater flowing via the surface streamlets at Glen Major has a high DO concentration and little interaction with surface peat. Surface flow on moderate slope gradients results in a limited water residence time of 0.5 to 1.0 h within the swamp (Hill, 1990a). In contrast regional groundwater reaching the outlet stream by subsurface paths is low in DO and has a longer residence time. Differences in the chemistry of these flow paths reflect several processes. Springs at the wetland perimeter have higher concentrations of NO_3–N and Cl than the streamlets entering the swamp outlet stream. This progressive decline in element concentration as the streamlets flow across the wetland surface results from the entry of regional groundwater with low NO_3–N and Cl levels rather than NO_3–N retention. Previous research has shown no nitrate removal from streamlets during nitrate enrichments and generally low denitrification rates in swamp soil (Warwick and Hill, 1988). However, local and regional groundwater had similar concentrations of NH_4–N (20 to 60 µg L^{-1}) and the low NH_4–N concentration (5 µg L^{-1}) in surface streamlet water entering the outlet stream resulted from microbial immobilization of NH_4–N in streamlet substrates (Hill and Warwick, 1987).

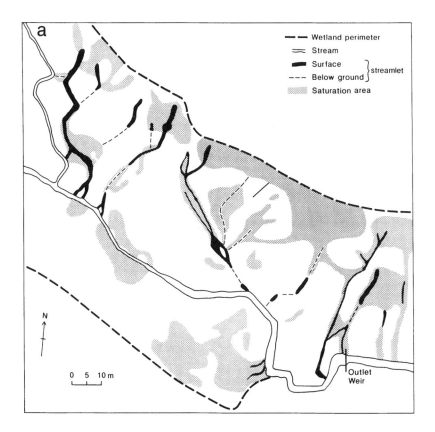

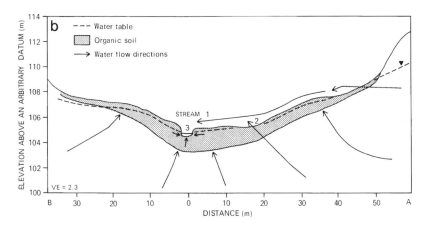

Figure 7 (a) Distribution of surface streamlets and permanently saturated areas in a portion of Glen Major swamp. (b) Cross-section of Glen Major swamp showing hydrological flow paths. 1. Surface streamlets. 2. Regional groundwater input to saturated areas within the swamp. 3. Regional groundwater entering outlet stream as subsurface bed and bank seepage.

Hydrological flow paths in Plastic and Harp showed similar seasonal patterns. During the fall and spring periods of high flow, the water table in both swamps was above the peat surface, and saturation overland flow (SOF) was the dominant water flow path (Devito, 1995). During the summer the water table dropped to or below the surface, and subsurface flow through the top 0.5 m of peat, which has higher hydraulic conductivity, became the major pathway. During periods of SOF in fall and spring, SO_4 concentrations in surface water in the swamps were similar to concentrations of stream and groundwater inputs. In summer as subsurface flow became dominant, surface water SO_4 concentrations declined below 2 mg L^{-1} during low flows (Devito, 1995). In Plastic swamp, NO_3–N concentrations in surface water varied from approximately 150 μg L^{-1} in late fall and spring to trace levels in early summer, whereas NH_4–N concentrations remained near detection limits through much of the year (Devito and Dillon, 1993). Nutrient flushing rates in Plastic swamp varied from 247 d in summer to 16 d in winter and spring (Devito and Dillon, 1993). During a period of SOF in the recession phase of snowmelt runoff in April 1987, an average residence time of 2 d was determined by LiBr tracing in Plastic swamp (Wels and Devito, 1989). Similar large seasonal contrasts in water residence time probably occur in Harp swamp.

Although the interaction of water and element inputs occurs primarily with the surface peat in Harp and Plastic swamp, upwelling groundwater through deeper peat is also an important flow path in Harp. Groundwater beneath the swamp contributed approximately 21 percent of the annual SO_4 input to Harp swamp in 1990 to 1992 (Devito, 1995). The passage of this water through deep anoxic peat resulted in the efficient reduction of SO_4 throughout the year.

PATTERNS OF ELEMENT RETENTION IN HEADWATER WETLANDS

Annual input-output budgets of nitrogen for 1982 to 1984 and SO_4 for 1990 to 1992 have been constructed for Plastic and Harp swamps (Devito et al., 1989; Devito, 1995). Devito and Dillon (1993) have also calculated a nitrogen budget for Plastic swamp for 1987 to 1988. Annual nitrogen inputs and outputs in groundwater fluxes have been previously reported for Glen Major swamp (Hill, 1991). During 1990 precipitation depth and nitrogen chemistry were measured at an open site adjacent to Glen Major swamp. Stream discharge at the swamp outlet has been monitored at a 53°V notch weir since 1986. In 1990 outlet stream samples were collected at 2 h intervals during spring snowmelt and at 20 to 30 min intervals during storms using a stage activated sampler (Hill, 1993). Stream samples were also collected manually at weekly intervals during low flows. Nitrogen chemistry was determined as outlined by Hill (1991). These data were used to calculate an annual nitrogen budget for the period January to December 1990.

Annual mass balances suggest that the three swamps had little effect on total nitrogen (TN) fluxes (Tables 1 and 2). Plastic and Harp swamps were nearly in balance, whereas the retention of 26 percent of TN input in Glen Major swamp was not much greater than estimated budget uncertainties. Plastic and Harp

Table 1   Glen Major Swamp Nitrogen Budget
(kg ha⁻¹) for 1990

	NO_3–N	NH_4–N	TON	TN
Groundwater	9.33	5.89	28.41	43.63
Precipitation	5.81	3.88	1.70	11.39
Total input	15.14	9.77	30.11	50.02
Total output	17.40	1.95	21.24	40.59
In-out	−2.26	7.82	8.87	14.43
% of input	−15	80	29	26

swamps transformed nitrogen by retaining NH_4–N (88 to 95 percent) and NO_3–N (43 to 70 percent) and exporting an equivalent amount of total organic nitrogen (TON). In contrast, only NH_4–N was retained and TON as a proportion of TN inputs and outputs was similar in Glen Major swamp.

Differences in catchment hydrogeology as reflected in upland-wetland hydrological connections and internal wetland hydrological pathways provide an explanation for these contrasts in the biogeochemical function of the three headwater swamps. In Glen Major swamp approximately 90 percent of the annual groundwater NO_3–N input was contributed by local groundwater emerging at the swamp perimeter (Hill, 1991). The swamp did not retain this NO_3–N because the groundwater flowed through the swamp in aerobic surface streamlets, which limited the water residence time and the interaction with the organic soils. Denitrification does not occur in the large volume of regional groundwater flowing through the peat within the swamp because this water contains only trace NO_3–N concentrations (Hill, 1991). Consequently, seasonal patterns of NO_3–N in the Glen Major swamp outlet stream are controlled by the mixing of local and regional groundwater with different NO_3–N concentrations and not by transformation processes in the swamp. Similarly during storm events, the close agreement between observed NO_3–N concentrations and values predicted from a chemical mixing model indicates that stream NO_3–N variations result from the mixing of precipitation with groundwater in surface storm runoff from saturated areas in the swamp (Hill, 1993).

Local and regional groundwater inputs of NH_4–N to Glen Major were retained as this water enters the surface swamp streamlets (Figure 7). Hill and Warwick (1987) found that short-term field enrichments of streamlets with NH_4–N produced rapid removal. Precipitation inputs of NH_4–N were also retained in surface storm runoff from the saturated areas (Hill, 1993). In contrast, little NH_4–N depletion occurred in regional groundwater, which flowed through anaerobic peat and entered the outlet stream as bed and bank seeps (Hill, 1990a).

The annual retention of inorganic nitrogen and the large export of TON in Harp and Plastic swamps are influenced by the considerable lateral dispersion of hillslope runoff and stream inflows, which results from low gradients and the absence of well-defined stream channels within the swamps. Although the dominant hydrological flow paths vary seasonally, both SOF and shallow subsurface

Table 2 Plastic and Harp Swamp Nitrogen Budget
(kg ha^{-1}) for 1982 to 1984

	NO$_3$–N	NH$_4$–N	TON	TN
Plastic swamp				
Stream inputs	0.11	0.04	2.97	3.12
Ungauged	0.06	0.05	2.11	2.22
Precipitation	5.15	3.15	1.49	9.79
Total input	5.33	3.24	6.57	15.14
Total output	1.58	0.17	13.30	15.05
In-out	3.74	3.07	−6.70	0.09
% of input	70	95	−102	1
Harp swamp				
Stream inputs	8.70	0.15	11.40	20.25
Ungauged	1.02	0.09	5.17	6.28
Precipitation	5.15	3.15	1.49	9.79
Total input	14.90	3.39	18.10	36.39
Total output	8.50	0.42	27.50	36.40
In-out	6.40	2.97	−9.40	−0.01
% of input	43	88	−52	<1

flow have greater residence times and more extensive contact with the surface peat and biota than occurs in Glen Major swamp.

Retention of NH$_4$–N and NO$_3$–N in Plastic and Harp swamps in the summer occurs during periods of low runoff from the catchment that result in a low water table and a dominance of subsurface flow maximizing the contact of inflowing water with swamp peat (Devito and Dillon, 1993). In the late fall to spring, periods of high runoff result in high water tables and dominance of saturated overland flow. The more limited interaction with swamp soils during these periods together with short residence times and aeration of surface water restricts nutrient retention. In Plastic swamp, seasonal NO$_3$–N retention was 100 percent in summer, 50 to 52 percent in fall to winter and only 28 percent in spring, whereas NH$_4$–N retention was 75 percent in spring and 90 to 100 percent throughout the rest of the year (Devito and Dillon, 1993). Retention of TON was only observed during the summer in Plastic and Harp swamps. Release of TON occurred during periods of standing water in the fall to spring period when SOF was important. Decomposition of peat and leaching of litter and plant tissues in the fall provide sources for increased export of organic N (Bowden, 1987).

Seasonal patterns of element retention were similar in the Plastic and Harp swamps for most years, but there were differences in sulfate dynamics in years with dry summers as a consequence of differences in upland-wetland hydrological links and water table fluctuation (Devito, 1995). Small groundwater volumes, which continuously discharged from deeper tills, were critical in maintaining saturated peat in Harp swamp, and SO$_4$ was retained (29 to 36 percent) in both 1990 to 1991 and 1991 to 1992 (Table 3). In contrast, Plastic swamp had large water table drawdowns during extended dry summers because it was disconnected

Table 3   Plastic and Harp Swamp Sulfate Budget (kg ha^{-1}) for 1990 to 1992

	Plastic swamp		Harp swamp	
	1990 to 1991	1991 to 1992	1990 to 1991	1991 to 1992
Total inputs	474	346	1524	1292
Total outputs	560	332	1079	825
In-out	−86	14	445	467
% of input	−18	4	29	36

from the uplands. Aeration of peat during these summer droughts was linked to elevated SO_4 concentrations in stream outflow during the fall. After the dry summer of 1990, Plastic swamp was a source of SO_4 during the fall and had an annual export of −18 percent in 1990 to 1991, whereas in 1991 to 1992, a year with a smaller summer water table decline, inputs and outputs were in balance (Table 3). Mass balance studies over 7 years (1983 to 1990) have also measured a net annual export of SO_4 (−8 to −32 percent) from Plastic swamp in 4 years with dry summers, whereas net SO_4 retention (20 to 29 percent) occurred during years with normal or greater summer rainfall (Lazerte, 1993). Export of elements following water table drawdown during drought conditions has also been observed in other northern wetlands (Bayley et al., 1987; Van Dam, 1988).

CONCLUSIONS

Standard wetland classifications identify these three Ontario wetlands as conifer swamps. However, the large variation in biogeochemical function which the swamps display indicates that traditional wetland classifications based on vegetation and substrate type (Zoltai and Pollett, 1973; Cowardin et al., 1979) are of limited use in evaluating the element dynamics of wetlands, nor can they be used to extrapolate studies from one wetland to another.

The process-based conceptual framework used to analyze these Ontario swamps shows how differences in upland-wetland linkages and wetland hydrological pathways as controlled by hydrogeological setting can provide an explanation for similarities and differences in the biogeochemistry of wetlands. The hydrological–chemical interaction perspective may also provide a framework for the extrapolation of results from individual wetlands to the landscape scale in some regions.

Analysis of Plastic and Harp swamps indicates that small differences in till depth from <1 m to 2 to 3 m between catchments are critical in determining the threshold response from continuous to discontinuous groundwater inputs and the biogeochemical behavior of the wetlands (Devito, 1995; Devito et al., 1995). It appears that till depth may be used as a surrogate variable to predict seasonal patterns of chemistry and the episodic export of elements such as SO_4 from headwater wetlands at the landscape scale in this region of the Precambrian Shield, which has a simple hydrogeology (Devito, 1994).

In southern Ontario landscapes covered by thick sequences of glacial deposits, which can contain several scales of groundwater flow, it may be impossible to generalize the biogeochemical behavior of wetlands with respect to physiographic setting. Site-specific characteristics of glacial stratigraphy and the difficulty of placing the wetland in relation to the larger hydrogeological system frequently necessitate detailed site studies.

Our comparison of three headwater swamps along a physiographic continuum that extends from glacial erosional landscapes with thin overburdens to areas of extensive glacial deposition suggests that a focus on interactions between hydrology and chemistry within the context of basin hydrogeology can explain differences in the role of wetlands as sources, sinks, and transformers of mineral elements. Further research using this conceptual framework would contribute to the development of reliable generalizations about the role of small wetlands in regulating water chemistry in headwater catchments.

REFERENCES

Bayley, S. E., Vitt, D. H., Newbury, R. W., Beaty, K. G., Behr, R., and Miller, C., Experimental acidification of a Sphagnum dominated peatland: first year results, *Can. J. Fish. Aquat. Sci.,* 44, 194, 1987.

Bowden, W. B., The biogeochemistry of nitrogen in freshwater wetlands, *Biogeochemistry,* 4, 313, 1987.

Brinson, M. M., Changes in the functioning of wetlands along environmental gradients, *Wetlands,* 13, 65, 1993.

Carter, V., An overview of the hydrologic concerns related to wetlands in the United States, *Can. J. Bot.,* 64, 364, 1986.

Carter, V. and Novitzki, R. P., Some comments on the relation between groundwater and wetlands, in *The Ecology and Management of Wetlands,* Hook, D. D. et al., Eds., Croom Helm, London, 1, pp. 68–86, 1988.

Cowardin, L. M., Carter, V., Golet, F. C., and LaRoe, E. T., Classification of Wetlands and Deepwater Habitats of the United States, U.S. Fish and Wildlife Service, U.S. Department of the Interior, Washington, D.C., 1979, 103 pp.

Devito, K. J., Hydrologic Control of Sulphur Dynamics in Headwater Wetlands of the Canadian Shield, Ph.D. thesis, York University, North York, Ontario, Canada, 1994, 210 pp.

Devito, K. J., Sulphate mass balances in headwater wetlands of the Canadian Shield: influence of catchment hydrogeology, *Can. J. Fish. Aquat. Sci.,* 52, 1750, 1995.

Devito, K. J. and Dillon, P. J., The influence of hydrologic condition and peat oxia on the phosphorous and nitrogen dynamics of a conifer swamp, *Water Resources Res.,* 29, 2675, 1993.

Devito, K., Dillon, P. J., and LaZerte, B., Phosphorus and nitrogen retention in five Precambrian Shield wetlands, *Biogeochemistry,* 8, 185, 1989.

Devito, K. J., Hill, A. R., and Roulet, N., Groundwater-surface water interactions in headwater forested wetlands of the Canadian Shield, *J. Hydrol.,* 1995, in press.

Hammer, D. E. and Kadlec, R. H., A model for wetland surface water dynamics, *Water Resources Res.,* 22, 1951, 1986.

Hemond, H. F., Biogeochemistry of Thoreau's Bog, Concord Massachusetts, *Ecol. Monogr.*, 50, 507, 1980.

Hill, A. R. Groundwater flow paths in relation to nitrogen chemistry in the near-stream zone, *Hydrobiologia*, 206, 39, 1990a.

Hill, A. R., Groundwater cation concentrations in the riparian zone of a forested headwater stream, *Hydrol. Proc.*, 4, 121, 1990b.

Hill, A. R., A groundwater nitrogen budget for a headwater swamp in an area of permanent groundwater discharge, *Biogeochemistry*, 14, 209, 1991.

Hill, A. R., Nitrogen dynamics of storm runoff in the riparian zone of a forested watershed, *Biogeochemistry*, 20, 19, 1993.

Hill, A. R. and Shackelton, M., Soil N mineralization and nitrification in relation to nitrogen solution chemistry in a small forested watershed, *Biogeochemistry*, 8, 167, 1989.

Hill, A. R. and Warwick, J., Ammonium transformations in springwater within the riparian zone of a small woodland stream, *Can. J. Fish. Aquat. Sci.*, 44, 1948, 1987.

Howard, K. W. F. and Beck, P., Hydrochemical interpretation of groundwater flow systems in Quaternary sediments of southern Ontario, *Can. J. Earth Sci.*, 23, 938, 1986.

Jacks, G., Joelsson, A., and Fleischer, S., Nitrogen retention in forest wetlands, *Ambio*, 23, 358, 1994.

LaZerte, B. D., The impact of drought and acidification on the chemical exports from a minerotrophic conifer swamp, *Biogeochemistry*, 18, 153, 1993.

Ogden, J. G., Seasonal mass balance of major ions in three small watersheds in a maritime environment, *Water Air Soil Pollut.*, 17, 119, 1982.

Riley, J. L., Peat and Peatland Resources of Southeastern Ontario, Mines and Minerals Div., Ontario Geological Survey, Ontario Ministry of Northern Development and Mines, Open File Report, 5633, 1988, 283 pp.

Roulet, N. T., Hydrology of a headwater basin wetland: groundwater discharge and wetland maintenance, *Hydrol. Proc.*, 4, 387, 1990.

Roulet, N. T., Storm runoff production in a headwater basin, *Nordic Hydrol.*, 22, 161, 1991.

Sparling, J. H., Studies on the relationship between water movement and water chemistry in mires, *Can. J. Bot.*, 44, 747, 1966.

Steele, D. W. and Buttle, J. M., Sulphate dynamics in a northern wetland catchment during snowmelt, *Biogeochemistry*, 27, 187, 1994.

Urban, N. R. and Bayley, S. E., The acid–base balance of peatlands: a short-term perspective, *Water Air Soil Pollut.*, 30, 791, 1986.

Urban, N. R. and Eisenreich, S. J., Nitrogen cycling in a forested Minnesota bog, *Can. J. Bot.*, 66, 435, 1988.

Urban, N. R., Eisenreich, S. J., and Grigal, D. F., Sulphur cycling in a forested Sphagnum bog in northern Minnesota, *Biogeochemistry*, 7, 81, 1989.

Van Dam, H., Acidification of three moorland pools in The Netherlands by acid precipitation and extreme drought periods over seven decades, *Freshwater Biol.*, 20, 157, 1988.

Verry, E. S. and Timmons, D. R., Water-borne nutrient flow through an upland-peatland watershed in Minnesota, *Ecology*, 63, 1456, 1982.

Waddington, J. M., Roulet, N. T., and Hill, A. R., Runoff mechanisms in a forested groundwater discharge wetland, *J. Hydrol.*, 147, 37, 1993.

Warwick, J. and Hill, A. R., Nitrate depletion in the riparian zone of a small woodland stream, *Hydrobiologia*, 157, 231, 1988.

Wels, C. and Devito, K. J., Tracing of flowpaths in a conifer swamp, Ontario Wetlands: Inertia or Momentum, Proceedings of a Conference of the Federation of Ontario Naturalists, Toronto, 1989, pp. 293–300.

Winter, T. C. and Woo, M.-K., Hydrology of lakes and wetlands, in *The Geology of North America*, Vol. 1, Wolman, M. G. and Riggs, H. C., Eds., Surface water hydrology: Boulder, Colorado, Geological Society of America, 1990, pp. 159–187.

Zoltai, S. C. and Pollett, F. C., Wetlands in Canada: their classification, distribution and use, in *Mires: Swamp, Bog, Fen and Moor*, Gore, A. J. P., Ed., Elsevier, Amsterdam, 1983, pp. 245–268.

CHAPTER 16

Application of Modeling Methods to Study Water Budgets in Forested Peatlands

J. Ruseckas

CONTENTS

INTRODUCTION

One method of analyzing the water budget of forest peatlands is modeling. The models must reflect main correlations between essential ecological factors of the peatland ecosystem. The main elements to consider for hydrological modeling of peatland are the lithosphere with water horizons and aeration zone, soil and peat deposition, vegetation cover, and atmosphere (Figure 1). Water content in the soil is the chief element joining all subsystems of the ecosystem.

EXPERIMENTAL APPROACH

Stationary complex investigations on the water budget were conducted on six experimental watersheds during the period 1974 to 1992. Evaporation from ground vegetation in harvested areas and forest plantations was determined using six hydraulic vaporizers (Ruseckas, 1990). The effect of the soil moisture on transpiration of trees was investigated in 1987 to 1989 in conjunction with studies conducted on the stability of young trees with moisture deficit. To determine the

1-56670-177-5/97/$0.00+$.50
© 1997 by CRC Press, Inc.

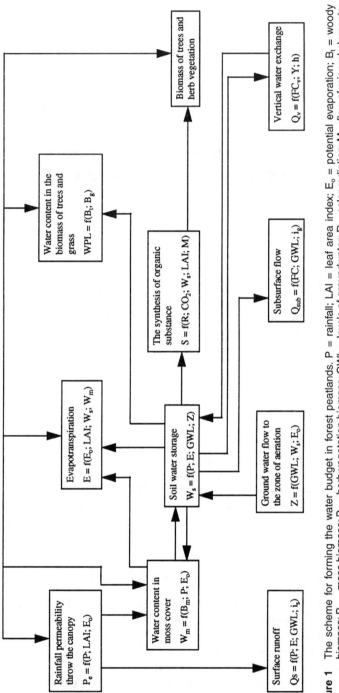

Figure 1 The scheme for forming the water budget in forest peatlands. P = rainfall; LAI = leaf area index; E₀ = potential evaporation; B_t = woody biomass; B_m = moss biomass; B_g = herb vegetation biomass; GWL = levels of groundwater; R = solar radiation; M = flow of mineral elements; FC = the filtration coefficient of peat; i_g = inclination of groundwater; F_cv = the filtration coefficient of the ground in the vertical direction; Y = water pressure; h = the thickness of the ground layer (calculated); i_s = the slope of the peatland surface.

transpiration from trees on these areas, 76 pine, spruce, black alder, and birch seedlings were sampled with monoliths of peat ($15 \times 15 \times 20$ cm). The productivity of individual trees and stands was determined according to the methods of Rodin, Remezov, and Bazilevich (1968).

A SCHEME FOR DEVELOPING THE WATER REGIME IN FOREST PEATLANDS

The general scheme of the water regime in forest peatlands is presented in Figure 1. In this scheme the vegetation cover is divided into the following layers: (1) woody vegetation, (2) grass-bushy, and (3) moss cover. This paper describes the development of empirical models for soil water storage only. Soil water storage was classified into four categories. The first soil water storage category of forest peatland included wilting point (WP). It has been explained that pine uses the reserve of soil moisture to the greatest extent, followed by spruce, birch, and black alder. The soil moisture of transitional and eutrophic peatland, which was found from the volume of peat monolith during wilting of pine, spruce, birch, and black alder, comprises 11.77 ± 1.11; 12.84 ± 1.62; 14.1 ± 1.23 and 16.06 ± 1.17 percent, respectively. Consequently, transpiration of the assimilation mass of the plant was equal to zero. The second category embraced the optimum moisture of forest peatland, which was necessary for the growth of trees. The limits of the optimum content of soil moisture ranged from 72 to 120 percent of the field capacity (FC). Spruce possessed the least range of optimum moisture (70 to 100 percent of FC), black alder the greatest (72 to 120 percent of FC), while pine and birch had intermediate ranges (67 to 102 percent and 65 to 101 percent of FC, respectively). Transpiration (E_t) per weight unit of the assimilative mass of plants was estimated according to the equation, $E_t = Z E_{ot}$, where E_{ot} refers to the evaporative capacity and Z to the coefficient of transpiration intensity by species. This coefficient depended upon the growth stock; birch and black alder averaged 3.5. The coefficient of transpiration intensity for spruce needles was least of all, $Z = 0.71$. Pine's intensity of transpiration was on average 2.5 times higher than that of spruce needles and 2.3 times less compared to the deciduous species (birch and black alder). For freely growing trees, this coefficient was nearly twice as high. The third category includes the content of soil moisture excluding WP and is less than the lower limit of the optimum content of soil moisture. Calculations showed the transpiration of trees within this soil moisture for spruce, pine, birch, and black alder was reflected best by a complex index $S = E_o (SM_t - WP)/(FC - WP)$, where SM_t refers to the content of soil moisture at time t (Figure 2). Concrete values of the coefficients of the equation for individual tree species are presented in Table 1.

It must be noted that these equations are effective if the content of soil moisture becomes stable during the process of significant drying. If the quantity of rainfall exceeds evaporation, the values of transpiration found according to the above equations should be multiplied by a correction coefficient $Q = D + LCM_t + Ft + JCM_t t$ for restoring the viability of plants. Coefficient (Q) depends upon

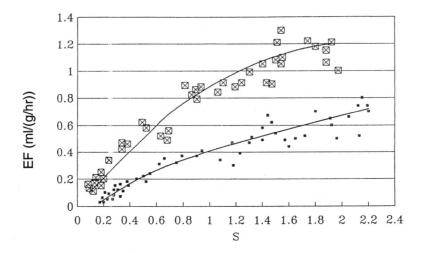

Figure 2   The loss of water per day by leaves of young black alder growing in the forest ⊠ and trees growing freely ☐ as a result of transpiration for a diapason WP < SM$_t$ 0.7 FC; S = E$_o$ (SM$_t$ – WP)/(FC – WP); E$_o$ = the evaporative capacity, WP = wilting point plants, FC = field capacity, SM$_t$ = actual moisture of the soil layer where roots are distributed.

the intensity of drought expressed by moisture content CM$_t$ of the soil and the number of days (t) after the drought. Other coefficients of the equation (D, L, F, and J) are constant. They are presented in Table 1. It was assumed that several

Table 1   The Values of the Empirical Coefficients in Mathematical Model

Coefficients	Tree species			
	Spruce	Pine	Black alder	Birch
A	0.061225	–0.04685	0.568758	0.380949
B	0.000	0.072165	–0.128146	0.000
C	–0.00107	0.038329	–0.04298	–0.041218
a	1.6785	4.021	8.467	8.7769
b	0.03508	0.0574	0.02775	0.10932
Z	0.712	1.488	3.512	3.52
K$_1$	0.70	0.67	0.72	0.65
K$_2$	1.00	1.00	1.2	1.00
N$_1$	6	7	16	8
N$_2$	24	29	41	31
N$_3$	12	14	21	15
D	–0.2681	–0.2691	–0.2807	–0.27961
L	0.0180	0.01794	0.01836	0.01927
F	0.1220	0.12164	0.1230	0.1241
A$_1$	1.1031	1.2112	1.4261	1.3131
B$_1$	0.0432	0.0426	0.0263	0.0398
C$_1$	0.2122	0.2231	0.3113	0.3061
J	0.00271	0.00268	0.00287	0.00294

days after the drought, the water regime and transpiration are still disturbed. The longer the drought, the longer this period is.

Soil moisture being higher than the upper limit of the optimum moisture belongs to the fourth category. The results of our experiment have shown that excess soil moisture lasting for 6, 7, 8, and 16 d (for spruce, pine, birch, and black alder, respectively) does not change the intensity of transpiration. (It is described by the same equation used for the optimum content of the soil.) In cases where the duration of flooding of the roots of young trees is longer, transpiration starts decreasing. If the period of root flooding exceeds a critical value (N_3) in days for the duration of excess soil moisture tolerated by plants, the physiological activity of the roots of trees and transpiration will increasingly diminish. For spruce, value N_3 is 12, for pine 14, for birch 15, and for black alder 21 d and nights (Table 1). The duration of flooding including the dispersion indexes of actual evaporation comprises 86 percent, while the maximum evaporation is only 4 percent. When the roots of trees are completely flooded by water in the period between the critical value (N_3) and the number of days tolerated by plants (N_1) and the number days of their die-off (N_2), the actual transpiration (E_t) is described by the following equation of regression:

$$E_t = E_o (a - bT_1) \tag{1}$$

where T_1 refers to the number of days of complete root flooding and E_o is total evaporation; a and b are constant coefficients (Table 1). It must be noted that N_2 depends upon saturation of water with oxygen and the season of the year. Value N_2 determined for spruce in July and August is 24, for pine 29, for birch 31, and for black alder 41 days and nights (Table 1). Several days (T_1) after flooding (if the duration of flooding does not exceed value N_2), the water regime of trees is still distributed. The longer the period of flooding, the larger the value T_1 is. The general mathematical model for calculating transpiration of trees growing on soils of different moisture is presented in Table 2. This model is recommended only for the calculation of transpiration intensity when rapid growth of the assimilation mass of trees stops. Calculation of the sum of evaporation per decade (using data not being used in constructing the model) has shown that the model error ranges from 5.3 to 9.7 percent. It has been determined that evaporation of rainfall absorbed by plants, leaves, and transpiration depends upon the same meteorological indexes. Consequently, if the moisture content of the soil is optimal, transpiration and evaporation of rainfall absorbed by the leaves of plants can be calculated according to the same empirical formula.

In a drained peatland, the moisture content of the soil layer where roots were distributed was calculated as a function of three values: (1) the quantity of precipitation (P), (2) groundwater level (GWL), and (3) the height of fringe water. It has been determined that GWL is high for oligotrophic swamps, up to 36 cm, and as high as 56 cm for the eutrophic swamps. The moisture evaporated from peat where roots are distributed is completely compensated for by the moisture

Table 2   Mathematical Model of Transpiration of Trees (mL/g of dry leaves per hour)

$$E_t = \begin{cases} O \dots \text{ when wi } CM_t < WP \text{ or wi } CM_t = SS; \; t > N_2 \\[6pt] Z\,E_{ot} \dots \text{ when wi } K_1 FC < CM_t < K_2\,FC \text{ or wi } CM_t = SS; \; t < N_1 \\[6pt] A\,E_o\,(SM_t - WP)/(FC - WP) + BE_o\,[(SM_t - WP)/(FC - WP)]^2 + c \\ \quad \text{when wi } WP < CM_t < K_1 FC; \; t = 0 \\[6pt] [A\,E_o\,(SM_t - WP)/(FC - WP) + BE_o\,[(SM_t - WP)/(FC - WP)]^2 + c] \\ \quad * (D + L\,CM_t + Ft + J\,CM_t\,t) \text{ when wi } WP < CM_t < K_1\,FC; \; t > 0 \\[6pt] Z\,E_{ot}\,(a - bT_1)\,(A_1 - B_1 T_1 + C_1 T_2) \text{ when wi } CM_t = SS; \; N_2 > t > N_1; \; T_2 > 0 \\[6pt] Z\,E_{ot}\,(a - bT_1) \text{ when wi } CM_t = SS; \; N_2 > t > N_1; \; T_2 = 0 \end{cases}$$

where E_t = transpiration; E_o = total evaporation (evaporative capacity); WP = wilting point; FC = field capacity of soils; CM_t = actual moisture of the soil layer where roots are distributed; SS = full field capacity; A; a; B; b; C; D; F; L; K; N; Z = the values of empirical coefficients. Limitation: If value $(a - bT_1)\,(A_1 - B_1 T_1 + C_1 T_2) > 1$, except when equal to 1; if value $[A\,E_o\,(SM_t - WP)/(FC - WP) + BE_o\,[(SM_t - WP)/(FC - WP)]^2 + c] * (D + L\,CM_t + Ft + J\,CM_t\,t) > ZE_{ot}$, except when to equal ZE_{ot}.

obtained from GWL through capillary action. It appeared that in the growing season for drained transitional swamps, the provision of water in the soil layer where roots were distributed was found to be a function of precipitation and its distribution through time. Therefore, for each growing period the integral coefficient of drought (KD) was calculated according to the formula:

$$KD = \sum_i^j m_1 T_1 + \sum_i^j m_2 T_2 + \dots + \sum_i^j m_j T_j \tag{2}$$

where $T_1 T_2 \dots T_j$ refers to the number of intervals without rain, which correspond to the above 5, 6 ... j days and nights in June to September of the calendar year; m_1, m_2 ... m_j are constant coefficients that are equal to 1; 1.1; $1 + 0.1\,(T_j - 5)$ for July to August, respectively, while for September they are equal to 0.5. Rainfall less than 2 mm/d was not considered and thus was equal to zero.

The complete evaporation (E_t) from pine stands of middle age can be determined according to the formula:

$$E_t = 123.967 E_o^{0.346544} KD^{-0.267436} + \left(8.9727 \sum g - 176.247\right) \quad (R = 0.9 \tag{3}$$

where $\sum g$ = stand basal area.

Groundwater flow was calculated as a function of three values: (1) GWL, (2) the coefficient of soil filtration (FK), and (3) slopes (i) of GWL, for modeling the water budget of forest peatlands. We suggest that the layer of water Q_v flowing through the thickness of layer H_n to H_{n-1} is not uniform (isotropic), and should be calculated according to Darcy's law using the following formula:

$$Q_v = \left((WY) / \int_{H_{n-1}}^{H_n} (1/FC_v) \, dh \right) \qquad (4)$$

where Y refers to the hydraulic gradient, W = the area of peatland (territory) on which similar hydraulic gradients are observed, FC_v is the coefficient of filtration of different ground layers.

For practical application of this formula, the function of distribution of value FC_v was calculated from depth H of peat. Ivanov (1957) showed that for oligotrophic swamps, the dependence of FC_v on the depth could be described by the following equation:

$$FC_v = a/(H + 1)^m \qquad (5)$$

where H refers to the depth of peat, and a and m are constant coefficients depending upon the kind of swamp. Combining Ivanov's (1957) with Equation 4, Q_v was calculated as:

$$Q_v = (WYa(m+1))/(H_n + 1)^{m+1} - (H_{n-1} + 1)^{m+1} \qquad (6)$$

Accordingly, the cycle of forming the water budget in forest peatlands can be described by local hydrological, meteorological, and biometric parameters. However, additional work is needed to test the validity of each model.

REFERENCES

Ivanov, K. E., *Basis of Peatlands Hydrology in Forestry Zone,* Gidrometeoizdat, Leningrad, 1957, 499 pp.

Rodin, L. E., Remezov, N. P., and Bazilevich, N. I., *Methodical Instructions on Studies on the Dynamics and Biological Exchange in Phytocoenoses,* L. Nauka, 1968, 143 pp.

Ruseckas, J. J., A Hydraulic Peat Evaporator, USSR patent Appl. # 4429894/30-15, 1988; USSR patent # 1566290, T.B. No. 19, 208, 1990.

CHAPTER **17**

Water Table Fluctuations Following Clearcutting and Thinning on Wally Creek Wetlands

Vincent Roy, John K. Jeglum, and André P. Plamondon

CONTENTS

INTRODUCTION

Forested wetlands (Lugo, 1990) on organic soils in Canada have low productivity resulting from the high water table, cold substrate temperature, and poorly aerated soil conditions. In spite of the slow growth on these wetlands, tree species such as black spruce (*Picea mariana* [Mill.] B.S.P.) provide material for the timber industry in Canada. It has been demonstrated in an extensive review of watershed studies by Bosch and Hewlett (1982) that tree felling reduces the evapotranspiration loss. The increased amount of water available on the site has considerable consequences in forested wetlands since the hydrological regime is the main environmental driving force (Verry, 1988; Lugo, 1990; Bridgham et al., 1991). The major change following clearcutting is the rise of the water table, referred to as "watering-up," which may delay regeneration (Bolghari, 1986) and reduce productivity of the site by decreasing the depth of the aerated zone for tree root exploitation (Heikurainen et al., 1964 in Mannerkoski, 1985). Respecting economic considerations, drainage after clearcutting can only be a solution on the better sites. Therefore, we must develop silvicultural systems for forested wetlands that minimize watering-up.

Hydrological studies to investigate watering-up have been conducted in many countries. An extensive review can be found in Dubé et al. (1995). The general conclusions from these studies are that decreases in the rate of transpiration and interception are responsible for the watering-up (Heikurainen, 1967). Also, the magnitude of the rise is directly related to the percentage of wood cut (Heikurainen and Päivänen, 1970; Päivänen, 1980), to the depth of the original water table (Heikurainen, 1967; Päivänen, 1980), and to the time water is available near the soil surface (Verry, 1980). Recently, Dubé et al. (1995) confirmed that clearcuttings, even on 1 ha, caused rises of the water table in the St. Lawrence Lowlands and that reduced interception appeared to be the most important parameter responsible. Intermediate cuttings such as thinning cause smaller rises on peat soils (Heikurainen, 1967; Heikurainen and Päivänen, 1970; Päivänen, 1982), and on mineral soils (Williams and Lipscomb, 1981). Therefore, it has been proposed that thinning, when applicable, could be effective to reduce watering-up by maintaining some degree of evapotranspiration (Dubé et al., 1995). Impairment of the natural drainage by roads and skidding trails is another cause of watering-up but is not treated in this study.

Extrapolation of the results of these studies from other countries to Canada requires some experiments to quantify the watering-up in relation to climate and soil characteristics. The objective of this project was to evaluate the effects of clearcutting and thinning on water table levels in the Clay Section of Northern Ontario.

SITE DESCRIPTION

The Wally Creek area is located 30 km east of Cochrane, Ontario, at 49°3'N, 80°40'W, in the Northern Clay Section of the Boreal forest (Rowe, 1972, in Berry and Jeglum, 1991) at an altitude of about 275 m above sea level (Figure 1B). Annual precipitation is 885 mm, and one-third of it falls as snow (Anon., 1982a,b, in Berry and Jeglum, 1991). Average June to September rainfall is 372 mm. Mean annual and July temperatures are 2 and 17°C, respectively. Potential evapotranspiration (Thornthwaite's method) in the area is estimated to be 490 mm per year

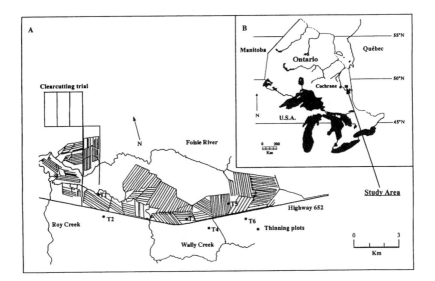

Figure 1 A — Wally Creek study area (clearcutting and thinning trials); B — location of the study area.

(Anon., 1985, in Berry and Jeglum, 1991). Hourly precipitation was measured in the study area with a tipping-bucket rain gauge.

The topography of the Wally Creek area is flat to gently rolling, with a general slope of 0.3 percent to the northwest. Organic soils have developed over heavy clays of lacustrine origin, with peats ranging up to 3 m in depth (Berry and Jeglum, 1991). Part of the forest was harvested in the 1930s by ground cutters using horses to skid trees to the roadside (Sundström, 1992). As a result, the area supports uneven-aged (50 to 140 years) 8 to 17 m black spruce stands that have been site-typed (Table 1) according to the Forest Ecosystem Classification for the Clay Belt of Ontario (Jones et al., 1983).

Table 1 Description of Site Types and the Related Operational Groups (OG) in the Study Area

Site type	OG	Depth of peat (cm)	Depth of fibric layer (cm)[a]	Bulk density (g/cm³) 0–20 cm	Bulk density (g/cm³) 20–40 cm	Rooting depth (cm)
Ledum	11	34–174	32	0.046	0.148	28
Alnus-herb poor	12	53–250	23	0.049	0.147	23
Chamaedaphne	14	100–300	30	0.046	0.114	33

[a] Fibric is defined as having von Post humification ratings from 1 to 4. (From Jones, R. K. et al., Field Guide to Forest Ecosystem Classification for the Clay Belt, Site Region 3E, Ontario Ministry of Natural Resources, Maple, Ontario, 1983. With permission.)

From Berry, G. J. and Jeglum, J. K., Hydrology of Drained and Undrained Black Spruce Peatlands: Groundwater Table Profiles and Fluctuations, COFDRA Report 3307, Forestry Canada, Ontario Region, Sault Ste. Marie, Ontario, 1991. With permission.

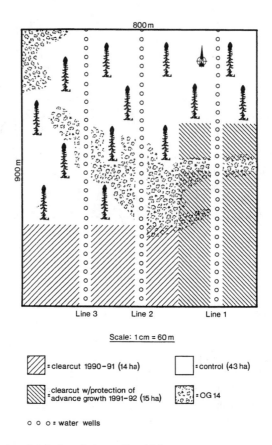

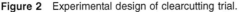

Figure 2 Experimental design of clearcutting trial.

METHODS

Clearcutting Trial

An 800 × 900 m forested block with a general slope of 0.2 percent to the southwest was selected as study site for the clearcutting trial (Figure 1A). The block was site-typed as a mixture of operational group 11 (OG 11) and OG 12 conifer swamp, with a central area classified as OG 14 poor-treed fen (Jones et al., 1983) (Figure 2). Peat depths ranged from 60 to 240 cm. It was divided into three sub-blocks, 267 m wide, to evaluate the effect of cutting in relation to soil wetness (east-west gradient) in the treatment area. In each sub-block, a north-south transect of 30 water wells, 30 m apart, was installed in the center axes of the east (line 1) and middle (line 2) sub-blocks. In the west (line 3) sub-block, the transect was located 77 m east of the center line to avoid an unharvestable area. Thin-walled steel pipes (1.5 m long and 2 cm diameter) were used to measure water levels. Four holes, 3 mm in diameter, were drilled around the pipe every 10 to 15 cm, up to about 30 cm of the well's top. Wells were installed in 1987

Table 2   Number of Water Table Measures During the Calibration and Treatment Periods for the Clearcutting Trial

Year	Period	Measures	Line 1	Line 2	Line 3
1988	Aug. 8–Oct. 15	5	Calibration	Calibration	Calibration
1989	May 25–Oct. 25	9	Calibration	Calibration	Calibration
1990	May 30–Sept. 25	16	Calibration	Calibration	Calibration
1991	June 25–Sept. 29	11	Calibration	Treated	Treated
1992	June 21–Sept. 26	12	Treated	Treated	Treated
1993	July 25–Sept. 20	8	Treated	—	—

with 30 to 60 cm of the pipe sticking above the ground. Measurements of water table levels were initiated in 1988 and were continued in 1989 and 1990 as the preharvest calibration period. In the winter of 1990 to 1991, 14 ha of the block was clearcut with full-tree logging of the southern part of sub-blocks 2 and 3 (Figure 2). The next winter, 15 ha of sub-block 1 was also full-tree harvested (Figure 2). However, in this case, careful logging with protection of advanced growth was carried out.

The wells were removed and replaced by a stake during harvesting in order to reinstall them at the same locations. Intensive level surveys were also done to determine the average ground levels within 1 m of the wells. In conducting the survey, sightings were taken relative to the top of the water well (pipe) and at 0.5 and 1.0 m on both sides of the well along the transect. All depths to water in each well were adjusted to reflect the mean distance to water from the ground surface (Berry and Jeglum, 1991). Water levels were measured periodically during the frost-free season of 1988 to 1993 (Table 2) with an electrical buzzer probe (Bodley et al., 1989).

The effect of clearcutting on the water table level was evaluated using calculated regression equations for the calibration period (1988 to 1990 for lines 2 and 3, 1988 to 1991 for line 1) and for 2 years after harvesting (1991 to 1992 for lines 2 and 3, 1992 to 1993 for line 1) (Table 2). The mean depth for control wells was used as independent variable (x) while the dependent variable (y) was the mean depth in wells in the clearcut area. Each pair of values represented, at a particular date of measurement, the mean depth to water for all pipes in the control and all pipes in the clearcut treatment. Large sample Z-tests (Kleinbaum and Kupper, 1978) were applied to test regression parameters. All analyses were carried out using the SAS statistical package (SAS Institute, 1990), except for Z-tests.

Thinning Trial

The Wally Creek thinning trial was established in July 1990 in black spruce stands with an average basal area of 22 m^2 ha^{-1}. Six plots were selected in OG 11, three each in drained (T1, T3, T5) and undrained (T2, T4, T6) forest (Figure 1A). In the drained forest, ditches were dug in 1984 with spacing of 40 m. In each plot, three 20 × 20 m subplots were established to represent unthinned (control), thinned from above (high thinning), and thinned from below (low

Table 3   Number of Water Table Measures
for the Thinning Trial

Year	Period	Measures
1991	July 2–Aug. 24	4
1992	June 25–Sept. 12	11
1993	July 13–Sept. 25	15

thinning) conditions. In high thinning, trees are removed from upper crown classes in order to open up the canopy, while in low thinning, trees are removed from lower crown classes (Smith, 1986). A 40 percent basal area reduction was chosen for both of the two thinnings in order to modify the stand structure significantly. The marked trees were cut and delimbed on site, and then the boles were removed to the periphery of the plots. Four wells per subplot were installed and depth to the water table was measured periodically during the frost-free season of 1991, 1992, and 1993 (Table 3).

Wells were analyzed either as drained or undrained, and as thinned from below, from above, or unthinned with three replicates. Since no measurements were made before thinning, thinned treatments were simply compared with controls to judge impacts.

RESULTS

Precipitation

Precipitation for the period of June to September was near normal during the 1988 to 1993 study period. It was above average in 1988 and 1990 by less than 12 percent, while below average by 3 to 18 percent during the other years (Figure 3). Water table fluctuations were directly related to the amount and frequency of precipitation (Figure 3).

Clearcutting Trial

The average depth to water table in the forested control part of each line ranged between 20 to 24 cm for both the calibration and treatment periods (Figure 3; Table 4). The precut depths to water table in the treated areas were 33, 25, and 14 for lines 1, 2, and 3, respectively. The water levels of the line 3 treatment wells were nearer the surface than their control wells, and both the treatment and control wells of lines 1 and 2. This corresponds to the lower ground level of the southwest corner of the study area. After cutting, the water table in the clearcut treatment on all lines rose in comparison to their respective controls (Figure 4).

The differences between predicted depth to water table for forested conditions and measured values indicated that mean water table levels were elevated by 5, 4, and 6 cm after clearcutting on lines 1, 2, and 3 respectively (Table 4). This is in agreement with Heikurainen and Päivänen (1970), who found a rise of 5 cm after clearcutting when the control water table was 25 cm deep. It is also in

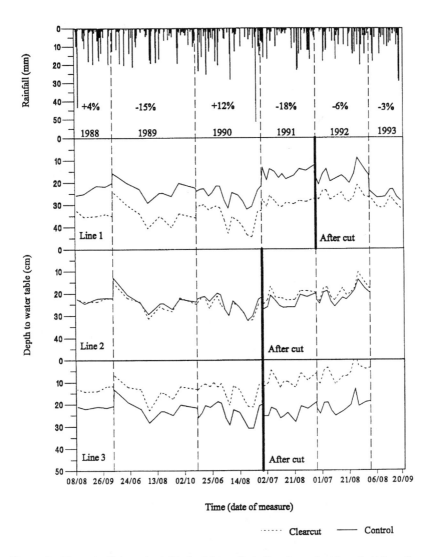

Figure 3   Mean depth to water table (cm) for wells in the clearcut and control along lines 1, 2, and 3, rainfall (mm), and percent of normal June to September rain from 1988 to 1993.

accordance with Dubé et al. (1995), who measured a rise of 4 to 7 cm on five forested wetlands on organic soils with depths to water table between 25 and 40 cm.

The linear regression of treatment against control before and after cutting had coefficients of determination (R^2) ranging between 0.77 and 0.90 (Table 5). Z-tests between slopes and intercepts before and after cut showed significant differences between the slopes of the regressions for line 1 (Table 5; Figure 4). These results indicate that the water table rose significantly due to cutting, and

Table 4   Mean[a] Depth to Water Table (cm) Before and After Clearcutting

| | | Before cut | After cut | | |
| | | Measured | Measured | Predicted[b] | |
	Well #	(x)	(x)	(y)	Difference
Line 1	#1–19 (Treated)	33	27	32	5
	#22–30 (Control)	22	20	—	—
Line 2	#1–9 (Treated)	25	20	24	4
	#11–30 (Control)	24	22	—	—
Line 3	#1–9 (Treated)	14	7	13	6
	#11–30 (Control)	23	22	—	—

[a] Means were the overall averages for all times of measurements for all years of measurements.
[b] Predicted: y calculated from the calibration equation in Table 5.

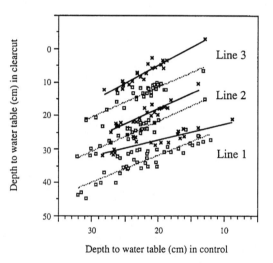

Figure 4   Linear relationship between clearcut and control depth to water table before (□) and after (x) clearcutting on lines 1, 2, and 3.

Table 5   Linear Relationship Between Clearcut and Control Depth to Water Table Before and After Clearcutting

| | | | | Z-tests | |
	Line #	Period	Function	R²	Slope	Intercept
1	1988–91	y = 0.813 x + 15.37	0.77	3.62[a]	1.13
	1992–93	y = 0.477 x + 17.68	0.79		
2	1988–90	y = 0.902 x + 3.443	0.90	−0.12	−1.82
	1991–92	y = 0.913 x − 0.366	0.89		
3	1988–90	y = 0.884 x − 6.302	0.83	−1.8	−4.10[a]
	1991–92	y = 1.087 x − 16.80	0.89		

[a] Statistical significance at 0.01 level.

Table 6   Differences Between the Smallest
and Greatest Depths to Water
Table (cm) for the Two Years After
Clearcutting

	Measured	Predicted	Change
Line 1	11	16	−5
Line 2	14	12	+2
Line 3	15	13	+2

Note: Predicted values are derived from the
calibration period regression equation
(Table 5).

it has a smaller effect during periods of high water table, as was found by Heikurainen and Päivänen (1970), Päivänen (1980), and Dubé et al. (1995). For line 2, z-tests on slopes and intercepts were found to be not significant, showing that the higher water level after clearcutting is not statistically significant (Table 5). In the case of line 3, however, z-tests indicated a significant difference between control and clearcut regression intercepts (Table 5; Figure 4). The cutting on this line has also raised the water table, but it would have the same effect regardless of the water levels between 10 and 30 cm below soil surface.

Water table fluctuations were reduced in the clearcut compared to the control after the cut on line 1 (Table 6). Dubé et al. (1995) also found reduced fluctuations ranging from 2 to 21 cm on five forested wetlands on organic soils after clearcutting. However, on lines 2 and 3, the differences between seasonal maximum and minimum water table were greater by 2 cm after clearcutting compared to predicted values (Table 6). This observation agrees with Verry (1980, 1988), who suggested that when the water table is within 30 cm of the surface, harvesting will increase the fluctuations of the water table. The proportion of net precipitation reaching the water table is probably responsible for these increased fluctuations.

Thinning Trial

In the thinning experiment, as expected, depths to water were greater in the drained than in undrained plots corresponding to the same treatments (Table 7). Measurements the first year after treatment did not indicate any clear relationships between thinning and water table levels (Figure 5). However, the second and third years showed clear trends.

In the undrained plots, the water levels after both thinnings were higher by 8 and 10 cm (Table 7), which is comparable to results from Heikurainen and Päivänen (1970) for the same level of thinning intensity. The water table was nearer the soil surface in the thinned-from-above plots than in the thinned-from-below plots. The thinning from above normally removes more canopy and thus reduces interception and allows more water to reach the ground.

In the drained plots, the ones thinned from above have higher water levels than control by 7 cm, while subplots thinned from below have lower water levels than control.

Table 7   Mean, Maximum, Minimum, and Differences of Depth to Water Table (cm) in the Subplots for the Three-Year Period After Thinning (1991 to 1993)

Subplots	Mean	Maximum	Minimum	Fluctuation
Drained and thin. below	51	58	39	19
Drained and thin. above	40	55	28	28
Drained, unthinned	47	65	34	31
Undrained and thin. below	31	42	21	21
Undrained and thin. above	29	45	21	24
Undrained, unthinned	39	51	25	26

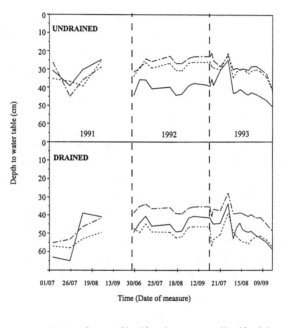

Figure 5   Mean depth to water table (cm) after thinning (1991 to 1993) for drained and undrained conditions. Same rainfall as for the clearcutting trial (Figure 3).

Fluctuations between seasonal lows and highs were greatest in controls, somewhat less for thinning from above, and least for thinning from below (Table 7). This is what would be expected from the influence of canopy on interception: canopy removal would tend to smooth out and reduce fluctuations, as was observed for the clearcut vs. forest areas with water table deeper than 30 cm.

DISCUSSION

In this study, the mean water table rise after clearcutting in the three sub-blocks is in the same range as other studies for similar water table depths

(Heikurainen and Päivänen, 1970; Dubé et al., 1995). Slight differences between the three lines may be due to the different treatment applied in the east sub-block (line 1) (Figure 2). We suggest that the clearcut with protection of advanced growth on line 1 reduced the impact of forest removal by intercepting part of the rainfall. Prévost and Plamondon (1987) have measured an interception rate of 32 percent 4 years after clearcutting with protection of advanced growth, while Roy, Dubé and Plamondon (unpublished) have measured 14 percent interception 3 years after clearcutting without advanced growth protection. Precise evaluation of the impact of protecting advanced growth on the hydrology of forested wetlands is lacking.

According to previous studies, the rise of the water table on a site is greater during dry, low water table periods (Heikurainen and Päivänen, 1970; Päivänen, 1980). We find this response pattern on line 1, but not on the other lines which have higher water tables (Figure 4). On lines 2 and 3, the water table fluctuates near the surface, and the slopes of the regression before and after cut are similar (Table 5). On a bog with a shallow water table, Dubé et al. (1995) found the same parallel response as our lines 2 and 3, with a decrease of the water table level. In the Clay Belt, however, the relationships of watering-up would require a detailed account of the underlying clay topography. This clay plain landscape is characterized by natural bowls and pockets covered by a mantle of peat. In our study area, the lines were linked in an overall hydrotopographic gradient from east to west (Figure 2), and we had naturally wet pockets, such as the basin filled with OG 14 Chamaedaphne and the lower basin in the clearcut treatment portion of line 3. It is possible that clearcutting induced an additional water movement to the lower area located in line 3 clearcut treatment. We suggest that lateral water movement be measured in this type of hydrogeomorphic landscape. The water table, being nearer the surface, is supplied directly by the net precipitation, while the expected increased evaporation may not occur for lack of exposed surface water or capillary rise. Evaluation of topsoil characteristics and evaporation from the surface should also be included in future studies.

The results from the thinning trial have to be interpreted with caution since there are no pretreatment measurements (calibration). For example, if we compared the depths to water table in the control and cut areas after the clearcutting treatment (without calibration) of lines 1, 2, and 3, the water level differences would be respectively −7, 2, and 15 cm (Table 4) instead of the 5-, 4-, and 6-cm rise attributed to the treatment. In the case of the thinning trial, however, the small subplots being located in the same plot are more amenable to an after-treatment comparison than the clearcutting treatments, which covered a wide area.

ACKNOWLEDGMENTS

The authors wish to thank Prof. Donald Siegel and an anonymous referee for their comments. Funding has been provided by Canadian Forest Service, NSERC and Fonds de coopération Québec-Provinces Canadiennes du Ministère de l'Enseignement Supérieur du Québec.

REFERENCES

Berry, G. J. and Jeglum, J. K., Hydrology of Drained and Undrained Black Spruce Peatlands: Groundwater Table Profiles and Fluctuations, COFRDA Report 3307, Forestry Canada, Ontario Region, Sault-Ste-Marie, Ontario, 1991.

Bodley, C. L., Jeglum, J. K., and Berry, G. J., A probe for measuring depth to water surface in wells, *Can. J. Soil Sci.,* 69, 683, 1989.

Bolghari, H., Perspectives du Drainage Forestier au Québec, Colloque sur le drainage forestier, Sept. 10–11, Ste-Foy, Québec, 1986, pp. 47–65.

Bosch, J. M. and Hewlett, J. D., A review of catchment experiments to determine the effect of vegetation changes on water yield and evapotranspiration, *J. Hydrol.,* 55, 3, 1982.

Bridgham, S. C., Faulkner, S. P., and Richardson, C. J., Steel rod oxidation as a hydrologic indicator in wetlands soils, *Soil Sci. Soc. Am. J.,* 55, 856, 1991.

Dubé, S., Plamondon, A. P., and Rothwell, R. L., Watering-up after clear-cutting on forested wetlands of the St. Lawrence Lowlands, *Water Res. Res.,* 31, 1741, 1995.

Heikurainen, L., Effect of cutting on the ground water level on drained peatlands, in *International Symposium on Forest Hydrology,* Sopper, W. E. and Corbett, E. S., Eds., Pergamon Press, Oxford, England, 1967, p. 345.

Heikurainen, L. and Päivänen, J., The effect of thinning, clearcutting, and fertilization on the hydrology of peatland drained for forestry, *Acta For. Fenn.,* 104, 1, 1970.

Jones, R. K., Pierpoint, G., Wickware, G. M., Jeglum, J. K., Arnup, R. W., and Bowles, J. M., Field Guide to Forest Ecosystem Classification for the Clay Belt, Site Region 3E, Ontario Ministry of National Resources, Maple, Ontario, 1983.

Kleinbaum, D. G. and Kupper, L. L., Applied Regression Analysis and Other Multivariable Methods, Service to Publisher, Inc., Duxbury Press, North Scituate, MS, 1978, 556.

Lugo, A. E., Introduction, in *Forested Wetlands, Ecosystems of the World,* Vol. 15, Regional Studies, Lugo, A. E., Ed., Elsevier, Amsterdam, 1990, chap. 1.

Mannerkoski, H., Effect of Water Table Fluctuation on the Ecology of Peat Soil, Department of Peatland Forestry, Publication No. 7, University of Helsinki, Finland, 1985.

Päivänen, J., The effects of silvicultural treatments on the ground water table in Norway spruce and Scots pine stands on peat, in Proc. 6th Int. Peat Congr., Duluth, MN, International Peat Society, Helsinki, Finland, 1980, p. 453.

Päivänen, J., The effect of cutting and fertilization on the hydrology of an old forest drainage area, *Folia For.,* 516, 1, 1982.

Prévost, M. and Plamondon, A. P., Interception de la pluie par les débris et la couverture végétale établie après la coupe à blanc, *Nat. Can.,* 114, 127, 1987.

SAS Institute Inc., SAS Language: References, Version 6.06, First Edition, Cary, NC: SAS Institute, Inc. 1990.

Smith, D. M., *The Practice of Silviculture,* John Wiley & Sons, New York, 1986, 527 pp.

Sundström, E., Five-year Growth Response in Drained and Fertilized Black Spruce Peatlands. Information Report O-X-417, Forestry Canada, Ontario region, Sault Ste. Marie, Ontario, 1992.

Verry, E. S., Water table and streamflow changes after stripcutting and clearcutting an undrained black spruce bog, in Proc. 6th Int. Peat Congr., Duluth, MN, International Peat Society, Helsinki, Finland, 1980, p. 493.

Verry, E. S., The hydrology of wetlands and man's influence on it, in *International Symposium on the Hydrology of Wetlands in Temperate and Cold Climates,* Joensuu, Finland, Vol. 2, Publ. of the Academy of Finland, Helsinki, 1988, p. 41.

Williams, T. H. and Lipscomb, D. J., Water table rise after cutting on Coastal Plain soils, *South J. Appl. For.,* 5, 46, 1981.

CHAPTER **18**

Effects of Engineered Drainage on Water Tables and Peat Subsidence in an Alberta Treed Fen

G. R. Hillman

CONTENTS

INTRODUCTION

Alberta's boreal forest contains extensive areas of wetlands that support stagnant stands of black spruce (*Picea mariana* [Mill.] B.S.P.) and tamarack

1-56670-177-5/97/$0.00+$.50

(*Larix laricina* (Du Roi) K. Koch). Very little is known about these stands or their potential for conversion to productive forest through groundwater table control. It is known, however, that peatland drainage has been practiced extensively and successfully in parts of Europe, resulting in wood volumes as high as 300 m^3 ha^{-1} and annual increments of over 10 m^3 ha^{-1} (Heikurainen, 1964). Finland has drained 5.7 million ha or 55 percent of its peatland to enhance growth of Scots pine (*Pinus sylvestris* L.) and Norway spruce (*Picea abies* [L.] Karst.) (Päivänen, 1991). The former Soviet Union has treated a similar amount of peatland (Vompersky, 1991) and Sweden has drained one million ha of peatland for forestry purposes (Hånell, 1991).

Given the potential for this kind of tree growth enhancement on suitable peatlands and also the withdrawal of forested public lands for uses other than forestry, the Alberta Forest Service (AFS) became interested in exploring the possibility of increasing forest productivity through drainage. Subsequently, in 1981, the first scientific forest drainage study in Alberta was initiated on a treed fen in the Saulteaux River drainage basin in central Alberta under the auspices of the Alberta Forest Service (Tóth and Gillard, 1984, 1988). The objective of the study was to develop techniques for designing drainage systems and measuring their performances.

The procedure used in the Saulteaux study entails the specification of certain design criteria, measurement of a number of environmental parameters, generation of synthetic hydrographs using a mathematical model that takes into account the variable site characteristics of the study area, comparison of synthetic hydrographs with field hydrographs, and eventually selection of appropriate ditch depths and spacings.

The synthetic hydrographs are obtained by solving a transient-flow drain spacing formula recommended by the U.S. Department of the Interior's Bureau of Reclamation (1984). Inputs required to solve the equation include measurements of hydraulic conductivity, water table buildup coefficient, depth to impermeable layer, rainfall distribution, ditch spacing, and ditch depth. A detailed description of the method is provided by Tóth and Gillard (1988).

In 1985, three additional experimental forest drainage studies were established in Alberta as part of the Wetlands Drainage and Improvement Program; this is a program initiated under the Canada–Alberta Forest Resource Development Agreement and designed to provide basic information on forested wetlands, the effects of water table control on the growth of existing forest stands, and the impact of water table control operations on the environment (Hillman et al., 1990). A number of studies were conducted on the Wolf Creek peatland, which is a treed fen and the smallest of the three experimental areas.

The objectives of the study described herein were (1) to determine the effects of different ditch spacings on groundwater table levels and (2) to assess the impact of drainage on peat subsidence. The relationship between ditch spacing and water table level is important because, for operational forest drainage, it is essential to use an optimum ditch spacing that is both biologically and economically suitable.

Peat subsidence is important because it relates to substrate compaction and loss, both of which have implications for forest growth. The purpose of this paper is to present results from the Wolf Creek study. In particular, it describes the effects of water table control and ditch spacings on groundwater levels and peat subsidence and reports on the effectiveness of the ditch network. In this paper, the term "drainage" is used to imply removal of excess water to obtain a desired average water table level.

SITE DESCRIPTION

The Wolf Creek experimental drainage area is located in the Whitecourt Forest about 30 km southeast of Edson, Alberta (53°25'N, 116°03'W) (Figure 1) at an elevation of 950 m. The legal description of the area is Sections 19, 30; Township 51, Range 14, west of fifth meridian.

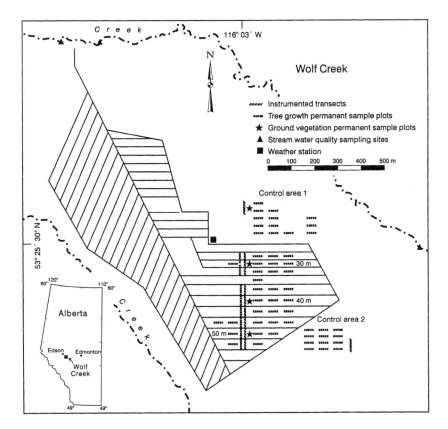

Figure 1   Wolf Creek experimental area — locations of instrumented transects and permanent sample plots.

Climate and Hydrology

The area experiences a subhumid continental climate with long, cold winters and moderately mild summers. It is subject to warm chinook winds during the winter (Dumanski et al., 1972). At Cold Creek Ranger Station, the station with long-term records and closest to the Wolf Creek site, the mean annual air temperature (1951 to 1980) is 1.1°C. The average January and July temperatures are −16.3 and 14.6°C, respectively (Atmospheric Environment Service, 1982a). The average annual precipitation is 536 mm, 361 mm of which falls during May through September. The average annual snowfall is 150 cm (Atmospheric Environment Service, 1982b). Similar temperature and precipitation averages were recorded for Edson, except that snowfall at Edson was slightly higher. Cold Creek Ranger Station averages about 1100 degree-days above 5°C annually (Atmospheric Environment Service, 1982c) and the area has about 70 frost-free days each year (Atmospheric Environment Service, 1982d).

The 132-ha experimental area is located within a confluence of Wolf Creek at the downstream end of a large (>2500 ha) treed fen. The fen is bounded on the northeast and southwest sides by tributaries of Wolf Creek (Figure 1). Groundwater within the fen flows in the direction of the topographic slope, from southeast to northwest, toward the north tributary of the confluence. Uplands with mineral soils located at the south end of the fen provide the fen with a steady water supply.

Vegetation

According to Rowe's (1972) forest classification, the study area falls within the Lower Foothills Section of the Boreal Forest Region. As is the case in much of Alberta's forested areas, the vegetation on the site originated after a fire. Black spruce (90 years) and tamarack (80 years) are the predominant species in the fen, but mineral islands within the fen support mature lodgepole pine (*Pinus contorta* Loudon) and white spruce (*Picea glauca* [Moench] Voss). The area has been subjected to intense geophysical exploration activity and, as a result, the forest is criss-crossed with a network of seismic lines, each about 7 m wide, that provide ready access to the area.

Before drainage, the dominant understory shrub was swamp birch (*Betula pumila* L.), with lesser amounts of bog willow (*Salix pedicellaris* Pursh) and Labrador tea (*Ledum groenlandicum* Oeder). Bog rosemary (*Andromeda polifolia* L.) was a conspicuous and constant dwarf shrub. The herb stratum was dominated by water sedge (*Carex aquatilis* Wahlenb.), prostrate sedge (*C. chordorrhiza* L.f.), and swamp horsetail (*Equisetum fluviatile* L.), with lesser amounts of two-stamened sedge (*Carex diandra* Schrank) and mud sedge (*C. limosa* L.). Labrador bedstraw (*Galium labradoricum* Wieg.), buck-bean (*Menyanthes trifoliata* L.), swamp cranberry (*Oxycoccus quadripetalus* Gilib.), marsh cinquefoil (*Potentilla palustris* [L.] Scop.), three-leaved Solomon's-seal (*Smilacina trifolia* [L.] Desf.), and seaside arrow-grass (*Triglochin maritima* L.) were constant and characteristic species, but with low cover. The predominant moss species were *Sphagnum warnstorfii* Russ. and *Tomenthypnum nitens* (Hedw.) Loeske, although *Sphagnum angustifolium* (C. Jens. ex Russ.) C. Jens. in Tolf was prominent in certain areas; *Drepanocladus vernicosus* (Mitt.) Warnst. occurred in wet hollows. *Aulacomnium palustre* (Hedw.) Schwaegr. and *Meesia triquetra* (Richt.) Aongstr. were constant species on hummocks and in hollows, respectively, but with low cover. (Ground vegetation information provided by D. Johnson, Canada Department of Natural Resources, Canadian Forest Service, personal communication.)

METHODS

Preliminary Work

A weather station (Figure 1) consisting of a recording precipitation gauge and a hygrothermograph was installed. Survey lines were cut, a topographic survey was conducted, and a topographic map was produced with a scale of 1:2000 and contour intervals of 0.5 m.

Preliminary peat, water, and vegetation surveys and sampling programs were completed to determine wetland, vegetation, and peat types, peat depths, nutrient status, and other site characteristics. A Macaulay-type sampler with an inside diameter of 4 cm was used to obtain fourteen peat cores from two transects running perpendicular to each other across the site. Core samples, 3 to 6 cm long, were taken at 15-cm intervals for physical and chemical analyses. Later, in the laboratory, they were oven-dried at $70 \pm 5°C$ and their moisture contents (volumetric and gravimetric) and bulk densities calculated. The depth and saturated hydraulic conductivity of peat were measured at 43 locations on a 200-m grid covering the site. Saturated hydraulic conductivity was measured in the top 1 m of soil using the piezometer method (U.S. Department of the Interior, 1984).

In 1986, prior to ditching, four transects were established, one on the control site (area 1) and three at different ditch spacings (30, 40, and 50 m) perpendicular to ditch lines, on the area to be drained (Figure 1). Each transect was instrumented, sampled, and surveyed to measure the effects of drainage on groundwater levels, peat subsidence, and other variables. In July 1987, a second control site (area 2) was established in the same manner as the first, south of the proposed ditch network (Figure 1).

Ditch Network: Design and Construction

A drainage ditch network design was prepared using field observations, a topographic map, and enlarged aerial photos. An adaptation of Tóth's synthetic hydraulic curve method (Tóth and Gillard, 1984, 1988) was used to find the optimum ditch spacings. The final design called for 35-m ditch spacings but allowed for evaluation of different ditch spacings (30-, 40-, and 50-m) on a homogeneous portion of the area (Figure 1).

The ditch network was marked, and 5-m rights-of-way were cleared with a D6 tractor during winter when the ground was frozen. In the fall of 1987 the ditch network (Figure 1) was constructed in unfrozen ground using a Lannen S10 ditcher. The ditches, parabolically shaped to maintain ditch stability, were 0.9 m deep and about 1.4 m wide. A total of 20 km of ditches were dug on 60 ha, resulting in a drainage ditch density of 333 m ha^{-1}. A large sediment pond was constructed near the downslope end of the main ditch to capture sediment originating primarily from the main ditch. A buffer strip was left between the main ditch and the watercourse; i.e., the main ditch terminated at the sediment pond before reaching the stream, and effluent water passed through the undisturbed stretch of ground (the buffer) between the sediment pond and stream before entering the watercourse. The purpose of the buffer was to filter out sediment particles that may escape the sediment pond. The work was completed October 28, 1987.

Evaluation of drainage network performance is an important part of this study. Results from drainage studies in European forests suggest that the optimum depth to water table for different tree species ranges from 0.18 to 0.50 m, measured at the midpoint between ditches (Heikurainen, 1964). The optimum depth is closely related to drainage norm, a term introduced by B. Meshechok of Norway to describe the effect of drainage on water table levels. Drainage norm is defined as the mean distance between the soil surface and the water table at the midpoint between ditches during the period June through September. It is dependent on climate, topography, hydraulic conductivity and vegetation, as well as on ditch depths and spacings (Päivänen and Wells, 1978).

Tóth and Gillard (1988) used a drainage norm (v_m) of 40 cm and a flood duration limit (λ_m) of 14 d as basic design criteria to evaluate drainage ditch system performance in their forest drainage experiment at Saulteaux. The flood duration limit (λ_m) is the specified maximum time interval during which the water table level at the midpoint between ditches is allowed to remain above a specified drainage depth, the drainage norm (v_m).

The same criteria were used to evaluate the effectiveness of the drainage ditches at Wolf Creek. The drainage system performance was rated acceptable if, during the growing season, the water table approximated, on average, a depth of 40 cm (v_m) without staying above this level for more than 14 d consecutively. On the other hand, the drainage system was rated overeffective if it held the average water table position consistently below 40 cm, and undereffective if the water table remained above 40 cm for more than 14 d any time during the growing season.

Groundwater Table Levels

Groundwater table configurations were monitored during May through October, 1986 to 1991 using 5-cm diameter wells installed along each of four transects in 1986 and, in the case of control area 2, in 1987. Two wells were installed on each of the control transects, six on the 30-m spacing transect, eight on the 40-m transect, and nine on the 50-m transect. Pressure transducers connected to battery-operated data recorders (Data recorders and pressure transducers were manufactured by Lakewood Systems Ltd., 9258 34A Avenue, Edmonton, Alberta, Canada T6E 5P4.) were inserted in ten wells to provide continuous records of changes in water levels with time; data were recorded at 90-min intervals. Two 15-cm diameter wells equipped with Leupold-Stevens F-type water level recorders were installed at the center between ditches on the 40- and 50-m transects. Depths to water table in all wells were measured once or twice a month with a carpenter's tape, using the top of each well as reference points. No corrections were made to compensate for peat subsidence; therefore, all water table measurements are referenced to the predrainage ground surface.

Peat Subsidence

Peat subsidence was measured by driving three 13-mm diameter steel reinforcing rods through the peat into mineral soil on control area 1 and on three transects for each ditch spacing, so that 15 cm of each rod projected above the ground surface. On each transect in the drained area, the three rods were placed 5 m from the ditch, at the center between ditches and at one-fourth the distance between ditches. The relative elevations of each rod were determined by referencing to a common bench mark. They were resurveyed once a year to determine if any of the rods had shifted vertically. The projection above ground was measured 1, 5, and 6 years after ditching.

RESULTS AND DISCUSSION

Peat Properties

The peat and groundwater chemistry data (not given) confirmed that the Wolf Creek experimental area was a relatively poor fen enriched by the local groundwater system. Data from four peat cores were selected to characterize peat properties across the peatland (Figure 2). They represent conditions before drainage when the depth to water table at the time of sampling and measurement (June 18–26, 1985) was between 20 and 55 cm. Fibric peat, consisting primarily of *Sphagnum* moss, formed the upper 30 to 40 cm layer of the profile (Figure 2). It was also the rooting zone, and the moisture content in this layer generally exceeded 90 percent. The surface was characteristically different from the rest of the fibric layer. Its surface bulk density ranged from 0.05 to 0.07 g cm^{-3}, and the von Post number was one, characteristic values for the loose, fibrous undecomposed (living) mat at the surface. Below the surface, within the fibric layer, the peat became more compact and decomposed with depth, as indicated by

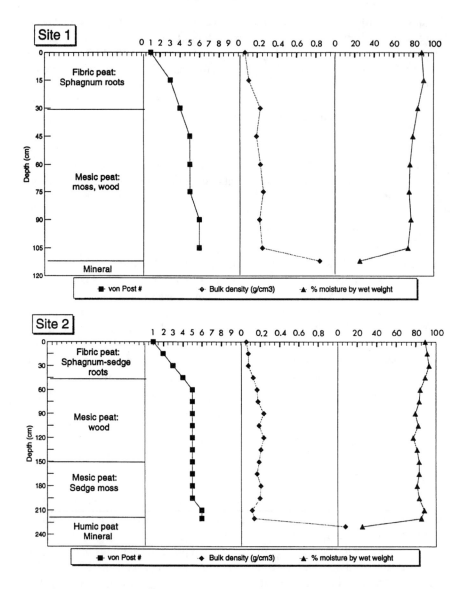

Figure 2 Properties of four peat profiles on the Wolf Creek experimental area.

increased bulk density (0.13 to 0.23 g cm⁻³) and increased von Post number (4) at the bottom of the fibric layer. In general, data for the fibric layer suggest that it readily transmits water.

The mesic peat horizon, located below the fibric layer (Figure 2), consisted of sedge moss remains and wood fragments in various stages of decomposition. It was denser (bulk density: 0.11 to 0.26 g cm⁻³) and more decomposed (von Post: 4-7) than the fibric layer, and the wood-dominated peat tended to be denser than the sedge-moss peat. These data signify that pore size in the mesic layer

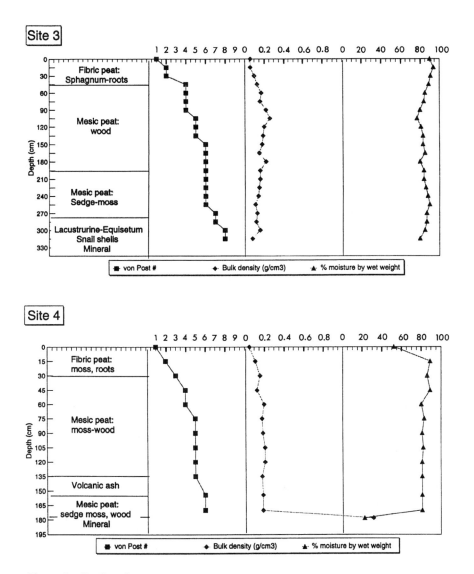

Figure 2 *Continued.*

was small compared to pore size in the fibric layer; thus the mesic layer will tend to retain water and transmit it less readily than the fibric layer. Moisture content in the mesic layer was generally in the 80 to 90 percent range; at site 1 it was in the 70 to 80 percent range. A noteworthy feature of site 4 (Figure 2) was the presence of a thin layer of Mazama volcanic ash, which is an important indicator in regional peat chronolgy.

The hydraulic conductivity in the top meter of the experimental area, based on measurements at 42 locations, ranged from 10^{-3} to 10^{-6} cm s^{-1} with an average of 2×10^{-3} cm s^{-1}. According to Päivänen (1973), peats with the stated average value are in the moderately rapid conductivity class.

As mentioned previously, the objective of the water table control treatment was to lower the average groundwater table of the site to 40 cm. The fibric layer at Wolf Creek had the necessary thickness and physical properties to facilitate drainage to this level. Pore size, hydraulic conductivity, and the water yield coefficient of peat increase as the bulk density decreases (Table 1) (The water yield coefficient describes the theoretical maximum quantity of water which can be removed from the topmost layers of a peat soil by means of draining alone.), and if a sufficient hydraulic gradient exists (e.g., after ditching) water will move rapidly out of the fibric peat. The tendency for the mesic peat to retain rather than transmit water (compared with fibric peat) may be considered an advantage in this situation because the mesic peat effectively serves as a reservoir for the root zone.

Before drainage, peat thickness at 42 locations on the experimental area averaged 1.70 m and ranged from 0.38 to 3.54 m, with a standard deviation of 0.88 m. The shallow peats occurred near the stream channels, and the deep peats were concentrated at the center of the confluence area away from the creeks. Peat thickness increased upslope and away from the confluence area.

Peat Subsidence

Removal of surface water through ditching resulted in a settling of the surface peat. On the unditched control, peat subsidence was minimal. Subsidence across the 30-m spacing transect tended to be greater than on the 40- and 50-m transects. Otherwise, ditch spacing did not appear to differentially affect the subsidence rate, and there was no clear pattern of subsidence as a function of distance from the nearest ditch. Consequently, the mean of all the drained areas is used to show changes in subsidence over time (Table 2). Although subsidence still continued 6 years after ditching, about half of the total subsidence occurred in the first year. Thereafter, peat subsided 1 to 2 cm each year (Table 2). Similar results were obtained in northern Alberta on the McLennan experimental drainage area where the peatland and vegetation resemble conditions at Wolf Creek. At McLennan, the peat subsided 11 cm during the first year after drainage and 1 to 2 cm during each of the following 6 years, for a total subsidence of 19.4 cm (Hillman, unpublished data). It was also noted during remeasurement of tree growth permanent sample plots at Wolf Creek in 1993 that there was a greater incidence of leaning trees among the larger trees in the drained plots. This implies that differential subsidence was taking place, and we can speculate that the "hollows" were sinking faster than the "hummocks" on which the trees were usually found. Alternatively, it may have been differences in peat water content and porosity that contributed to the greater incidence of leaning trees.

No attempt was made to partition subsidence into settlement due to drainage and peat loss due to decomposition. It is suggested, however, that most of the subsidence during the first year can be attributed to settlement following water removal and that subsidence in the following years was a combination of addi-

Table 1 Water Yield Coefficients, Hydraulic Conductivity, and Average Pore Size Distribution of Peats for Different Bulk Densities (Percent of the Volume of Water at Saturation)

Bulk density (gm cm⁻³)	Water yield coeff. (cm³ cm⁻³)	Hydraulic conductivity × 10⁻⁶ (cm s⁻¹)	Pore size, μm						
			>300	300–100	100–30	30–3	3–0.3	0.3–0.2	<0.2
0.05	0.6	1632.3	21.3	22.3	20.2	13.7	10.5	3.6	8.4
0.10	0.36	282.9	3.9	16.7	18.9	27.2	11.1	6.6	15.6
0.15	0.22	49.0	3.3	5.2	16.7	31.3	13.3	8.7	21.5
0.20	0.18	8.5	3.0	7.2	11.6	26.2	16.5	9.9	25.6

From Päivänen, J., *Acta For. Fenn.*, 129, 1, 1973. With permission.

Table 2   Change in Peat Surface Elevation for the Mean of All Drained Areas

Date	Drained (cm)	Drained change (cm)	Change interval (years)	Annual subsidence (cm year⁻¹)
Sept. 1987	0	—	—	—
Sept. 1988	−7.9	−7.9	1	−7.9
July 1992	−13.3	−5.4	4	−1.4
Sept. 1993	−15.1	−1.8	1	−1.8

tional settlement and organic decomposition. The results suggest that more research needs to be done on peat subsidence. In particular, more information is required on the rate of subsidence due to drainage and the rate due to microbial decomposition. As Walmsley (1977) pointed out, once the water table is lowered, the balance between peat accumulation and decomposition is so altered that decomposition is favored. If decomposition is allowed to continue, the organic material will disappear until the subsiding surface is close to the water table again.

An indication of the long-term effects of forest drainage on subsidence is provided through Finnish experience. According to Laine et al. (1994), average subsidence of peat layers on Finnish sites 60 years after drainage varied from 12 to 33 cm, increasing from nutrient-poor to nutrient-rich site types. They also cited other studies on Finnish peatlands in which subsidence ranged from 17 to 41 cm, 14 to 36 years after drainage.

Groundwater Table Levels

Two years of pretreatment (1986 to 1987) and four years of posttreatment (1988 to 1991) groundwater data were obtained for Wolf Creek. The average groundwater table profiles derived from manually obtained (discrete) data for each ditch spacing for the two periods (Figure 3) and the overall average depths to water table across these profiles (Figure 4) indicate that, on all sites before ditching, the water table was parallel to and less than 20 cm below the ground surface. The slopes of the predrainage profiles (Figure 3) indicate that the direction of groundwater flow is from south to north.

After drainage, the overall average depths to water table were 77, 52, and 58 cm for the 30-, 40-, and 50-m ditch spacings, respectively (Figure 4). These depths are considerably lower than the corresponding predrainage depths at all ditch spacings. Before drainage, the overall depth to water table was greater for Control 2 than for Control 1. After drainage, the reverse was true (Figure 4). The greater change on Control 1 may be attributed to its location downstream from the ditch network (Figure 1). The ditches to the south likely prevented some portion of the total groundwater flow from reaching the Control 1 area. Control 2, located upstream from the ditch network, was not affected by the ditches.

After ditching, the water table profiles were significantly lower for all ditch spacings, and each showed the characteristic postdrainage "mounding" effect, with the greatest depth to water table near the ditches and the least at the midpoint between ditches (Figure 3). In each case, the average depth to water table at or near the midpoint fell below the drainage norm (v_m) of 40 cm, to 61, 46, and 49

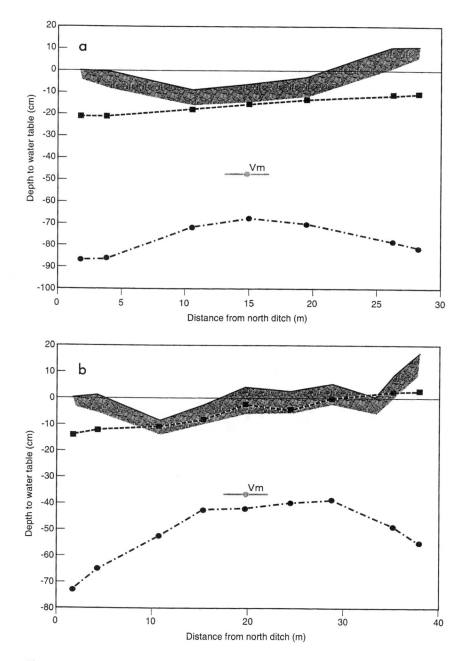

Figure 3   Average groundwater table profiles between ditches for the 30-m (a), 40-m (b), and 50-m (c) ditch spacings: hatched (upper) line = peat surface at time of drainage (Sept. 1987); ■ 2-year (1986 to 1987) predrainage means; ● 4-year (1988 to 1991) postdrainage means; ν_m drainage norm.

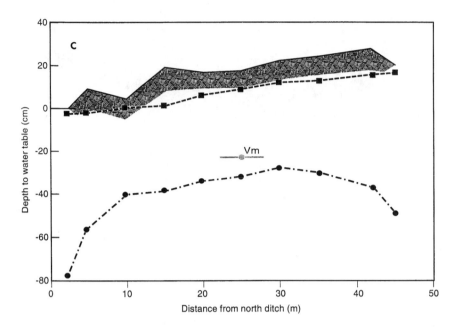

Figure 3 *Continued.*

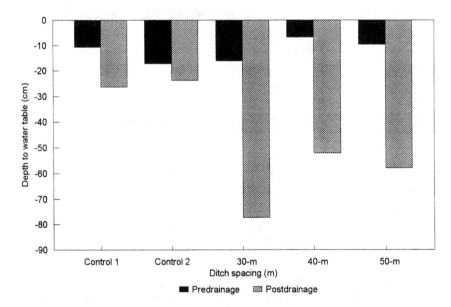

Figure 4 Overall average depths to water table across profiles: 2-year (1986 to 1987) predrainage mean; 4-year (1988 to 1991) postdrainage mean.

cm for the 30-, 40-, and 50-m spacings, respectively. The greatest deviation from the norm (21 cm) occurred on the 30-m spacing.

Continuous traces of water table levels at the midpoints between ditches obtained for the predrainage (Figure 5) and postdrainage (Figure 6) periods show the water table fluctuations above and below the norm during the 6-year interval. At the midpoints, the average depth to water table before ditching was 11, 5, and 12 cm for the 30-, 40-, and 50-m spacings, respectively. At the beginning of the predrainage measurement period, in July and August 1986, water was ponded

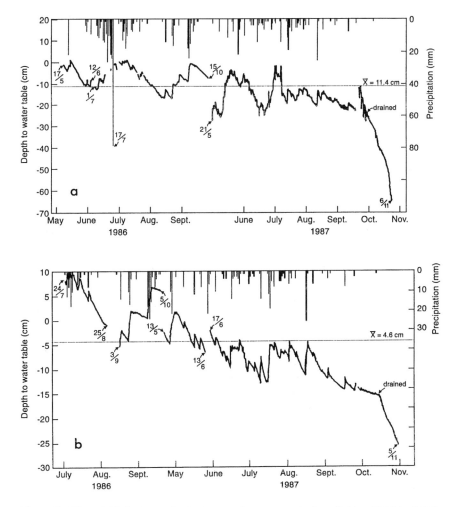

Figure 5 Water table hydrographs for two growing seasons before ditching (1986 to 1987) recorded at the midpoint between ditches on the 30-m (a), 40-m (b), and 50-m (c) ditch spacing areas. The dates (day/month) of major peaks and lows are shown together with precipitation for the same period. x̄ = mean depth to water table.

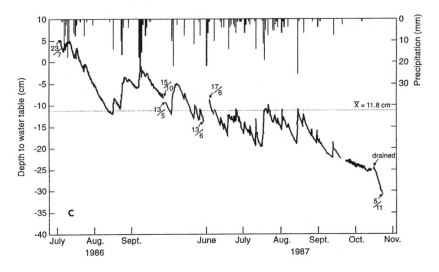

Figure 5 *Continued.*

above the ground surface around the midpoint wells on the 40- and 50-m spacings, the depth of water fluctuating between 0 and +10 cm, and 0 and +5 cm, respectively. Successive rainstorms in early to mid-July, culminating in a 79-mm storm on July 17 (Figure 5a), contributed to the above-surface groundwater levels. During the predrainage period, the water table for each spacing was above average in 1986 and below average in 1987 (Figure 5). This response was due to the total annual precipitation in 1986 (734 mm) being almost twice that of 1987 (370 mm).

When the ditches were dug near the instrumented transects in October 1987, the water levels dropped rapidly by 42, 10, and 6 cm for the 30-, 40-, and 50-m spacings, respectively (Figure 5). Because data for the period mid-October to early November 1987 represent the transition from an undrained to a drained condition, they were not used to calculate predrainage or postdrainage mean values.

After drainage the average depths to water table at the midpoints over 4 years were 64, 47, and 53 cm for the 30-, 40-, and 50-m spacings, respectively (Figure 6). These values compare favorably with the corresponding midpoint averages (61, 46, and 49 cm) obtained using discrete measurements (Figure 3). For each spacing, the mean was lower than the norm, and the difference between the mean and the norm was least for the 40-m spacing. The average postdrainage midpoint depths to water table for all ditch spacings were weighted low by the dry period extending from July to October, 1990 when water table levels approached or fell below the 1-m depth. The postdrainage midpoint water table levels remained below the 20-cm depth for all spacings for the entire 4-year period, except on one occasion in 1988 on the 40-m spacing (Figure 6b). The closest the water table came to the surface was at the 25-, 18-, and 23-cm depths for the 30-, 40-, and 50-m spacings, respectively (Figure 6).

When the concepts of flood duration limit and drainage effectiveness were applied to the data, it was obvious that the 30-m ditch spacing was overeffective.

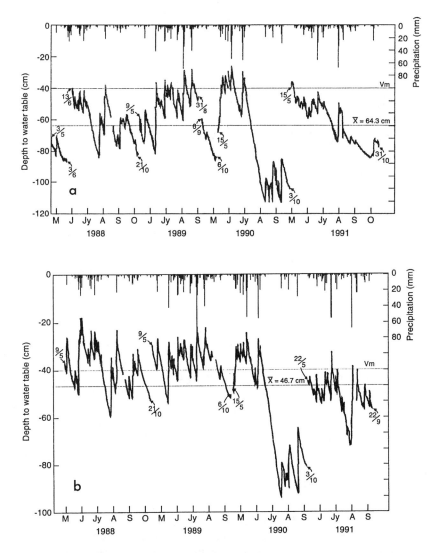

Figure 6 Water table hydrographs for four growing seasons after ditching (1988–1991), recorded at the midpoint between ditches on the 30-m (a), 40-m (b), and 50-m (c) ditch spacing areas. The dates (day/month) of major peaks and lows are shown together with precipitation for the same period. $\bar{x}$ = mean depth to water table; v_m = drainage norm.

The water level remained below the drainage norm for almost the entire 4-year period (Figure 6a), with the mean depth (64 cm) well below the drainage norm.

The hydrographs for the 40- and 50-m spacings required more detailed analysis because they both fluctuated above and below the drainage norm. Inspection of the numerical data rather than the compressed graphical data was undertaken to determine the number of times the flood duration limit of 14 d was exceeded. On the 40-m spacing, the water level rose above the drainage norm 18 times

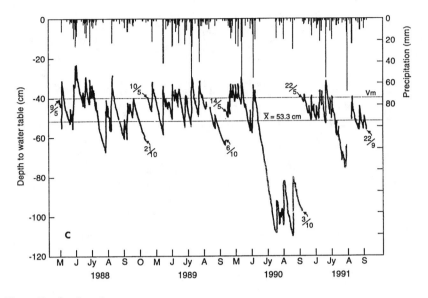

Figure 6   *Continued.*

(Figure 6b) and remained above that level for less than 14 d on 13 occasions. The flood duration limit was far exceeded on three occasions, once for 46 d, once for 45 d, and once for 32 d.

In contrast, on the 50-m spacing, the water level rose above the norm on 27 occasions, and only once did it exceed the flood duration limit (Figure 6c). The 50-m spacing is, therefore, hydrologically speaking, the most suitable of the three evaluated that could be applied to the Wolf Creek experimental drainage area and similar sites. It is possible, however, that a spacing greater than 50 m will suffice, thereby making a proposed drainage operation with this wider spacing more economical than a 50-m spacing operation.

In this attempt to assess ditch spacing performance at Wolf Creek, it was clear that the method used had some limitations. The targeted norm of 40 cm is a reasonable one, but, in light of the results, the flood duration limit of 14 d appeared to be too stringent and likely could not be met over a prolonged wet period. The postdrainage hydrographs (Figure 6) show the tendency for water tables to rise above the norm during major rainfall events and fall below it between these events.

CONCLUSIONS

The draining of a portion of a treed fen resulted in the peat subsiding 15 cm in 6 years. Half of this amount occurred during the first year. The consequences of peat subsidence include decreased depths to water table, loss of substrate through organic decomposition, and possible reduced windfirmness of trees. The results indicate that there is a need for additional research into the effects of forest drainage on subsidence, particularly on the rates of organic decomposition and on physical settlement.

The results show that even though the difference between the mean water table level and the drainage norm was less for the 40-m spacing than for the 50-m spacing, on the basis of flood duration limit criteria, the 50-m spacing was considered the most suitable for the Wolf Creek drainage site and similar areas. It is possible, of course, that a ditch-spacing greater than 50-m would be adequate and, for economic reasons, would be preferred. Although it is helpful to use hydrographs to design ditch networks and appraise network performances, ultimately, for enhancing tree growth, it is better to relate tree growth to site moisture, temperature, and nutrient conditions directly for this purpose. In addition, pre-drainage mean water table levels can be quite variable over an area, which must be considered when trying to attain a specified drainage norm for that area.

ACKNOWLEDGMENTS

This study was a cooperative effort between the Canada Department of Natural Resources and the Alberta Forest Service and was funded under the Canada-Alberta Forest Resource Development Agreement. The author is much indebted to the foresters and rangers of the Whitecourt Forest and AFS headquarters for assistance in the field. S. Takyi, the AFS coordinator for the study, and T. Nash, who supervised the construction of the drainage ditches, deserve special mention in this regard. Sincere thanks are due to B. Robson who, with assistance from many summer students and part-time help, was largely responsible for collecting the data for this study. The author is also grateful to D. Johnson for identifying the ground vegetation. The very helpful comments and suggestions of reviewers E. Verry and T. Williams are gratefully acknowledged.

REFERENCES

Atmospheric Environment Service, Canadian Climate Normals (1951–1980), Vol. 2, Temperature, Environment Canada, Downsview, Ontario, 1982a, 82 pp.

Atmospheric Environment Service, Canadian Climate Normals (1951–1980), Vol. 3, Precipitation, Environment Canada, Downsview, Ontario, 1982b, 100 pp.

Atmospheric Environment Service, Canadian Climate Normals (1951–1980), Vol. 4, Degree Days, Environment Canada, Downsview, Ontario, 1982c, 114 pp.

Atmospheric Environment Service, Canadian Climate Normals (1951–1980), Vol. 6, Frost, Environment Canada, Downsview, Ontario, 1982d, 113 pp.

Dumanski, J., Macyk, T. M., Veauvy, C. F., and Lindsay, J. D., Soil Survey and Land Evaluation of the Hinton-Edson Area, Alberta, Alberta Institute of Pedology Report No. S-72-31, University of Alberta, Edmonton, Alberta, 1972, 20 pp.

Hånell, B., Peatland forestry in Sweden, in Proc. Symp. '89. *Peat and Peatlands Diversification and Innovation,* Vol. 1, *Peatland Forestry,* Jeglum, J. K. and Overend, R. P., Eds., The Canadian Society for Peat and Peatlands, 1991, p. 19.

Heikurainen, L., Improvement of forest growth on poorly drained peat soils, *Int. Rev. For. Res.*, 1, 39, 1964.

Hillman, G. R., Johnson, J. D., and Takyi, S. K., The Canada-Alberta Wetlands Drainage and Improvement for Forestry Program, Canada-Alberta Forest Resource Development Agreement Report, Edmonton, 1990, 1.

Laine, J., Minkkinen, K., Puhalainen, A., and Jauhiainen, S., Effect of forest drainage on the carbon balance of peatland ecosystems, in *The Finnish Research Program on Climate Change, Second Progress Report*, Kanninen, M. and Heikinheimo, P., Eds., Publications of the Academy of Finland 1/94, Painatuskeskus, Helsinki, 1994, p. 303.

Päivänen, J., Hydraulic conductivity and water retention in peat soils, *Acta For. Fenn.*, 129, 1, 1973.

Päivänen, J., Peatland forestry in Finland: present status and prospects, in Proc. Symp. '89. *Peat and Peatlands Diversification and Innovation*, Vol. 1, *Peatland Forestry*, Jeglum, J. K. and Overend, R. P., Eds., The Canadian Society for Peat and Peatlands, 1991, p. 3.

Päivänen, J. and Wells, E. D., Guidelines for the Development of Peatland Drainage Sytems for Forestry in Newfoundland, Information Report N-X-156, Fisheries and Environment Canada, Newfoundland Forest Research Centre, St. John's, Newfoundland, 1978, 17 pp.

Rowe, J. S., Forest Regions of Canada, Environment Canada, Canadian Forestry Service Publication 1300, Ottawa, Ontario, 1972, 38 pp.

Tóth, J. and Gillard, D., Design of Drainage System in Forested Peatland with Special Reference to Water Regime Near Saulteaux River, Slave Lake Forest: Phase II and III, Interim Report III, University of Alberta, Edmonton, Alberta, 1984, 1.

Tóth, J. and Gillard, D., Experimental design and evaluation of a peatland drainage system for forestry by optimization of synthetic hydrographs, *Can. J. For. Res.*, 18, 353, 1988.

U.S. Department of the Interior, Drainage Manual. A Water Resources Technical Publication. Bureau of Reclamation, Denver, CO, 1984, 143 pp.

Vompersky, S. E., Current status of forest drainage in the USSR and problems in research, in Proc. Symp. '89. *Peat and Peatlands Diversification and Innovation*, Vol. 1, *Peatland Forestry*, Jeglum, J. K. and Overend, R. P., Eds., The Canadian Society for Peat and Peatlands, 1991, p. 13.

Walmsley, M. E., Physical and chemical properties of peat, in Muskeg and the Northern Environment in Canada, 15th Muskeg Research Conf., Radforth, N. W. and Brawner, C. O., Eds., University of Toronto, 1977, p. 82.

CHAPTER 19

Wetland Effects on Hydrological and Water Quality Characteristics of a Mid-Michigan River System

T. M. Tompkins, W. W. Whipps, L. J. Manor, M. J. Wiley,
C. W. Radcliffe, and D. M. Majewski

CONTENTS

INTRODUCTION

It is often asserted that wetlands contribute both to improved water quality and flow stability in river systems (van der Valk et al., 1979; Brinson et al., 1984; Brown, 1985; Kadlec, 1987; Johnston, 1991; Hammer, 1992). Bottomland hardwood swamps, for example, have been found to modify both discharge patterns and water quality in downstream river channels (Hammer, 1992). Wetlands are thought to contribute to flow stability by both desynchronizing peak flows (Novitzki, 1981) and by slowly releasing stored water during dry periods (Hammer, 1992). Others contend, however, that the usual effect of wetlands is to reduce low flow. They argue that although wetland soils may be capable of storing large amounts of water during the growing season, the extensive consumption of available water by evapotranspiration from plants tends to reduce base flow (Vecchioli et al., 1962; Miller, 1965; Verry and Boelter, 1981; Novitzki, 1981). Similarly, some researchers question the generalization that wetlands serve as nutrient sinks and sediment traps (Kraus, 1987; Richardson, 1989), since studies have shown that under certain conditions, nutrients may be rapidly lost from wetland ecosystems (Richardson et al., 1978; Simpson et al., 1978; Peverly, 1982; Elder, 1985; Kadlec and Bevis, 1990; Chescheir et al., 1991).

Studies of the ecological role of wetlands have principally focused on peatland systems in the northern boreal zone (e.g., Verry and Bolter, 1981; Berry and Jeglum, 1991), or bottomland forests in the southern United States (Vecchioli et al., 1962; Miller, 1965; Elder, 1985; but see, O'Brien, 1977; Novitzky, 1981). Several studies of wetlands occurring on glacial drift deposits in the northern United States and Canada (Devito et al., 1989; Dillon et al., 1991; Hill, 1991, 1993; Seitzinger, 1994) are well known, but focus on exports from relatively small catchments.

In this chapter we present the results of a study of a large wetland complex located in the upper Tittabawassee River basin of Michigan's lower peninsula. We examine both the water discharge and water quality characteristics of a series of small to medium size catchments (7 to 123 km^2) with varying degrees of wetland cover. Our objective was to determine the degree to which wetlands contribute to improved water quality and enhanced flow stability in a larger stream system.

METHODS

Study Site

The Tittabawassee River, the largest tributary of the Saginaw River system, drains roughly 6500 km^2, and eventually empties into Lake Huron at Saginaw Bay. Stream water quality and discharge were measured at 25 stream sites in a wetland complex in the upper portion of the Tittabawassee River Watershed (located in the east-central region of the lower peninsula of Michigan). This region is composed of more or less contiguous patches of predominantly rheotrophic wetlands covering an area of roughly 1300 km^2.

Patches of agricultural development in this region occur on clay and loamy soils of the more well-drained outwash and moraine features. The wetlands are diverse and include lowland deciduous forest, hardwood and shrub swamp, open marsh, and bog communities. The most common vegetation types include

Chamaedaphne calyculata in the more open sites; *Alnus rugosa* and *Salix* spp. in the shrub swamps, and mixed forest covers of *Populus* spp., *Betula papyrifera*, *Acer rubrum*, and *Fraxinus nigra*.

Analytical Methods

GIS and Statistical Methods

Spatial analysis of landscape features present in the sample watersheds was performed using a raster-based geographic information system developed with Erdas GIS software (Erdas Corporation, Atlanta, GA). Geographic data was obtained from the Michigan Rivers Inventory Database (Fisheries Division, Michigan Department of Natural Resources (MDNR)).

Land cover was based on 1978 aerial photo interpretations prepared by the Michigan Resource Information System (MIRIS, Land and Water Management Division, MDNR). We further classified wetland cover by subdividing the level 1 MIRIS wetland classifications into two categories: swamp and lowland deciduous forest. This was done by digitizing areas denoted as wetlands on 30 × 60 min U.S. Geological Survey (USGS) maps, and overlaying these on the MIRIS data. Land units identified as forested wetland on both covers were classified as swamp, these being the wettest areas. The wetland units on the MIRIS cover that did not include areas denoted as wetland on the USGS maps were classified as lowland deciduous forest.

Geology data used in the GIS were based upon the Quaternary Geology of Southern Michigan map compiled by Farrand and Bell (1983). Soil data were based on county soil surveys issued by the U.S. Department of Agriculture Soil Conservation Service. The original raster soil cover was produced by the Michigan State Center for Remote Sensing (MSCRS), as a part of the MSCRS Michigan Geographic Information System. Watersheds of each of the 25 sample sites were delineated using USGS 7.5 min topographic maps.

Basin composition in terms of land cover/land use, soil, and surficial geology was determined for the catchment basin of each sample site using standard GIS techniques. For the purposes of analysis, sample watersheds were initially divided into two groups: those composed predominantly of wetlands (*high wetland* sites, >60 percent wetland, n = 12), and those that were not (*low wetland* sites, <60 percent wetland, n = 13). To further facilitate analysis, we identified 15 sample sites that were relatively homogeneous in terms of catchment composition. Based on predominant land cover, they were grouped into five categories: *agriculture* sites (>60 percent agriculture cover), *upland forest* sites (>50 percent upland forest cover: mesic to dry mesic mixed hardwoods), *lowland forest* sites (>60 percent lowland forest cover: wet-mesic mixed hardwoods), and *swamp* sites (>85 percent swamp cover: mostly *Alder* wet shrub/scrub), respectively. A fifth category, designated *wetland mix* sites, comprised a combination of swamp and lowland deciduous forest (60 to 74 percent swamp, 10 to 17 percent lowland forest).

Data were examined statistically using standard ANOVA and ANCOVA techniques. A posteriori contrasts for land use categories were evaluated using Tukey's

HSD test (Wilkenson, 1990). Data distributions for hydrological and chemical parameters were summarized for various land use categories using box plots (McGill et al., 1978).

Hydrological and Chemical Measurements

To index hydrological and chemical characteristics, 25 stream sites were sampled twice during both spring and late summer periods (i.e., on four dates): May 9 and 16, and August 22 and 29, 1992. Sites were chosen arbitrarily based on road access and a desire to include sites draining catchments influenced by either forest, wetlands, or agriculture. Sampling in May began just after peak annual flow from snowmelt. The cumulative 10-d precipitation preceding both the spring and summer sampling periods was less than 5 mm. Streamflow during the August 29 sample period included the response to a 40-h rain event on August 27 and 28, which yielded 47.5 mm of rain in the study area (National Oceanic and Atmospheric Administration weather data collected at the Gladwin, MI station, 1992).

Stream discharge was measured using standard cross-section techniques, standardized for drainage area, and reported as water yield (discharge in $m^3 ha^{-1} d^{-1}$). Flow variability was characterized for each site by examining the difference between highest and lowest measured flows (ΔQ_{season}); and between low and high August flows (ΔQ_{event}). Water samples were collected midstream in acid-rinsed nalgene bottles, at the same times that discharge was measured. Stream water samples were tested chemically following standard methods (American Public Health Association, 1981) for: alkalinity (titrametric), conductivity, gilvin (absorbance at 440 nm, a measure of absorbance due to DOC), nitrate-N (nitrate-N + nitrite-N; ultraviolet spectrophotometric screening method, cadmium reduction), pH, soluble reactive phosphate (SRP; ascorbic acid method), total phosphorous (TP; ascorbic acid method, with digestion), total suspended sediments (TSS), and turbidity (photometric, FTU). We analyzed water quality data both in terms of concentration and net yield (in $kg ha^{-1} d^{-1}$) to the river from the catchment.

RESULTS

Hydrology

Water yields from the high wetland catchments were significantly greater than those from the low wetland catchments (Table 1). Mean water yields were generally highest for wetland mix, followed by lowland forest and upland forest sites (Figure 1a). Water yields were lowest for the agriculture and swamp sites. Mean yields for both wetland mix and lowland forest sites were significantly higher (Tukey's HSD test; $\alpha < .05$) than those for sites dominated by agricultural use, upland deciduous forest, or shrub/scrub swamp.

Seasonal variation in flow ($\Delta Q_{seasonal}$) tended to be larger for the high wetland sites than for the low wetland sites (Table 1). Among the land cover categories, median $\Delta Q_{seasonal}$ was highest in wetland mix, followed by lowland forest and

Table 1   Hydrology and Water Quality Parameter Means, and t-Statistics Comparing Low Wetland (<60 Percent Wetland Cover) and High Wetland (>60 Percent Wetland Cover) Sites

Parameter	Low wetland	High wetland	t-Value
Mean water yield (m^3 ha^{-1} d^{-1})	1.037	2.765	2.42*
$\Delta Q_{seasonal}$ (m^3 ha^{-1} d^{-1})	3.673	5.764	1.41[n.s.]
ΔQ_{event} (m^3 ha^{-1} d^{-1})	0.680	1.753	1.94[n.s.]
Alkalinity (mg $CaCO_3$ L^{-1})	171.53	108.02	6.30***
Conductivity (microsiemens cm^{-1})	573.98	270.58	8.20***
Gilvin (g440)	0.0235	0.0379	−4.51***
Nitrate (μg L^{-1})	254.60	231.5	1.14[n.s.]
Nitrate yield (kg ha^{-1} d^{-1})	0.00037	0.00062	0.460[n.s.]
pH ($-$log [H^+])	7.70	7.45	2.75**
TP (μg L^{-1})	105.20	81.36	1.30[n.s.]
TP yield (kg ha^{-1} d^{-1})	0.00009	0.00026	2.49*
SRP (μg L^{-1})	18.94	9.11	13.25*
SRP yield (kg ha^{-1} d^{-1})	0.00001	0.00003	0.016[n.s.]
TSS (μg L^{-1})	25.23	21.23	0.976[n.s.]
TSS yield (kg ha^{-1} d^{-1})	0.0440	0.1000	2.40*
Inorganic sed (mg L^{-1})	6.53	6.01	0.2280[n.s.]
Organic sed (mg L^{-1})	16.23	10.35	1.38[n.s.]
Turbidity (PTU)	15.34	11.72	1.73[n.s.]

Note: Degrees of significance: * = $p < 0.05$, ** = $p < 0.01$, *** = $p < 0.001$; n.s. = nonsignificant

then upland forest (Figure 1b) catchments. $\Delta Q_{seasonal}$ was lowest for agriculture and swamp watersheds. Event flow responses (ΔQ_{event}) also tended to be greater for the high wetland group than for the low wetland group (Table 1), but due to higher variability within the wetland watersheds, the difference was not statistically significant. Comparing land cover categories, ΔQ_{event} was significantly higher for lowland deciduous forest-dominated sites than for swamp sites (Figure 1c). ΔQ_{event} for the other three categories, while not statistically different, tended to be intermediate.

Water Quality

Alkalinity and conductivity were significantly greater, and turbidity tended to be greater, in the low wetland catchments (Table 1). A similar pattern was observed for alkalinity and conductivity with respect to the five land cover categories (Figures 1d and 1e). As expected, gilvin was significantly greater in the high wetland watersheds (Table 1; Figure 1f). There was no statistically significant difference between mean nitrate concentrations associated with extent of wetland cover, although mean nitrate concentration tended to be higher for the low wetland group (Table 1). Mean nitrate concentrations also varied little among the five land cover categories (Figure 1g), although concentrations generally were lower in swamp and wetland mix sites and higher in the agriculture sites.

Examination of nitrate yields (in kg ha^{-1} d^{-1}) indicated a trend toward higher yields from the predominantly wetland watersheds, although there was no significant difference in the mean yields (Table 1). Lowland forest and wetland mix

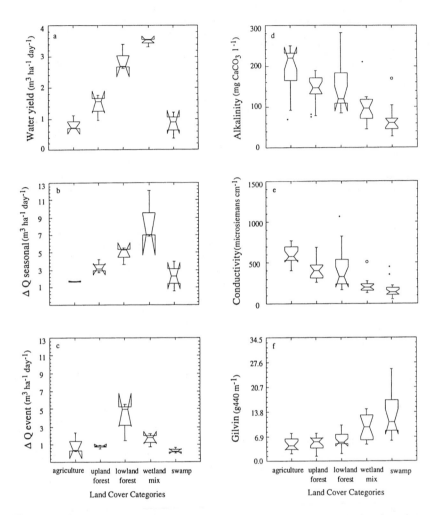

Figure 1 Hydrological and chemical parameters across the five land cover categories. Hydrological parameters: (a) relationship between land cover categories and mean water yield; (b) seasonal change in stream flow; (c) precipitation even effect on stream flow. Chemical parameters: (d) patterns in alkalinity across land cover categories; (e) patterns in conductivity across land cover categories; (f) patterns in gilvin across land cover categories; (g) patterns in nitrate concentration across land cover categories; (h) relationship between land cover and nitrate yield; (i) relationship between land cover and soluble reactive phosphate yield; (j) total suspended sediment concentration as a function of land cover; (k) Relationship between land cover and total suspended sediment yield.

sites also tended to have a higher mean nitrate yield (Figure 1h), presumably due to their higher water yields. Since dissolved load and water yield are necessarily correlated, we employed analysis of covariance (ANCOVA) to test for differences in nitrate yield while controlling for discharge effects. We found a barely significant difference in slopes, and a highly significant difference between the y-intercepts of the regression models for the high wetland and low wetland catch-

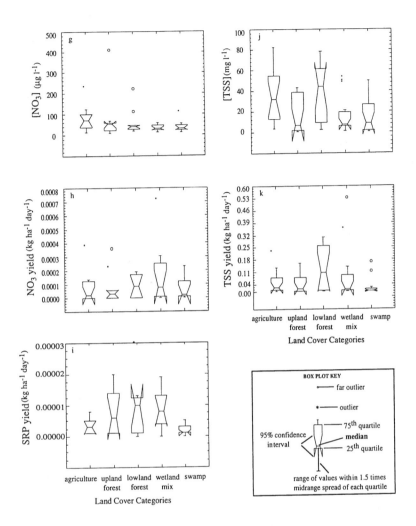

Figure 1 *Continued.*

ments (Table 2). The y-intercept for the high wetland sites was lower than that of the low wetland sites. This result indicates that, given the same discharge, nitrate yield from catchments comprised mostly of wetlands was significantly lower. However, since wetland-influenced watersheds had significantly higher water yields, average nitrate yields were similar. Based on the mean water yields of the low wetland and high wetland groups (1.04 and 2.76 m³ ha⁻¹ d⁻¹, respectively), the observed nitrate yield was 50 mg ha⁻¹ d⁻¹ for both groups.

Mean TP concentrations tended to be lower in wetland-dominated sites, but mean TP yield (kg ha⁻¹ d⁻¹) was significantly greater in the high wetland sites (Table 1). However, no significant differences were found between the high wetland and low wetland sites in the ANCOVA model of TP yield. Mean SRP concentrations were significantly lower in the high wetland sites, while mean

Table 2 ANCOVA Tables for Nitrate Yield, Soluble Reactive Phosphorus Yield, and Total Suspended Sediment Yield. Model: $\ln Y = a + b(\ln Q)$, where $Y = $ kg ha^{-1} d^{-1}, and $Q = $ water yield in m^3 ha^{-1} d^{-1}

Treatment	n	y-Intercept	Slope	ANOVA F	ANCOVA F
				Regression	
Nitrate yield					
Pooled	81	−135.30	27.98	64.2***	slope = 4.378*
Low wetland	38	−250.62	50.49	13.5***	intercept = 64.158***
High wetland	43	−137.58	28.44	120.0***	
Soluble reactive phosphorous yield					
Pooled	69	−118.00	23.80	27.1***	slope = 27.378***
Low wetland	37	−532.20	116.50	105.0***	intercept = 27.167***
High wetland	32	−98.83	19.47	20.6***	
Total suspended sediment yield					
Pooled	99	−127.0	27.58	116.0***	slope = 13.197***
Low wetland	49	−284.0	62.74	114.0***	intercept = 18.045***
High wetland	50	−120.4	26.08	71.7***	

Note: Degrees of significance: * = p <0.05, ** = p <0.01, *** = p <0.001; n.s. = non-significant

SRP yields tended to be greater in high wetland sites (Table 1). SRP yields were quite variable, depending upon the type of wetland occurring in the catchment. Lowland forest sites had the highest yields and swamp sites the lowest (Figure 1i). In contrast to the TP yield ANCOVA, the SRP yield ANCOVA model displayed significant differences in the regression slopes of the low wetland and high wetland sites (Table 2). SRP yield increased much more rapidly with increasing discharge in the low wetland sites.

There was no significant difference in sediment concentration between the low wetland and high wetland sites. However, high wetland catchments did have a significantly greater mean yield (kg ha^{-1} d^{-1}) of sediment (Table 1). Although no statistically significant differences were found, total suspended sediment concentration exhibited a great deal of variability within each land cover category. Agriculture and lowland forest sites had the highest median sediment concentrations (Figure 1j). Lowland forest sites also had higher and more variable sediment yields than the other land cover categories (Figure 1k). An ANCOVA of total suspended sediment yield, with discharge as the covariate, indicated significantly different slopes for the low wetland and high wetland sites (Table 2). The slope of the regression line for the low wetland sites was steeper, indicating sediment yield increased more rapidly with discharge in the low wetland compared to the high wetland sites. At very high water yields, the low wetland sites appear to export greater amounts of sediment. However, given the significantly higher water yield of the high wetland sites, average sediment yields during our observations were greater from these than from the low wetland catchments by approximately a factor of 2 (0.040 and 0.024 g ha^{-1} d^{-1} for high wetland and low wetland catchments, respectively).

DISCUSSION

While differences in water yield might be expected from different types of wetland catchments, the large variations in water yield and flow variability observed between our swamp and wetland mix categories was surprising, because the dominant land cover in both watersheds was shrub/scrub swamp. Analyses did not reveal any correlation with other measured catchment characteristics (e.g., differences in topography, soils, geology, or watershed size). A review of forest compartment maps indicated a higher incidence of recently clearcut acreage in lowland forest and wetland mix than in swamp catchments (MDNR, Au Sable State Forest compartment maps, 1992). Because clear-cutting is often associated with increased water yield (see Brooks et al., 1991), logging might in part explain the relatively high water yields observed in lowland forest and wetland mix sites.

The spatial location of wetlands within catchments, as well as the amount and kinds of wetlands present, may have influenced the observed variations in water yield. For example, in many of the swamp catchments, wetlands were located adjacent to the upstream riparian corridor and therefore relatively near the stream sampling site. In contrast, many of the lowland forest sites had substantial palustrine wetlands located away from the stream channel. It would be preferable in future studies to control for spatial location and hydrological delivery times, although this is not possible in many landscape settings. Another potential problem in our analysis is the lack of detailed data on vegetation types within our landcover characterizations. Areas we broadly characterized as being swamp cover (based on the 1:24,000 USGS topographic maps) probably had variations in water table elevation due to evapotranspirational differences associated with vegetational composition.

Stability of Stream Flow

The results of our flow stability analysis do not support earlier studies suggesting that wetland cover is related to stream stability (Novitzki, 1981; Verry and Bolter, 1981; Hammer, 1992). In August most sites exhibited zero or very low water yield. As a result, $\Delta Q_{seasonal}$ values follow substantially the same pattern as the seasonal high flow yields, that is, flow variation is greater in the predominantly wetland catchments. It seems clear, however, that this result is primarily due to the wetland mix and lowland forest categories. The predominantly swamp watersheds had a significantly lower $\Delta Q_{seasonal}$ value than the other wetland types. Responses to the summer storm event followed a similar pattern: lowland forest and wetland mix systems were least stable, and swamp systems most stable.

Water Quality

As expected, basic water quality characteristics varied with extent of wetland cover. The mineral soils characteristic of the low wetland catchments (including agricultural and upland forest sites) resulted in higher alkalinity and conductivity.

Conversely, lower conductivity and alkalinity and higher gilvin values were associated with the highly organic soils of the predominantly wetland watersheds. The generally higher turbidity found at agricultural sites is likely related to the soil exposure associated with such a land use.

Predominantly wetland watersheds exported significantly lower amounts of nitrate at any given discharge. Nitrate has been widely reported to be reduced in waters as they pass through wetland ecosystems (Bartlett et al., 1979; Davis et al., 1981; Peterjohn and Correll, 1984; Jacobs and Gilliam, 1985; Brodrick et al., 1988; Hill, 1991, 1993), with assimilation, denitrification, and interrupted nitrification being the relevant mechanisms (Armstrong, 1975; Howard-Williams, 1985; Seitzinger, 1994). Lower concentrations of nitrate in the swamp vs. lowland forest watersheds may be due to longer periods of soil saturation. Natural wetlands are believed to be most effective as nutrient sinks when they receive nutrient inputs at low concentrations delivered by a constant, slow flow of water (Howard-Williams, 1985).

Given that the high wetland sites exhibited lower TP concentrations, the tendency of the high wetland watersheds in this study to generate higher TP yields is likely due to the higher water yields of those catchments. While Dillon et al. (1991) found high TP exports correlated with a measure of the percent wetland in a catchment, one must be cautious in comparing results from a single season (instantaneous) study, such as this one, and one involving long-term annual nutrient budgets, such as that of Dillon et al. (1991). However, our results over the course of a single spring-summer sampling are in agreement with TP retention means reported in Devito et al (1989). In that study, TP retention is lowest (hence export is highest) in the spring when water yields are presumably highest, and TP retention is highest during mid to late summer (i.e., during periods of presumed lower water yields).

Since a majority of the agricultural land in our study area has been artificially drained, our results are also consistent with those of Yarbro et al. (1984), suggesting that channelized streams may facilitate nutrient export. Even within predominantly wetland catchments, many of the extreme values for nitrate we observed were measured at heavily channelized sites. Additionally, wetland soil disturbance associated with logging increases the rate of nitrification and thus the amount of nitrate available to be transported to the stream (Kovacic et al., 1990).

The effect of wetlands on river hydrology and water quality is clearly complex. Our results support the assertion of Glooschenko et al. (1987) that generalizations about the ecological function of all wetlands may be difficult to make. Even within a river basin, and on a relatively homogeneous geology (glacial lake basin), we found substantial differences in hydrology and chemistry between the different types of wetland-dominated catchments identified in our study. Water quality and hydrological characteristics are also clearly linked. Since observed concentrations are the net result of ratios of material to water exports (i.e., yields) from the watershed, understanding the potential implications of wetland management necessarily will require detailed information on the hydrological function and land-cover characteristics of individual wetland units.

REFERENCES

American Public Health Association, Standard Methods for the Examination of Water and Wastewater, 15th ed., American Public Health Association, Washington, D.C., 1981.

Armstrong, W., Waterlogged soils, in *Environment and Plant Ecology*, Etherington, J. R., Ed., John Wiley & Sons, New York, 1975, p. 181.

Bartlett, M. S., Brown, L. C., Hanes, N. B., and Nickerson, N. H., Denitrification in freshwater wetland soil, *J. Environ. Qual.*, 8(4), 460, 1979.

Berry, G. J. and Jeglum, J. K., Hydrology of Drained and Undrained Black Spruce Peatlands: Surface Water and Groundwater Quality, Forestry Canada, Ontario Region, Sault Ste. Marie, Ontario, COFDRA Report 3316, 1991.

Brinson, M. M., Bradshaw, H. D., and Kane, E. S., Nutrient assimilative capacity of an alluvial floodplain swamp, *J. Appl. Ecol.*, 21(3), 1041, 1984.

Brodrick, S. J., Cullen, P., and Maher, W., Denitrification in a natural wetland receiving secondary treated effluent, *Water Res.*, 22(4), 431, 1988.

Brooks, K. N., Folliot, P. F., Gregersen, H. M., and Thames, J. L., *Hydrology and Management of Watersheds*, Iowa State University Press, Ames, 1991.

Brown, R. G., Effects of an urban wetland on sediment and nutrient loads in runoff, *Wetlands*, 4, 147, 1985.

Chescheir, G. M., Gilliam, J. W., Skaggs, R. W., and Broadhead, R. G., Nutrient and sediment removal in forested wetlands receiving pumped agricultural waters, *Wetlands*, 11 (1), 87, 1991.

Davis, C. B., Baker, J. L., van der Valk, A. G., and Beer, C. E., Prairie pothole marshes as traps for nitrogen and phosphorous in agricultural runoff, in Selected Proc. of the Midwest Conf. on Wetland Values and Management, Richardson, B., Ed., Minnesota Water Planning Board, St. Paul, MN, 1981, p. 153.

Devito, K. J., Dillon, P. J., and Lazerte, B. D., Phosphorus and nitrogen retention in five Precambrian Shield wetlands, *Biogeochemistry*, 8, 185, 1989.

Dillon, P. J., Molot, L. A., and Scheider, W. A., Phosphorus and nitrogen export from forested stream catchments in central Ontario, *J. Environ. Qual.*, 20, 857, 1991.

Elder, J. F., Nitrogen and phosphorous speciation and flux in a large Florida river-wetland system, *Water Resources Res.*, 21, 724, 1985.

Farrand, W. R. and Bell, D. L., Quaternary Geology of Southern Michigan, Michigan Department of Natural Resources, Land and Water Management Division, Lansing, 1982.

Glooschenko, W. A., Archbold, J. H., and Weninger, J. M., The importance of swamp wetland habitat in southern Ontario, in *Symposium '87 Wetlands/Peatlands*, Environment Canada, Edmonton Convention Center, Edmonton, Alberta, Canada, 1987.

Hammer, D., *Creating Freshwater Wetlands*, Lewis Publishers, Chelsea, MI, 1992.

Hill, A. R., A ground water nitrogen budget for a headwater swamp in an area of permanent ground water discharge, *Biogeochemistry*, 14, 209, 1991.

Hill, A. R., Nitrogen dynamics of storm runoff in the riparian zone of a forested watershed, *Biogeochemistry*, 20, 19, 1993.

Howard-Williams, C., Cycling and retention of nitrogen and phosphorous in wetlands: theoretical and applied perspective, *Freshwater Biol.*, 15, 391, 1985.

Jacobs, T. C. and Gilliam, J. W., Headwater stream losses of nitrogen from two coastal plain watersheds, *J. Environ. Qual.*, 14 (4), 467, 1985.

Johnston, C. A., Sediment and nutrient retention by freshwater wetlands: effects on surface water quality, *Crit. Rev. Environ. Control*, 21, 491, 1991.

Kadlec, R. H., Northern natural wetland water treatment systems, in *Aquatic Plants for Water Treatment and Resource Recovery*, Reddy, K. R. and Smith, W. H., Eds., Magnolia Publishing Inc., Orlando, FL, 1987, p. 83.

Kadlec, R. H. and Bevis, F. B., Wetlands and wastewater: Kinross, Michigan, *Wetlands*, 10 (1), 77, 1990.

Kovacic, D. A., Ciravolo, T. G., McLeod, K. W., and Erwin, J. S., Potential nitrate leaching losses and nitrogen mineralization in an Atlantic coastal plain watershed following disturbance: preliminary results, in *Ecological Process and Cumulative Impacts: Illustrated by Bottomland Hardwood Wetland Ecosystems*, Gosselink, J. G., Lee, L. C., and Muir, T. A., Eds., Lewis Publishers, Inc., Chelsea, MI, 1990, p. 103.

Kraus, M. L., Wetlands: toxicant sinks or reservoirs?, in *Proceedings of the National Wetland Symposium: Wetland Hydrology*, Kusler, J. A. and Brooks, G., Eds., Omnipress, Madison, WI, 1987.

McGill, R., Tukey, J. W., and Larsen, W. A., Variations of box plots, *Am. Stat.*, 32, 12, 1978.

Miller, E. G., Effect of Great Swamp, New Jersey, on Streamflow During Base-Flow Periods, U.S.G.S. Research Paper 525-B, 1965.

Novitzki, R. P., The hydrologic characteristics of Wisconsin's wetlands and their influence on floods, streamflow, and sediment, in Selected Proceedings of the Midwest Conference on Wetland Values and Management, Richardson, B., Ed., Minnesota Water Planning Board, St. Paul, MN, 1981, p. 111.

O'Brien, A., Hydrology of two small wetland basins in eastern Massachusetts, *Water Resources Bull.*, 13 (2), 325, 1977.

Peterjohn, W. T. and Correll, D. L., Nutrient dynamics in agricultural watersheds: observations on the role of a riparian forest, *Ecology*, 65, 1466, 1984.

Peverly, J. H., Stream transport of nutrients through a wetland, *J. Environ. Qual.*, 11 (1), 38, 1982.

Richardson, C. J., Freshwater wetlands: transformers, filter, or sink?, in Freshwater Wetlands and Wildlife, Proceedings of a Symposium, Sharitz, R. R. and Gibbons, J. W., Eds., U.S. Department of Energy, 1989, p. 25.

Richardson, C. J., Tilton, D. L., Kadlec, J. A., Chamie, J. P. M., and Wentz, W. A., Nutrient dynamics of northern wetland ecosystems, in *Freshwater Wetlands: Ecological Processes and Management Potential*, Good, R. E., Whigham, D. F., and Simpson, R. L., Eds., Academic Press, New York, 1978, p. 217.

Seitzinger, S. P., Linkages between organic matter mineralization and denitrification in eight riparian wetlands, *Biogeochemistry*, 25, 19, 1994.

Simpson, R. L., Whigham, D. F., and Walker, R., Seasonal patterns of nutrient movement in a freshwater tidal marsh, in *Freshwater Wetlands: Ecological Processes and Management Potential*, Good, R. E., Whigham, D. F., and Simpson, R. L., Eds., Academic Press, New York, 1978, p. 243.

van der Valk, A. G., Davis, C. B., Baker, J. L., and Beer, C. E., Natural freshwater wetlands as nitrogen and phosphorous traps for land runoff, in *Wetland Functions and Values: The State of Our Understanding*, Greeson, P. E., Clark, J. R., and Clark, J. E., Eds., American Water Resources Association, Minneapolis, MN, 1979.

Vecchioli, J., Gill, H. E., and Lang, S. M., Hydrologic role of the Great Swamp and other marshland in the Upper Passaic River Basin, *J. Am. Waterworks Assoc.*, 54, 695, 1962.

Verry, E. S. and Boelter, D., Peatland hydrology, in Selected Proceedings of the Midwest Conference on Wetland Values and Management, Richardson, B., Ed., Minnesota Water Planning Board, St. Paul, MN, 1981, p. 121.

Wilkenson, L., SYSTAT: The System for Statistics, SYSTAT, Inc., Evanston, IL, 1990.
Yarbro, L. A., Kuenzler, E. J., Mulholland, P. J.. and Sniffen, R. P., Effects of stream channelization on exports of nitrogen and phosphorous from a North Carolina coastal plain watershed, *Environ. Manage.,* 8 (2), 151, 1984.

CHAPTER **20**

Relationships Between Groundwater Level and Temperature in Peat

Hannu Hökkä, Timo Penttilä, and Matti Siipola

CONTENTS

INTRODUCTION

Soil temperature directly influences tree root growth and water uptake. In addition, soil temperature affects, for example, microbial activity (Silvola et al., 1985), nitrogen mineralization (Kaunisto and Norlamo, 1976), and decomposition processes. The rates of these processes indirectly affect tree growth. Changes in soil temperature can thus have a significant effect on tree growth.

The soil temperature regime is determined by bulk density, water content, and soil composition (Campbell, 1977). The soil's volumetric heat capacity, as well as its thermal conductivity, are known to vary considerably with changes in water content. Thus, decreasing the water content through drainage decreases the soil's heat capacity and thermal conductivity. Thermal conductivity and heat capacity, in turn, determine the soil's heat storage, heat transfer, and temperature (Stathers and Spittlehouse, 1990).

1-56670-177-5/97/$0.00+$.50
© 1997 by CRC Press, Inc.

Peat is a substrate with high porosity and naturally high water retention capacity (Päivänen, 1973). Wet peat has a heat capacity almost equal to that of water. When dried, the heat capacity of peat decreases to a level close to that of mineral soils, and its thermal conductivity decreases as well (Stathers and Spittlehouse, 1990). After drainage, the temperature of peat is also influenced by decreased evaporation from soil surface and by compacting of the peat (Heikurainen and Seppälä, 1963).

In several Finnish studies, increased drainage intensity has been observed to decrease soil temperature in the surface layer of peatland sites drained for forestry (Heikurainen and Seppälä, 1963; Mannerkoski, 1988; Hytönen and Silfverberg, 1991). The greatest differences in temperature between dry and wet peat have been observed in early summer (Heikurainen and Seppälä, 1963). Peat temperature during the growing season has been shown to correlate with water table depth, which, in turn, is dependent on drainage intensity (Mannerkoski, 1988; Hytönen and Silfverberg, 1991). Spatial variation of peat temperature has been studied by Heikurainen and Seppälä (1963). Although significant variation between and within sites existed, the general conclusion was that drainage decreases surface peat temperature. This was assumed to be due to the different heat capacity and thermal conductivity of peat under varying moisture conditions.

On an intermediate fen in north-central Alberta, Canada, maximum summer temperatures, as well as the temperature sum of the surface peat, were higher on the drained site than on the undrained control site (Lieffers and Rothwell, 1987). Earlier flowering and budflush of tamarack (*Larix laricina* [Du Roi] K. Koch) and dwarf birch (*Betula pumila* L.) also occurred on the drained area. In spring, wet peat warmed up more quickly than dry peat. This was concluded to be due to the wet peat's higher thermal conductivity; in midsummer, dry peat needed less heat than wet peat to warm up. On the other hand, it is known that dry peat has poor heat storage properties. The same conclusions were also reached by Pessi (1958). In his study on a treeless flark fen in northern Finland, the peat (5 to 15 cm depth) on a drained site was warmer than that on the undrained control site during the summer months.

The objective of the present study was to investigate the relationship between the groundwater table (GWT) depth and the peat temperature and its variation during the growing season on a drained peatland site. To test the relationship in extreme circumstances, the study was conducted under harsh climatic conditions with respect to forest drainage. The aim was also to provide information about the spatial variation of temperature in the surface peat.

MATERIAL AND METHODS

The study area is located in northern Finland, ca. 80 km north of the Arctic Circle (Figure 1). The average temperature sum (with threshold value 5°C) on the site is 800 dd, and the precipitation sum from May 1 to September 30 is 300 mm. The site was drained for forestry in 1969 with 50 m ditch spacing. Ditch

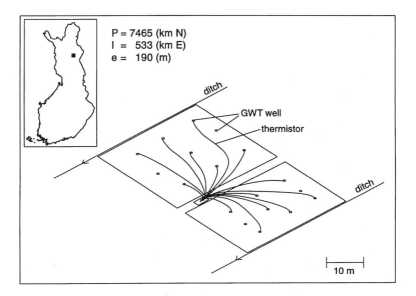

Figure 1 Location of the study area and the experimental layout.

cleaning was carried out in 1984. The site is occupied by a vigorous pole-stage stand of Scots pine (*Pinus sylvestris* L.) with a volume of about 50 m³ ha⁻¹. Due to thinning treatments carried out in 1989, the stand density was higher on one side of the drainage strip. The original site type was a meso-eutrophic fen, with a peat depth of more than 1 m. The average bulk densities of the SC-type peat in the 0 to 5, 5 to 10, and 10 to 20 cm surface layers were 0.123, 0.138, and 0.132 g cm⁻³, respectively. Bulk density did not depend on the distance away from the ditch. The site was rather even in terms of topography and dominated by lawns with only minor hummocks.

In order to monitor the GWT depth, a systematic network of wells was established across a drainage strip in 1991. Perforated plastic pipes (diameter 3.5 cm, length 1 m) were inserted into the peat in a gridlike layout. The GWT depth was expected to be influenced by the distance to the nearest ditch and, to some extent, by stand density. Peat water content was not measured directly. Adjacent to the 14 wells, thermistors were set at depths of 10 cm (Figure 1). Both the pipes and the thermistors were set up in similar microsites representing the lawn surface.

The study was conducted during the 1993 growing season. Peat temperature data were collected at 3-h intervals using a data logger (Grant 1200 Series Squirrel Meter, Model 1206). The temperature sum of the peat was calculated as the cumulative sum of the daily mean temperatures. In addition to the commonly used threshold value of 5°C, the peat temperature sum was also calculated using 2°C as the threshold value, because considerable soil respiration also occurs in drained peat soils at temperatures under 5°C (Silvola et. al., 1985). Air temperature data were collected in a similar way using two sensors at a height of 30 cm above the soil surface.

The GWT depth in the wells was monitored manually from July 8 to October 1 at 2-week intervals, seven times in all, in order to cover the temporal variation during the period, which is the most important with respect to tree root growth. GWT depths were expressed as the difference between the soil surface and the measured GWT level. For each measurement, correlations between the daily minimum, mean, and maximum temperatures of the peat and the GWT depth were then calculated. Correlations between the temperature sum of the peat (with threshold value 2°C) and the mean GWT depth in the adjacent well were also calculated. The effect of distance away from the ditch on GWT levels and the temperature sum was studied.

RESULTS

The 1993 growing season was colder than normal. The temperature sum (with a threshold value of 5°C) was 607 dd°C. There were 89 d when the temperature exceeded 5°C. The air temperature peaked in late July (Figure 2). The temperature sum (>5°C) of the peat was 456 dd°C, calculated as the mean of readings provided by 14 sensors, and the average number of days with peat temperature exceeding 5°C was 97. The range in soil temperatures was at its minimum (0.9°C) on October 1 and at its maximum (2.1°C) on July 22. Although the temperature sum in the air was higher than in the peat, the daily mean temperatures during August and September were higher in the peat.

The GWT levels were highest in early midsummer and in the autumn (Figure 2). The mean GWT depth, calculated on the basis of all readings, was 36.5 cm

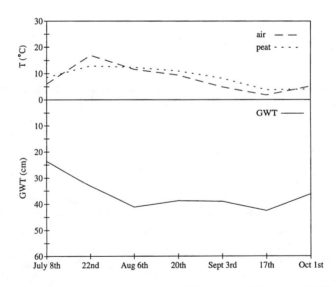

Figure 2   Mean temperatures of the air and of the peat, and the mean GWT depth on different measurement occasions.

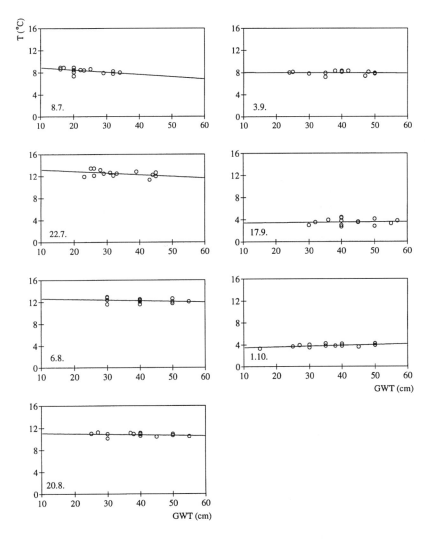

Figure 3   Relationship between the daily minimum peat temperature and GWT depth in adjacent wells and the temporal variation of this relationship.

below the surface. The range in GWT depths was at its minimum (18 cm) on July 8 and at its maximum (35 cm) on October 1.

Since there was practically no difference in the correlations between GWT depths and daily minimum, mean, and maximum temperatures, the dependence of peat temperature on GWT depth was tested using only the daily minimum temperatures. The correlations were significant only for the first and last measurements, with a significance level of 0.05 (Figure 3; Table 1). The slope was negative for the first measurement and positive for the last measurement. Only the last measurement differed from the others in terms of parallelism (test of

Table 1   Correlation Coefficients of the Peat Temperature Sum at 10 cm Below Surface (DD2, Threshold Value 2°C), Bulk Density at 0 to 5 cm Below the Surface (ρ), GWT Levels at Different Measurements (GWT1 to GWT7), and Daily Minimum Temperatures at Different Measurements (T1 to T7)

	DD2	ρ	GWT1	GWT2	GWT3	GWT4	GWT5	GWT6	GWT7	T1	T2	T3	T4	T5	T6
ρ	0.149														
GWT1	-0.655*	-0.213													
GWT2	-0.340	-0.014	0.870***												
GWT3	-0.321	0.276	0.584*	0.692**											
GWT4	-0.595*	0.182	0.625*	0.517	0.858***										
GWT5	-0.689*	0.161	0.710**	0.598*	0.824***	0.935***									
GWT6	-0.509	0.249	0.528*	0.476	0.889***	0.911***	0.913***								
GWT7	-0.347	0.194	0.659**	0.632*	0.873***	0.896***	0.841***	0.808***							
T1	0.794**	0.170	-0.545*	-0.315	-0.264	-0.425	-0.565*	-0.517	-0.156						
T2	0.980***	0.092	-0.714***	-0.412	-0.383	-0.642**	-0.714**	-0.538	-0.442	0.721**					
T3	0.948***	0.095	-0.570*	-0.270	-0.282	-0.517*	-0.620*	-0.515	-0.238	0.880***	0.916***				
T4	0.637*	-0.033	-0.310	-0.118	-0.162	-0.274	-0.403	-0.440	0.002	0.905***	0.570*	0.831***			
T5	0.126	-0.064	0.022	0.032	-0.016	0.040	-0.057	-0.233	0.253	0.672**	0.023	0.378	0.799***		
T6	-0.430	-0.138	0.370	0.177	0.153	0.393	0.302	0.053	0.399	0.152	-0.508	-0.207	0.344	0.768***	
T7	-0.270	-0.111	0.442	0.336	0.349	0.475	0.346	0.149	0.532*	0.199	-0.362	-0.048	0.412	0.712**	0.898***

Note: Statistical significance level indicated by asterisks.

Table 2 Regression Between the Daily Minimum Temperatures (T1, T7) of Peat and GWT Depths (GWT1, GWT7), and Between the Peat Temperature Sum (DD2, Threshold Value 2°C) and Mean GWT Depth (GWTm)

Dependent variable	Constant (std. err.)	Independent variable	Coefficient (std. err.)	R^2	Standard error of estimate
T1	9.273 (0.430)	GWT1	−0.04 (0.013)	29.7	0.397
T7	3.361 (0.226)	GWT7	0.013 (0.006)	28.3	0.226
DD2	782.220 (47.646)	GWTm	−2.858 (1.281)	29.3	34.882

equal slopes, p = 0.041). The results of pairwise comparisons indicated that the first (p = 0.004) and second (p = 0.029) measurements differed from the last measurement. The slopes for the other measurements were parallel. All in all, the GWT depth did influence the minimum daily temperatures of peat to some extent, mostly in early midsummer.

For the first and last measurements, regression equations were calculated for the dependence of soil temperature on GWT depth (Table 2). The coefficient of GWT1 was of the same magnitude as that presented by Mannerkoski (1988) for a longer measurement period in a study of an afforested bog in southern Finland.

Throughout the whole study period, the GWT observations at almost all measurement occasions were significantly correlated with all the others (Table 1). This means that the spatial pattern of the GWT level was relatively stable during the study period. For the daily minimum temperatures, the first measurement correlated with the second, third, fourth, and fifth measurements, and the second measurement correlated with the third and fourth. All the other measurements correlated only with the preceding one. This indicated that the spatial distribution of soil temperature at a depth of 10 cm was changing during the period; i.e., the warmest wells in early summer were not the warmest in autumn.

With threshold value 2°C, the mean temperature sum of the peat was 678 dd°C (SD 39.9). The temperature sum of the peat decreased significantly with the drawdown of the mean GWT level (r = −0.542, p = 0.045). Calculations with the equation in Table 2 showed that the drawdown of the water level from 25 to 50 cm below the surface would decrease the temperature sum at a depth of 10 cm by 70 degree-days, i.e. by 10 percent. The spatial variation of the temperature sum was considerable; the maximum was 22 percent higher than the minimum. According to results obtained with ANOVA, distance away from the ditch did not explain the variation of the temperature sum (F = 2.274, df = 2, p = 0.149), but it was highly significant (F = 17.137, df = 2, p = 0.000) in explaining the mean GWT depth in the wells. The bulk density of the surface peat was not correlated with the temperature sum nor with the minimum temperatures of the peat (Table 1).

Correlations between the daily minimum temperatures of the peat at each measurement and the temperature sum (>2°C) were also calculated. Peat temperatures for the second measurement (T2, July 22) were highly correlated with the temperature sum (Table 1). This was the time of the highest peat temperature and the largest range of temperature during the study period (Figure 3). T3 and T1 also correlated significantly with the temperature sum.

DISCUSSION

The decreasing temperature sum of the surface peat with the drawdown of the groundwater level is in accordance with the results given by Hytönen and Silfverberg (1991). However, the relationship was rather weak in the present study. Also, the daily minimum temperatures were only slightly influenced by variation in the GWT depth (cf. Heikurainen and Seppälä, 1963; Mannerkoski 1988; Hytönen and Silfverberg, 1991). Some influence occurred in early mid-summer when the temperature of the peat had not yet peaked. At that time, the drawdown of the GWT resulted in lower surface peat temperatures than when the GWT was closer to the peat surface. During the rest of the growing period, the level of the GWT did not affect surface peat temperatures. The positive correlations of GWT depth and peat temperature in the autumn observed in this study can be explained by the cooling of the peat surface due to frequent frosts. Wet surface peat is affected more by cold air temperatures than drier peat. There is also a lag with depth in the phase of the yearly temperature wave (Steenhuis and Walter, 1986). In the autumn, peat at a depth of 10 cm may be cooling off, while deeper down it is still warming.

In the site examined in this study, the spatial variation of the mean GWT depth was mainly influenced by the distance away from the ditch, unlike the spatial variation of the temperature sum of the peat. The spatial variation of the temperature sum was not dependent on the bulk density of the surface peat, either. With respect to the lawn surface, similar sites for the thermistors were selected, but variation in the peat composition, as well as in the ground vegetation and stand density, may have accounted for the spatial variation of the temperature sum of the peat.

Although the study area was located almost 100 km further north than the pine mire studied by Hytönen and Silfverberg (1991), it seemed to be a more appropriate site for tree growth with respect to soil temperature. The average temperature sum of the air was 300 dd lower than the 9-year average presented by Hytönen and Silfverberg, but the mean temperature sum of the peat in 1993 was over 84 dd higher than in the site studied by Hytönen and Silfverberg (1991). The ratio of the annual peat temperature sum to air temperature sum was 0.75 in this study, whereas in their case it varied between 0.36 and 0.67. This supports the important effect of site, previously emphasized by Heikurainen and Seppälä (1963).

The results of this study and those of Heikurainen and Seppälä (1963) and Hytönen and Silfverberg (1991) concerning the effect of drainage intensity on the thermal regime of the surface peat were partly inconsistent with those reported by Pessi (1958) in Finland and Lieffers and Rothwell (1987) in Alberta, Canada. The reasons for the contrary changes in temperature behavior due to drainage may lie in climatic and site differences. Lieffers and Rothwell (1987) obtained their results in a more continental and southern climate than that in Finland. Pessi (1958) made his comparisons on a treeless flark fen with no variation in temperature caused by tree stands and only minimum variation caused by the ground vegetation and the composition of the peat. In the case of mires with a sparse tree cover, Heikurainen and Seppälä (1963) also noticed higher temperatures in dry peat. The data of the present study represent cold and humid climatic conditions. In such conditions, Hytönen and Silfverberg (1991) also observed considerable cooling of peat due to intensive drainage. The weak relationship between groundwater level and peat temperature in the data obtained in the present study could be due to the lower heat load reaching the ground compared to that of the other studies in more southern locations; the heat transfer into the soil profile would have been correspondingly weaker.

The studies of Lieffers and Rothwell (1987) and Pessi (1958) differ from the others in that they were conducted within 0 to 3 years since drainage, while in other studies the time since drainage was 5 to 7 years (Mannerkoski, 1988), 10 to 40 years (Heikurainen and Seppälä, 1963), 24 years (this study), and 30 to 60 years (Hytönen and Silfverberg, 1991). Because of drainage, the water content of surface peat decreases in the growth period immediately following ditching. Within a considerably longer period, drainage also causes peat subsidence and subsequently increases peat density. The water retention capacity of peat increases with increasing density (Päivänen, 1973), due to which the water content of the surface peat may vary even if the depth of the groundwater table is the same. The capillary connection between the water table and dry surface peat with low density may be weak. Peat densities were not observed in other studies, but within 0 to 3 years, great deviations from the undrained state may not be expected. The average bulk densities for some undrained peatland sites (0.032 to 0.064 g cm^3) given by Westman (1981) are considerably lower than those included in the data for this study at 24 years after drainage (0.123 g cm^3).

Although variation of water content does not affect peat's diffusivity (i.e., how rapidly temperature changes are transmitted through the soil profile), diffusivity is influenced by bulk density (Campbell, 1977; Stathers and Spittlehouse, 1990). In the studies of Lieffers and Rothwell (1987) and Pessi (1958), the low diffusivity of dry surface peat may have caused the warming of the surface peat. In older drainage areas, the increased density of the surface peat with subsequent changes in diffusivity, thermal conductivity, and volumetric heat capacity could contribute to cooler temperatures following the drawdown of the water level. On older drained sites, the decreased thermal conductivity of dry peat could be the major agent for the low temperatures observed in several studies. All in all, air temperature and humidity, together with the variation in the heat load reaching the soil surface and the physical properties of the peat, probably mainly determine the changes in surface peat temperatures after drainage. If these factors are constant, peat temperature is influenced by the peat water content to some extent.

ACKNOWLEDGMENTS

We would like to express our thanks to Jean-Louis Belair, who provided us with valuable information concerning the problem. We are also grateful for the valuable comments provided by Richard L. Rothwell and an anonymous referee concerning an earlier version of the manuscript. The English was revised by Michael Hurd from the University of Lapland.

REFERENCES

Campbell, G. S., *An Introduction to Environmental Biophysics,* Springer-Verlag, New York, 1977, 159 pp.

Heikurainen, L. and Seppälä, K., The effect of drainage degree on temperature conditions of peat, *Acta For. Fenn.,* 76, 33, 1963 (in Finnish with English summary).

Hytönen, J. and Silfverberg, K., Effect of drainage on thermal conditions in peat soils, *Fol. For.* (Helsinki), 780, 24, 1991 (in Finnish with English summary).

Kaunisto, S. and Norlamo, M., On nitrogen mobilization in peat. I. Effect of liming and rotavation in different incubation temperatures, *Comm. Inst. For. Fenn.,* 88 (2), 27, 1976.

Lieffers, V. J. and Rothwell, R. L., Effects of drainage on substrate temperature and phenology of some trees and shrubs in an Alberta peatland, *Can. J. For. Res.,* 17, 97, 1987.

Mannerkoski, H., Effect of drainage on the temperature of surface peat, in *Symposium on the Hydrology of Wetlands in Temperate and Cold Regions,* Vol. 1, The Publications of the Academy of Finland, Joensuu, Finland, 4/1988, p. 60.

Päivänen, J., Hydraulic conductivity and water retention in peat soils, *Acta For. Fenn.,* 129, 70, 1973.

Pessi, Y., On the influence of bog draining upon thermal conditions in the soil and air near the ground, *Acta Agric. Scand.,* 24, 359, 1958.

Silvola, J., Välijoki, J., and Aaltonen, H., Effect of draining and fertilization on soil respiration at three ameliorated peatland sites, *Acta For. Fenn.,* 191, 32, 1985.

Stathers, R. J. and Spittlehouse, D. L., Forest Soil Temperature Manual, FRDA report 130, 1990, 47 pp.

Steenhuis, T. S. and Walter, M. F., Will drainage increase spring soil temperatures in cool and humid climates?, *Am. Soc. Agric. Eng.,* 29, 1641, 1986.

Westman, C. J., Fertility of surface peat in relation to site type and potential stand growth, *Acta For. Fenn.,* 172, 77, 1981.

CHAPTER 21

Factors Affecting Sediment Accumulation in Sedimentation Ponds

Samuli Joensuu

CONTENTS

INTRODUCTION

Forest drainage is used to change the hydrological conditions in the soil into a state favorable for forest growth. During the past few decades over 5 million ha of peat and mineral soils with a poor hydrological status and low stand growth have been converted into productive forest land in Finland (Paavilainen and Tiihonen, 1989; Päivänen, 1990).

The erosion and sedimentation problems caused by ditching have been recognized throughout the 60 year history of forest drainage in Finland. The first ditching recommendations emphasized the serious nature of these problems from the point of view of ditch maintenance, the main attention being paid only to channeling the water out of the drainage area (Lukkala, 1940, 1948; Tanttu, 1943).

By the beginning of the 1970s practical forest ditching recommendations and textbooks started to mention the role of environmental protection structures such as sedimentation ponds. More attention also began to be paid to the harmful environmental impacts of drainage schemes (Heikurainen, 1971). The ditch maintenance problems caused by of the transport of sediment in the runoff from forest drainage areas have been covered in a number of early publications (Multamäki, 1934; Heikurainen, 1957, 1959; Antola and Sopo, 1966; Timonen, 1971). Corresponding studies concerning ditch maintenance are only in their infancy (Isoaho et al., 1993).

The first studies on the negative environmental impacts of forest drainage were carried out at the end of the 1970s. Monitoring data on this topic were not collected until the beginning of the 1980s (Heikurainen et al., 1978; Kenttämies, 1980). The Nurmes Project, which was started during the 1980s and involved monitoring a number of drainage basins is, for the present, the longest and most comprehensive study on the harmful effects of forestry on the watercourses (Ahtiainen, 1988, 1990). The METVE Project (The harmful effects of forestry on watercourse quality, and the abation of these impacts) was started in Finland in 1990, and was continued to the end of 1995 (Saukkonen and Kenttämies, 1993). One part of the project deals with the effects of sedimentation ponds on the transport of solid material and nutrients from ditch maintenance areas (Ahti et al., 1993).

Problems Caused by Ditch Erosion

Forest ditching clearly increases erosion, the transport of solid matter by the outflowing water. The soil is loosened and mixed into the water during ditching work. The sides of the ditches and ditching soil are susceptible to the eroding effects of flowing water for a considerable period of time.

The discharge from lateral ditches is usually small; hence the erosion and transport of solid material are low. In contrast, large amounts of water are carried by the main ditches, which are therefore frequently the sites of considerable erosion. A vegetation cover gradually develops in the ditch network; this increases either the fixation or the release of certain nutrients and helps to retain some of the solid matter carried by the water. It has been estimated that the transport of solid matter from ditch maintenance areas can sometimes be as great as that associated with pioneer drainage work, and in some cases even greater.

The surface runoff resulting from heavy rainfall and melting snow often causes erosion (meteorological factor in Figure 1). On the other hand, the degree of erosion depends on the type of ground surface, gradient of the terrain, length of the slopes, water permeability of the soil, etc. Precipitation, soil properties, and the gradient of the terrain are examples of local factors that cannot be regulated (Tiainen and Puustinen, 1989).

The gradient of the terrain determines the flow rate of the water moving down the ditch channels. The particle size distribution of mineral soil and the degree

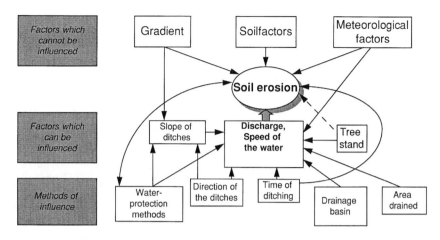

Figure 1 Factors affecting erosion in the forest drainage areas.

of humification of peat determine the extent to which the flowing water can detach and transport material from the sides of the ditches. Precipitation intensity, time, and amount may have a decisive effect on the occurrence of solid matter peaks. In particular, the water equivalent and melting rate of the snowpack have a considerable effect on the magnitude of flood peaks.

The amount of runoff and its flow rate are important factors from the point of view of erosion. The location of the ditch lines in drainage projects can be used to affect the flow rate of the ditch water by regulating the gradient of the ditches. The tree stand also has an effect on the amount and timing of runoff (Ahti, 1975).

The size of the drainage basin has a decisive effect on the magnitude of the discharge and subsequently on the extent of possible erosion. Erosion can be minimized by regulating the size of drainage basin. Ditching often results in the exposure of surfaces sensitive to water erosion. Increasing the number of ditches also increases the magnitude of the flood peaks. Watercourse protection methods can be used to regulate the amount of runoff and rate of flow of the water. On the other hand, the erosion risk also affects the choice of watercourse protection method.

Sedimentation Ponds as a Watercourse Protection Measure

Sedimentation ponds are excavated to retain sediment from large drainage basins. Their effective functioning presupposes a suffcent reduction in the water flow rate for the waterborne solid matter to sediment on the bottom of the pond (Selin and Koskinen, 1985; Ihme et al., 1991). Ponds are usually constructed in feeder or main ditches.

Despite the common practice of constructing sedimentation ponds, little research has been carried out on their effectiveness in reducing the solids load caused by forest drainage (Sallantaus and Pätilä, 1983; Ahtiainen et al., 1988;

Joensuu, 1990, 1992; Ylimannila et al., 1993). In contrast, the role of sedimentation ponds in reducing the solids load in peat production areas has been investigated considerably more thoroughly (Hynninen and Sepponen, 1983; Selin and Koskinen, 1985; Ihme et al., 1991). Aho and Kantola (1985) have, for instance, used laboratory experiments to investigate the functioning of sedimentation ponds and the particle size distribution in the runoff water.

The efficiency of sedimentation ponds to retain solid material has been monitored for a considerable period of time in peat production areas in Ireland (Hannon and Coffey, 1984; Harkins and Wynne, 1991; Wynne, 1992). According to their studies, sufficiently large and correctly dimensioned sedimentation ponds are capable of retaining over 90 percent of the solid material entering the pond.

Experience about the efficiency of sedimentation ponds is also available from the former Soviet Union, primarily from peat production areas. In Largin's (1983) studies sedimentation ponds have retained as much as 80 percent of the solid material. Largin and Lavrov (1983) state that the amount and flow rate of the water entering the pond have an effect on the functioning of sedimentation ponds.

MATERIAL AND METHODS

Sedimentation Pond Material

The efficiency of sedimentation ponds in six forestry board districts was investigated during summer 1988 (Joensuu, 1990). A total of 97 pools were studied (Figure 2). The parameters needed for calculating the original volume of the pools and that at the measuring instant were determined in the field. The excavated depth and depth of the sediment layer in the pools were determined using a measuring stick. The measurements were made with respect to the water level at the measuring instant, the water surface also being measured with respect to the ground surface. These measurements were made along lateral transects at ca. 2-m intervals along a line running along the length of the pond. At least three measurements were made along each transect. The length and breadth of the ponds were also measured for the volume calculations.

The location and pond type were also noted. A map (1:100 to 1:200) was prepared of each pond, as well as a cross-sectional drawing based on the depth and amount of sediment in the pond. The excavation date and information about the excavation work and emptying of the pond were obtained from the records. Missing information was added during 1989 to 1990. The sedimentation pond information also includes ponds that were emptied prior to the field measurements. However, no information was available about the amount of sediment removed during emptying.

The particle size distribution of the mineral material and the degree of humification of the peat in the sediment that had accumulated in the ponds were estimated by visual observation. The forest site type distribution in the drainage area above the pond was surveyed, and the soil type in the ditches estimated

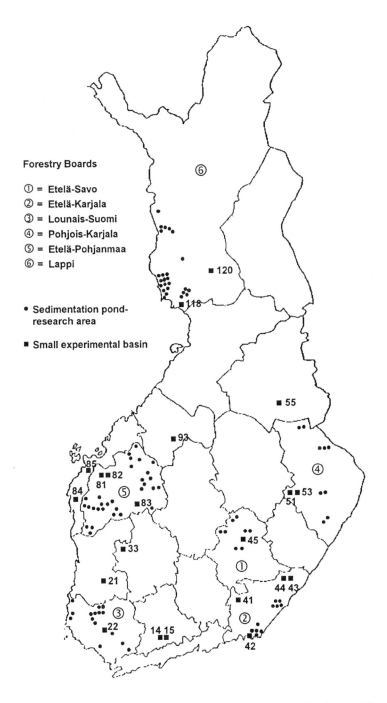

Figure 2 Location of the sedimentation ponds and the experimental basins used for estimating runoff. The estimates are based on runoff observations from nearby experimental basins measured by the National Board of Waters and the Environment.

visually. The forest compartment data in the ditching plan were also used. Information about the tree stand was obtained from the forestry plans. Special attention was paid to the subsoil type in the main ditches.

Information was obtained about the size of the ditching area and number of ditches from the drainage project records. The area of the drainage basin above the pond was measured from maps (1:10,000) using a planimeter. The area covered by the drainage basins was determined on the basis of the height contours. The mean gradient in the direction of the main ditch was also measured from maps.

Discharge Measurements

Seasonal discharge values for each pond were calculated from routine measurements of the water volume of so-called small drainage basins carried out by the National Board of Waters and the Environment. Small drainage basins located as close as possible to the sedimentation ponds were selected (Figure 2). The mean spring flood discharge values during the period when the ponds were functioning were calculated for each area on the basis of the small drainage basin data.

Statistical Treatment of the Material

The study material was depicted using means, standard deviations, and percentage distributions. Correlation analysis (Pearson correlation) and regression analysis were used in attempts to explain the interdependency between the amount of sediment in the ponds and background variables.

Residence time values corresponding to the spring mean flood discharge were determined for the sedimentation ponds. The residence time refers in this instance to the relationship between the maximum water volume in the pond and the spring flood discharge. Similarly, the surface load and water flow rate corresponding to the spring mean flood discharge were calculated for each pond. The surface load was determined as the ratio between the discharge and the surface area of the sedimentation pond. These parameters were used as independent variables in the regression analyses.

Owing to the skewed distribution of the material, the dependent and most important independent variables were logarithmically transformed for regression analysis. Degrees of statistical significance were $p < 0.05$, $p < 0.01$, and $p < 0.001$ (Snedecor and Cochran, 1967). The SYSTAT computer program (Systat, 1989) was used in the statistical treatment of the material.

The variation in the sediment accumulation of the sedimentation ponds was investigated by means of regression analysis. The annual accumulation of sediment per drainage basin unit was used as the dependent variable. The factors associated with the sedimentation pond parameters or the discharge above the pond proved to be the best explaining variables. Analysis of the residuals of the regression models showed that the models functioned well with very small sed-

Table 1   Correlations Between the Mean Annual Sediment Accumulation and
Dimension Parameters of the Sedimentation Ponds and Some Areal Factors

Variable	Whole material	Class 1	Class 2	Class 3	Class 4
Residence time of the pond	0.714***	0.874**	0.793***	0.461*	0.606*
Sediment volume in the pond	0.534***	0.777*	0.478**	0.456*	0.835**
Surface load to the pond	−0.334**	−0.328	−0.243	0.542**	−0.639*
Water flow rate	−0.383**	−0.342	−0.412**	−0.368	−0.512
Gradient of the main ditch	0.226*	0.242	0.229	0.227	0.170
Age of the sedimentation pond	−0.251*	−0.361	−0.179	−0.209	−0.434
n	82	8	41	22	11

Note: Classes: 1 = compacted soils, 2 = fine-textured soils, 3 = coarse-textured soils, 4 = peat.
Degrees of significance: * = p <0.05, ** = p <0.01, *** = p <0.001.

iment accumulation values, but that the estimates given by the model became less reliable as the sediment accumulation increased. The small number of observations in the compacted soil and peat soil classes especially increased the unreliability of the results. The regression models are presented in Tables 2 to 4. The overall form of the model was:

$$y = e^{a + b*\ln X1 + c*\ln X2 + d*\ln X3 + \sigma^2/2}$$

where y = the mean annual sediment accumulation, m^3 ha^{-1} yr^{-1}; X1 = the residence time of the pond; X2 = the sediment volume index (100 * sv/ev), sv = sediment volume/drainage basin ha/year; ev = excavated volume; X3 = the water flow rate in the pond or the surface load in the pond or the drainage basin size or the spring mean flood discharge.

The sediment accumulation values estimated using the models are presented in Figures 3 to 5.

Table 2   Factors Affecting the Filling Up of Sedimentation Ponds

Independent variable	Whole material	Class 1	Class 2	Class 3	Class 4
Constant	−2.396***	−3.842*	−2.580***	−1.399*	−2.212*
ln (X1)	0.597***	0.487*	0.502***	0.898***	0.806***
ln (X2)	0.842***	0.876***	0.765***	0.965***	1.020***
ln (X3)	−0.302***	−0.469*	−0.321***	0.009 ns	−0.235ns
n	82	8	41	22	11
R	0.88	0.99	0.84	0.96	0.97
d.f.	3, 78	3, 4	3, 37	3, 18	3, 7
F	188.8***	103.6***	64.8***	130.3***	67.0***
S_e	0.409	0.193	0.461	0.240	0.301

Note: Y = ln (sediment accumulation m^3/drainage basin ha/year). Independent variables: residence time of the pond (X1), sediment volume index (X2) and water flow rate in the pond (X3). Classes: 1 = compacted soils, 2 = fine-textured soils, 3 = coarse-textured soils, 4 = peat. Degrees of significance: * = p <0.05, ** = p <0.01, *** = p <0.001.

Table 3   Factors Affecting the Filling Up of the Sedimentation Ponds

Independent variable	Whole material	Class 1	Class 2	Class 3	Class 4
Constant	−1.303***	−1.511**	−1.243***	−1.443***	−1.234***
ln (X1)	0.531***	0.507*	0.476***	0.899***	0.581*
ln (X2)	0.888***	0.908**	0.858***	0.965***	0.939***
ln (X3)	−0.378***	−0.476 ns	−0.391***	0.008 ns	−0.511 ns
n	82	8	41	22	11
R^2	0.89	0.97	0.84	0.96	0.97
d.f.	3, 78	3, 4	3, 37	3, 18	3, 7
F	215.2***	49.0**	66.9***	130.3***	73.9***
S_e	0.386	0.278	0.454	0.240	0.287

Note: Y = ln (sediment accumulation m³/drainage basin ha/year). Independent variables: residence time of the pond (X1), sediment volume index (X2), and surface load to the pond (X3). Classes and degrees of significance the same as in Table 1.

Table 4   Factors Affecting the Filling Up of Sedimentation Ponds

Independent variable	Whole material	Class 1	Class 2	Class 3	Class 4
Constant	−0.198 ns	−0.228 ns	0.042 ns	−0.998*	−0.746*
ln (X1)	0.641***	0.612***	0.551***	0.852***	0.840***
ln (X2)	0.844***	0.983***	0.754***	0.945***	1.030***
ln (X3)	−0.290***	−0.367*	−0.311***	0.108 ns	−0.181*
n	82	8	41	22	11
R^2	0.91	0.99	0.88	0.96	0.97
d.f.	3, 78	3, 4	3, 37	3, 18	3, 7
F	249.3***	160.8***	88.0***	148.3***	82.3***
S_e	0.362	0.155	0.404	0.225	0.273

Note: Y = ln (sediment accumulation m³/drainage basin ha/year). Independent variables: residence time of the pond (X1), sediment volume index (X2), and the drainage basin size (X3). Classes and degrees of significance the same as in Table 2.

RESULTS

In about 80 percent of the cases the sediment that had collected in the sedimentation ponds was predominantly mineral material. The forest drainage projects included in the sample were mainly new drainage works carried ouin thin peated areas. In ditch maintenance an ever increasing proportion of the digging operation penetrates down into the underlying mineral soil owing to the compaction of the peat layer. The mean thickness of the peat in the drainage areas was less than 0.5 m.

The predominant particle type in the sediment in about one-quarter of the ponds was fine sand (0.2 to 2.0 mm), and in around one-fifth coarse sand (2.0 to 20.0 mm). The proportion of very fine sand (0.02 to 0.2 mm) and silt (0.002 to 0.02 mm) was also relatively high in the ponds. There were also many different gravel fractions in the ponds, as well as some clay, but these fractions were not predominant in any of the ponds. Peat was the predominant form of sediment in about one-fifth of the ponds.

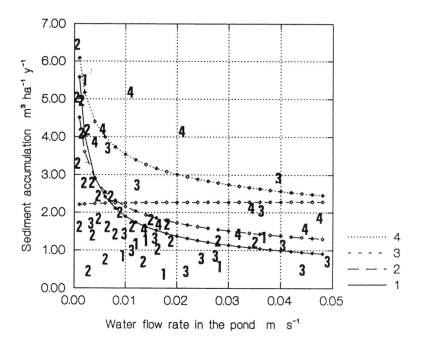

Figure 3 Sediment accumulation estimated as a function of the pond discharge water flow. Standardized variables: residence time of the water in the pond = 1 h. Sediment volume index = 10. Classes: 1 = compacted soils, 2 = fine-textured soils, 3 = coarse-textured soils, 4 = peat.

The ponds appeared to have filled up rapidly in those areas where the dominant soil type in the ditches was humified peat. The amount of solids carried by the water was also high in cases where the dominant soil type in the ditches was silt, very fine sand, or fine sand. Less sediment appeared to accumulate in those ponds where the subsoil type in the ditches was clay, compacted glacial till containing clay, or unhumified peat.

Filling up of the sedimentation ponds was at its greatest immediately after ditching and during flooding. Of the ponds excavated during the inventory year, on the average about half of them had become filled up by the time the measurements were performed. Unemptied ponds over 5 years old had, in practice, already become so filled up that their sedimentation capacity had essentially deteriorated.

The material of this study was divided into four classes on the basis of the predominant soil type in the ditches (Table 1). Class 1 included those areas where the subsoil type ranged from clay to compacted glacial till. In class 2, silt and very fine sand were predominant, and in class 3 sand and gravel fractions. Class 4 consisted of peat soils.

The mean annual sediment accumulation correlated statistically highly significant ($p < 0.001$) with the dimension parameters of the ponds, residence time, and sediment volume. The positive correlation between the sediment volume and the residence time with the annual sediment accumulation was statistically highly

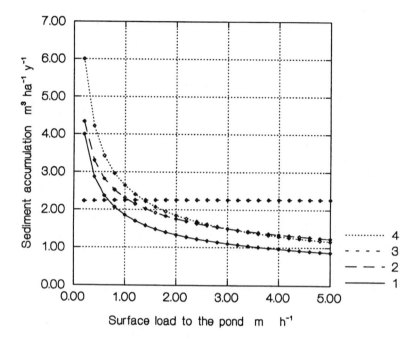

Figure 4   Sediment accumulation estimated as a function of the surface load on the pond. Standardized variables and the classes as in Figure 3.

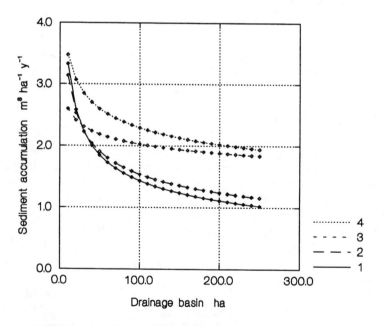

Figure 5   Sediment accumulation estimated as a function of the area of drainage basin above the sedimentation pond. Standardized variables and the classes as in Figure 3.

significant (p < 0.001), especially in those ditching areas where the dominant fraction was silt or very fine sand (Table 1). The increase in the statistical significance of the correlation on fine-textured soils suggests that the sedimentation rate of fine particles is slow. For this reason, the water flow rate should be sufficiently reduced in sedimentation ponds that the finest fractions would have time to be sedimented. Peat particles require a considerably longer sedimentation time than the heavy mineral soil particles (Aho and Kantola, 1985).

There was positive correlation between the gradient of the main ditch and sediment accumulation in the whole material. The correlation was statistically almost significant (p < 0.05). The positive correlation between the gradient and sediment accumulation in the fine-particle class (Class 2) was also rather strong but not statistically significant. There was negative correlation between the age of the sedimentation pond and sediment accumulation. The correlation was statistically almost significant (p < 0.05) in the whole material. The annual sediment accumulation decreased along with increasing time since the sedimentation pond had been excavated.

DISCUSSION AND CONCLUSIONS

The rate at which sedimentation ponds become filled up depends on the pond dimensions and local factors like discharge and predominant soil type. Correlation and regression analyses of the independent variables supported the hypothesis that an increase in the pond residence time especially increases the amount of sediment accumulating in the ponds. An increase in the sediment volume in the sedimentation ponds has a similar effect as the increase in the residence time on sediment accumulation. Large amounts of relatively heavy particles are transported by the water, especially in the early stages after ditching and during high water periods. These particles sediment on the bottom of the ponds as the water flow rate decreases. When the residence time and sediment volume remain constant, an increase in the amount of water, flow rate, surface load, and drainage basin in turn reduce sediment accumulation in the ponds.

The strong correlation between the parameters depicting the amount and flow rate of the water and the correlation with the residence time reduces the explaining power of these variables in the model. On the other hand, they show that the variation in the sediment accumulation is strongly connected to the discharge. The water flow rate in the ponds depends on the variation in the discharge. On the other hand, the pond parameters have an effect on the water flow rate. The surface load varies according to the water volume. An increase in the discharge probably also increases the amount of solid matter flowing through the ponds, which is a factor that was not measured in this study. Settling of the sediment in the ponds depends on the particle size distribution of the solids and the water flow rate. Accumulation of sediment in the sedimentation ponds can be depicted using a simplified equation (Figure 6).

The weakness of this material is the skewed distribution caused by its heterogeneity. The material does not follow a normal distribution. The proportion of

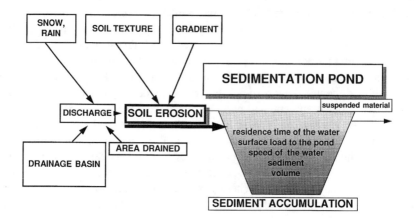

Figure 6   Factors affecting sediment accumulation in the sedimentation ponds.

small ponds and small annual sediment accumulation values in the material is high. This causes problems in the statistical analysis of the material and a need for transforming the variables. Linear regression analysis, the only method used in this study, may not in every case be the best method to estimate the filling of the ponds. In future studies we must take account that the nature of runoff, channel flow, and sedimentation pond design should offer some reasons for the nonlinear relationships.

The material is based on only one measurement instant. One problem has therefore been the lack of data about the development history of the ponds. Under what sort of conditions were the ponds excavated, how heavy were the first rains in the drained area above the ponds, what were the weather conditions when the ditching work was carried out, what is the real impact of flood timing on the filling up of the ponds, etc.?

Owing to the small number of measurement times, we do not know how the sediment accumulating in the ponds really behaves during the spring flood and periods of heavy rain. It would appear that only coarser particles are deposited in the ponds. Only small amounts of very fine material (i.e., clay and fine silt) were present in the ponds together with the coarser material. The variation in the particle size distribution depends on the specific water flow conditions in each pond (Aho and Kantola, 1985). During periods with a low flow rate, fine material accumulates in the ponds, and during fast flow rates, coarse particles only are sedimented. Part of the sedimented fine material may be transported out of the ponds if the flow rate is sufficiently high. However, this phenomenon could not be studied here. Neither does this study provide any information about the amount of material passing through the ponds as suspended material.

The material of this study also included ponds that had been emptied before the field measurements of the study. It was not possible to estimate the amount of sediment removed from the ponds during these operations. This introduces some error when estimating the sediment accumulation. Another drawback to the study is the fact that no basin-specific information was available about the variation in the water volume. The basin-specific discharge values calculated on the basis of discharge observations for small drainage basins are at their best only indicative.

This study shows that more attention should be paid to the design of sedimentation ponds in forest drainage areas. The ponds included in this study were so small that the solids that accumulated only included particles larger than a certain size. Visual examination indicated that particles smaller than silt had not accumulated in the ponds. In most cases the ponds filled up with sand and fine sand species. This presumably means that the flow rate in these ponds is so high that particles smaller than silt had no possibility to be deposited.

According to the models constructed using regression analysis, it is extremely important to take the discharge rate into account in designing sedimentation ponds. Additional important factors in pond design appear to be a sufficient residence time, the sediment volume, flow rate in the ponds, and the surface load. The size of the drainage basin and ditching area and the spring mean flood discharge can be used, in addition to the above-mentioned factors, to predict sediment accumulation.

According to study, the residence time explained the sediment accumulation better than surface load value. Ponds are able to accumulate more solids carried by the water the longer the residence time of the pond. Because the reach of the excavating machinery determined the width of the pond and the volume of the pond, the residence time could be increased by increasing the length of the pond. However, this is partly in contradiction with the surface load theory. According to the theory, a shallow, wide pond is the most effective solution in ideal flow conditions (Aho and Kantola, 1985). In such a case the flow rate slows down as the water spreads out over a wide area, and the particles therefore rapidly settle to the bottom in a shallow pond. However, shallow, broad ponds do not function for long in practice.

We can conclude on the basis of this study that the planning of sedimentation ponds should be made more effective in forest ditching areas. Design based on water volume calculations for sedimentation ponds is especially important on fine-textured soils. In such cases we can assume that not all of the solid matter will be sedimented in the ponds during peak flow rate periods. On the other hand, repeated emptying of the ponds will most likely be necessary in areas susceptible to erosion. On compacted soils and especially on clay soils, other watercourse protection measures should be considered instead of expensive sedimentation ponds.

On coarse soils (i.e., those comprising fine sand and sand) a small reduction in the water flow rate in the ponds is almost always sufficient to deposit the solid material. The shape of the pond can, for this reason, vary more than on fine-textured soils. On coarse soils the ponds can, for instance, be built as ditch-widening types without reducing their sedimentation properties. On fine-textured soils the water flow rate in ditch-widening ponds of this sort may be too strong for the particles to sediment to the bottom of the ponds.

REFERENCES

Aho, M. and Kantola, K., Kiintoaineen sedimentoituminen turvetuotannon vesissä (Summary: Sedimentation of solid matter in peat production waters), *Tech. Res. Cent. Finl. Res. Rep.*, 345, 56, 1985.

Ahti, E., Ajatuksia metsäojituksen tulvavaikutusten arvioimisesta (Summary: on the estimation of the effect of forest drainage on flooding), *Suo*, 26, 9, 1975.

Ahti, E., Joensuu, S., and Vuollekoski, M., Laskeutusaltaiden vaikutus kiintoaineen huuhtoutumiseen kunnostusojitusalueilta (Leaching of suspended solids after ditch maintenance as affected by sedimentation ponds) in publ.: Metsätalouden vesistöhaitat ja niiden torjunta — METVE — projektin väliraportti, Saukkonen, S. and Kenttämies, K., Eds., National Board of Waters and the Environment Report 455, 111, 1993.

Ahtiainen, M., Effects of clear-cutting and forestry drainage on water quality in the Nurmes-study, *Symposium on the Hydrology of Wetlands in Temperate and Cold Regions*, Vol. 1, Joensuu, Finland, 6–8 June 1988, The Publications of the Academy of Finland, 4, 206, 1988.

Ahtiainen, M., Avohakkuun ja metsäojituksen vaikutukset purovesien laatuun (Summary: the effects of clearcutting and forestry drainage on water quality of forest brooks), Publications of the Water and Environment Administration — Series A, 45, 1990, 1.

Antola, A. and Sopo, R., Tutkimus 1930-luvulla kaivettujen metsäojien kunnosta ja perkaustarpeesta Helsingin, Mikkelin ja Seinäjoen metsänparannuspiireissä (Summary: cleaning forested swamp drains), *Suo*, 3, 39, 1966.

Hannon, G. P. and Coffey, P. J., Control of silt in peatland waters, Proceedings of 7th Int. Peat Congr., Dublin, 2, 237, 1984.

Harkins, J. and Wynne, D., Control of Silt in Run-Off from Peatland Bogs, 1991.

Heikurainen, L., Metsäojien syvyyden ja pintaleveyden muuttuminen sekä ojien kunnon säilyminen (Summary: changes in depth and top width of forest ditches and the maintaining of their repair), *Acta For. Fenn.*, 65, 1, 1957.

Heikurainen, L., Tutkimus metsäojitusalueiden tilasta ja puustosta (Referat: uber waldbaulich entwässerte Flächen und ihre Waldbestände in Finland), *Acta For. Fenn.*, 69, 1, 1959.

Heikurainen, L., *Metsäojituksen alkeet*, Gaudeamus, Helsinki, 1971, 281 pp.

Heikurainen, L., Kenttämies, K., and Laine, J., The environmental effects of forest drainage, *Suo*, 29, 49, 1978.

Hynninen, P. and Sepponen, P., Erään suoalueen ojituksen vaikutus purovesien laatuun Kiiminkijoen alueella, Pohjois-Suomessa (Summary: the effect of drainage on the quality of brook waters in the Kiiminkijoki river basin, northern Finland), *Silva Fenn.*, 17, 23, 1983.

Ihme, R., Heikkinen, K., and Lakso, E., Laskeutusaltaiden toimivuuden parantaminen turvetuotantoalueiden valumavesien käsittelyssä (Summary: development of the sedimentation basin structures for purifying runoff water from peat production areas), Publications of the Water and Environment Administration — Series A, 77, 117, 1991.

Isoaho, P., Lauhanen, R., and Saarinen, M., Metsäojien jatkuvan kunnossapidon vaikutus ojien tilaan Keski-Pohjanmaalla (Summary: effects of continuous ditch netvork maintenance on the condition of forest drainage areas in Central Ostrobothnia district), *Suo*, 44, 33, 1993.

Joensuu, S., Protection of watercourses from peatlands utilized for forestry: sedimentation ponds, Proceedings of the International Seminar in Umea, Sweden, September 3–7, 1990 on Biomass Production and Element Fluxes in Forested Peatland Ecosystems, 157, 1990.

Joensuu, S., Peatland forest ditch maintenance and runoff water quality: the effect of sedimentation ponds, Proceedings of the 9th International Peat Congress, Uppsala, Sweden 22-26 June 1992, 1992, p. 423.

Kenttämies, K., The effects on water quality of forest drainage and fertilization in peatlands. The influence of man on the hydrological regime with special reference to representative and experimental basins, Proceedings of the Helsinki Symposium, June 1980, IAHS — AISH Publication 130, 277, 1980.

Largin, I. F., Tutkimus seisottamalla ja suodattamalla tapahtuvasta turvesoiden valumavesien puhdistuksesta, Kalin Peat Institute — Report, 1983, 8 pp.

Largin, I. F. and Lavrov, N. V., Turvetyömaan vesien puhdistus selkeytysaltailla, *Torfnaja promylennost,* 8, 21, 1983.

Lukkala, O. J., *Metsämiehen suo-oppi,* Keskusmetsäseura Tapio, Helsinki, 1940, 192 pp.

Lukkala, O. J., *Metsän ojitus sekä ojien kunnossapito ja suometsien hoito,* Keskusmetsäseura Tapio, Helsinki, 1948, 54 pp.

Multamäki, S. E., Metsäojien mittojen ja muodon muuttumisesta (Deutches Referat: Uber die Grössen- und Formveränderungen der Waldgräben), *Acta For. Fenn.,* 40, 1, 1934.

Paavilainen, E. and Tiihonen, P., Suomen suometsät vuosina 1951–1984 (Summary: peatland forests in Finland in 1951–1984), *Folia For.,* 714, 1, 1989.

Päivänen, J., *Suometsät ja niiden hoito,* Kirjayhtymä, Helsinki, 1990, 231 pp.

Sallantaus, T. and Pätilä, A., Runoff and water quality in peatland drainage areas, Proc. Int. Symp. of Forest Drainage, Tallin 1983, 1983, p. 183.

Saukkonen, S. and Kenttämies, K., Eds., Metsätalouden vesistöhaitat ja niiden torjunta — METVE — projektin väliraportti, National Board of Waters and the Environment — Report, 455, 1, 1993.

Selin, P. and Koskinen, K., Laskeutusaltaiden vaikutus turvetuotantoalueiden vesistökuormitukseen (Summary: the effects of the sedimentation ponds on the load coming from the peat production areas to the watercourses), National Board of Waters, Finland — Report 262, 1985, 1.

Snedecor, G. W. and Cochran, W. G., *Statistical Methods,* 6th ed., The Iowa State University Press, Ames, IA, 1967.

Systat, *The System for Statistics for the PC,* 2nd ed., 1989.

Tanttu, A., *Metsäojittajan opas,* Keskusmetsäseura Tapio, 1943, 221 pp.

Tiainen, I. and Puustinen, M., Peltoviljelyn vaikutus eroosioon, National Board of Waters and the Environment — Report 190, 1989, 1.

Timonen, E., Auraus-ja kaivuriojien koon ja muodon muutoksista (Summary: on the changes of the size and shape of ditches made by plows and tractor diggers), *Silva Fenn.,* 5, 70, 1971.

Wynne, D., Control of peat silt in drainage waters from milled peat production bogs in Bord na Mona, Proceedings of the 9th International Peat Congress, Vol 2, Uppsala, Sweden 22–26 June 1992, 1992, p. 470.

Ylimannila, S., Härkönen, P., and Lakso, E., Metsätalouden aiheuttaman kiintoainekuormituksen vähentäminen, National Board of Waters and the Environment — Report 476, 1993, 1.

CHAPTER **22**

The Dynamics of Peat Accumulation by Mires of Northern Eurasia During the Last 3000 Years

Vladimir A. Klimanov and Andrej A. Sirin

CONTENTS

INTRODUCTION

A major part of the biosphere's carbon can be found in mires; estimates range up to 12 percent. Mires play a key role in the temperate belt of the Northern Hemisphere, which contains about 95 percent of the world's surveyed peat reserves. Mires are the only subaerial ecosystems that cause an overall, continuing net sequestering of atmospheric carbon into peat (Vompersky, 1994). The rate of peat accumulation varies somewhat, but it is the basis for estimating the deposition of carbon in mires. Different peats have a very similar proportion of carbon (about

1-56670-177-5/97/$0.00+$.50
© 1997 by CRC Press, Inc.

50 percent of dry mass), and their dry bulk density does not differ greatly (Vompersky, 1994).

Previous peat accumulation rates can be correlated with past climatic conditions. Such a study can help estimate current peat accumulation rates more accurately and suggest trends for the future. It can allow us to explore how natural and anthropogenic climate change may affect peat accumulation rates. For these reasons we analyzed the dynamics of peat accumulation, mainly during the last 3000 years, using a selection of mires previously studied in Russia and adjacent countries.

METHODS

The correlation of modern pollen spectra with present vegetation and present climate allowed us to use pollen spectra to calculate the average yearly temperature and the average July temperature to an accuracy of ±0.6°C, the average January temperature to ±1.0°C, and the average yearly total precipitation to ±25 mm (Klimanov, 1976, 1984). This method of reconstruction has been widely used (Velichko et al., 1991).

Periods of rapidly changing temperature regime in northern Eurasia were nearly synchronous, within the limits of errors in the absolute stratigraphy (Klimanov, 1994a). Synchronism is revealed in both longitudinal (Figure 1) and latitudinal (Figure 2) directions. The paleotemperature curves have different degrees of detail that are associated with variable detail in spore-pollen diagrams

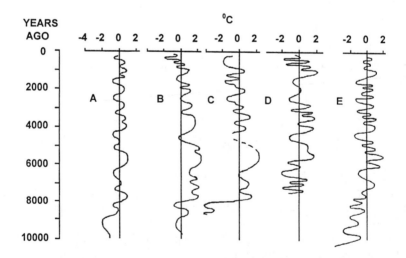

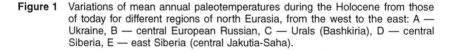

Figure 1   Variations of mean annual paleotemperatures during the Holocene from those of today for different regions of north Eurasia, from the west to the east: A — Ukraine, B — central European Russian, C — Urals (Bashkiria), D — central Siberia, E — east Siberia (central Jakutia-Saha).

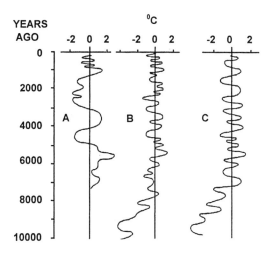

Figure 2   Variations of mean annual paleotemperatures as above for different regions of north Eurasia, from the north to the south: A — north Urals, B — central West Siberia, C — Kazakhstan.

over the range of hundreds to a thousand years. Local changes in paleoclimatic indices are characterized by much finer detail. Some extreme values of cooling and warming can last for only decades. This makes it possible to construct detailed summary temporal paleotemperature curves for each region. Individual C^{14} dates for a region are usually scattered among many profiles, with few dates in any one profile. Detailed paleotemperature curves for such profiles allow cross-correlation of climate and dates from different profiles.

In the present work we used mainly the reconstructed average yearly temperature. Paleotemperature curves show the peaks and troughs in temperature, the rates of warming and cooling, and the changing from a cold to a warm period (or vice versa). Both the change of sign in the curve and the peak values are usually coincident in a particular region. The paleoclimatic curves have been dated by some combination of C^{14}, dendroclimatology, the results of lacustrine sediment studies, glaciological studies, historical information for the last two millennia, and instrumental meteorology for the most recent two or three centuries.

We used the general paleoclimatic curves (changes and peaks) for particular peatlands, tying them to the particular profile by its own C^{14} dates and paleotemperature curve. The dated paleoclimatic curves were used to define zones with different rates of peat accumulation in their local stratigraphy. A time resolution of 50 years was used, as that is the dating precision of the fastest cooling and warming episodes. Crude peat accumulation rates (mm/year) for these zones were calculated without any allowance for differing extents of continuing decay. The true rate of accumulation for older peats will be more underestimated than that of recent peat.

For each of the climatically defined zones, both the climate and the crude accumulation rate are known. Twenty-eight such peat profiles associated with pollen-climate curves were used. The profiles came from peatlands situated in different regions of Russia: the Kola Peninsula, Karelia, the north and central parts of the European territory, the north and south Ural mountains, the north, central, and south parts of west Siberia, central Siberia, and the north and south parts of east Siberia. Profiles from peatlands in Belarus, Ukraine, Estonia, and Kazakhstan were also used. These sites span the vast territory from 50 to 71° North and from 24 to 136° East.

The peatlands that have been studied occur in various natural regions: in tundra, taiga, broad-leaved forests, and steppe. Their nature is diverse: from the palsa mires of the permafrost zone to river valley fens. But raised bogs predominate because their deep peats are most often chosen for paleobotanical analysis.

Graphs of crude peat accumulation rates were made for the full thickness of each profile. Mean values for three separate climatic periods within the Holocene were calculated: Subatlantic (2600 years B.P. to present), Subboreal (4600–2650 years B.P.), and Atlantic (8000–4650 years B.P.). We used the time limits within the Holocene periods that are most frequently used in Russia (Hotinsky, 1977). The Holocene (HOL) is usually considered to be the most recent 10,300 years, but we used only the most recent 8000 thousand years here because this covers the main period of peat formation for most of the studied territory. Peat profiles often lack any peat formed in the Preboreal and Boreal periods, which makes overall comparison difficult.

RESULTS AND DISCUSSION

Pre-Holocene

Thick ice covered most of the vast territories of Russia's northern plains and adjacent countries from the beginning of the Middle Pleistocene (about 400,000 years B.P). The earliest residual peat deposits before the Holocene are from the Mikulino (Eemian) last interglacial optimum (125,000 to 110,000 years B.P.) and the Valdai (Wurm) glaciation interstages (50,000 to 44,000, 42,000 to 33,000, 30,000 to 24,000 years B.P.) (Markhov et al., 1965).

Holocene

The first centers of mire formation in Russia and adjacent countries date from the period of warming, which began 8000 to 10,000 years B.P. and reached its optimum 5000 to 6000 years B.P. Simultaneous cooling in all areas began in the late Holocene (i.e., the last 4600 years — the Subboreal and the Subatlantic). During these periods, forests receded southward, and the present tundra was formed. The southern boundary of permafrost also moved south. The late Holocene cooling slowed peat accumulation as a whole. In some regions (in southern Yakutija, for example), it nearly stopped about 3000 years B.P. as permafrost expanded.

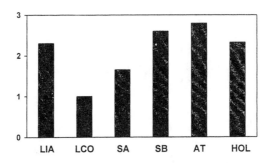

Figure 3   Relative values of mean apparent peat accumulation rate for different periods of the Holocene as compared with the Little Climatic Optimum (= 1). HOL — Holocene (8000 years B.P. to present), AT — Atlantic (8000 to 4650 years B.P.), SB — Subboreal (4600 to 2650 years B.P.), SA — Subatlantic (2650 years B.P. to present), LCO — Little Climatic Optimum (800 to 1100 A.D.), LIA — Little Ice Age (1550 to 1850 A.D.)

Currently the surfaces of permafrost peatlands (palsa mires) in these regions thaw to only a few tens of centimeters in summer, and there is no significant accumulation of new peat.

Temporal trends in peat accumulation during the three main periods of the Holocene and for the whole Holocene period are summarized in Figure 3 as the Holocene as a whole (HOL, 8000 years B.P. to present), the Atlantic (AT, 8000 to 4650 years B.P.), the Subboreal (SB, 4600 to 2650 years B.P.), and the Subatlantic (SA, 2600 B.P. to present) periods. Additionally two shorter periods within the late Subatlantic period are shown: the Little Climatic Optimum (LCO, 800 to 1100 A.D.) and the Little Ice Age (LIA, 1550 to 1850 A.D.). For this graph we did not use the mean value of peat accumulation rate for these time periods, since the rate showed much spatial variation (e.g., from very low values in Kazakhstan to high values in Estonia). For example, the mean value of peat accumulation rate among the profiles for the Subatlantic period of the Holocene differs among the profiles under consideration by 15 times (from 0.07 to 1.1 mm yr⁻¹). Thus, we showed the mean values for all periods under consideration relative to the Little Climatic Optimum, because it has the lowest accumulation rate among the intervals studied.

There is a decrease in peat accumulation rate from the Atlantic to the Subboreal and then to the Subatlantic period. In addition to this general trend, each of these periods contained various phases of warming and cooling, which influenced peat accumulation in a complicated way. Also, we did not take into consideration the pattern of precipitation, which changed more intricately and ambiguously.

Subatlantic Period

The average peat accumulation rate for the Subatlantic period (2600 years B.P. to present) in all the peatlands studied are shown in Figure 4. Based on the results shown above, we decided to consider the last 3000 years in finer detail,

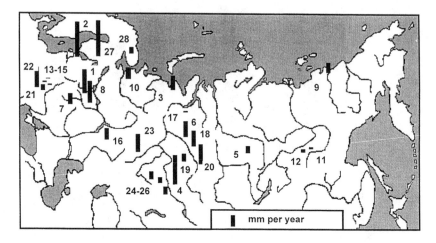

Figure 4   Mean annual rate of apparent peat accumulation during the Subatlantic period of the Holocene (2600 years B.P to present) over Russia and adjacent countries. (Numbers are the number of profiles averaged; bar heights are proportional to rate of peat accumulation.)

which primarily encompasses the Subatlantic period of the Holocene. Twenty-eight profiles are not sufficient to be sure of fine details in the geographic pattern of accumulation rate over such a vast area, but the zones of highest peat accumulation rate are quite clear. These are the regions of raised bog peatlands, which are in the northwest and central parts of Russia, in Estonia, and in West Siberia. The average rate of peat accumulation occurs both north and south of these central zones. Lower values are also found in Fenno-Scandinavian countries: in Finland 0.4 to 0.6 mm yr[-1] (El-Daoushy et al., 1982; Ruuhiijarvi, 1983), and in southern Sweden 0.53 mm yr[-1] (Franzen, 1992).

The temporal patterns in peat accumulation rates are interesting, since they appear independent of their geographic location and peatland type (Figures 5 and 6). These patterns are also independent of the average rate of peat accumulation during the Subatlantic, which ranged from 0.07 to 1.1 mm yr[-1] in the profiles we studied.

There were several peaks of warming and troughs of cooling lasting a century or longer during the Subatlantic period (Figures 1 and 2). The main peaks of warming were about 2300, 1800, 1600, 1300, 1000, 600, 300, and 150 years ago, and at the present time. The troughs of cooling were about 2500, 2200, 1700, 1500, 1200, 700, 500, 250, and 100 years ago. As a whole, longer cold phases occurred before 300 B.C., from 100 to 500 A.D., and from 1100 A.D. to the middle of the last century. Longer warm phases were from 300 B.C. to 100 B.C., from 500 to 100 A.D., and from the middle or end of the last century until the present time.

Peaks and troughs in average yearly total precipitation were asynchronous. In general, the amount of precipitation increased in the north during warming and decreased during cooling. In the middle latitudes, the peaks of precipitation variation were more often between the extreme values of warming and cooling;

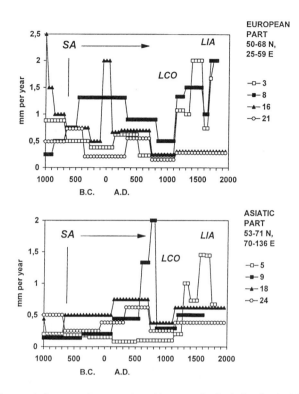

Figure 5   Temporal changes of apparent peat increment rate during the last 3000 years for mires of different regions of Russia and adjacent countries. (Numbers are the number of profiles studied.) SA — Subatlantic (2600 B.P. to present), LCO — Little Climatic Optimum (800 to 1100 A.D.), LIA — Little Ice Age (1550 to 1850 A.D.)

in southern regions precipitation usually decreased during warming and increased during cooling.

When we look at variation in the peat accumulation rate (Figures 5 and 6), the increase in rate can be correlated with cool thermal phases and the decrease with warm thermal phases. This does not mean, however, that we can infer a direct causal dependence between the intensity of peat formation and thermal conditions. The processes of peat formation are very complicated and depend upon many factors (Moore, 1991).

Moore's (1991) findings about the periods of intensive growth in northern Europe's bogs and other information on the temporal course of peat accumulation rate (Figure 7) do not contradict our findings reported here. Our findings also agree with highly detailed C[14] datings profiles of Finnish peatlands (Tolonen, 1987), which differ from the pattern of temporal change in accumulation rate recorded for Sweden by Franzen (1992).

The rate of peat accumulation in the warm phases of the last one-third of the Subatlantic period changed sharply. Its rate decreased precipitously at the end of the first millennium A.D. (Figures 5 and 6), while its rate began to increase during 1200 to 1400 A.D., especially during the most recent 500 years (Neishtadt, 1985).

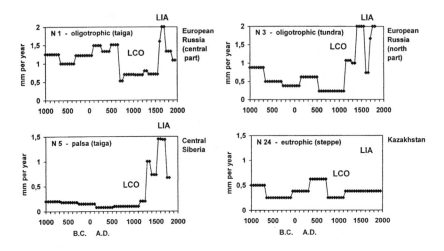

Figure 6   Temporal changes of apparent peat increment rate during the last 3000 years for mires of different nature of Russia and adjacent countries. (Numbers are the number of profiles studied.) LCO — Little Climatic Optimum (800 to 1100 A.D.), LIA — Little Ice Age (1550 to 1850 A.D.)

The high values of peat accumulation in the upper layers of peatlands cannot be explained only by their being as yet uncompacted. For the most part, they are out of the zone of aeration (the acrotelm), and the period of rapid decomposition is already over. Subsequent peat compaction under pressure in the permanently saturated zone (the catotelm) is not great.

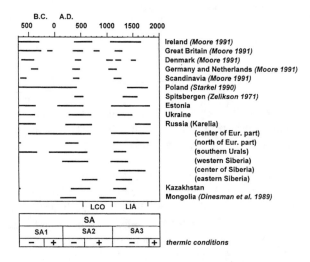

Figure 7   Periods of intensive peat accumulation in different regions of Eurasia. SA — Subatlantic (2600 B.P. to present) and its subperiods (SA1, SA2, SA3), LCO — Little Climatic Optimum (800 to 1100 A.D.), LIA — Little Ice Age (1550 to 1850 A.D.)

The period from 800 to 1100 A.D. was the period of the Little Climatic Optimum of the Middle Ages, which is clearly shown in both historical and paleoclimatic records. Although its thermal conditions did not reach the heights of maximum warming in the Subboreal period, it was a noticeable event in the climate regime of the Holocene (Klimanov, 1994a). At that time precipitation was high only in northern regions. When that warming ended about 800 years ago, the Little Ice Age began (Imbrie and Imbrie, 1986; Anonymous, 1988). The maximum cooling during that period occurred at different times in the various regions of our study area, but most occurred 200 to 250 years ago. There was a significant cooling in all regions in the last century as well, particularly in the north. The mean temperatures were lower by as much as 2°C during the Little Ice Age. This is equal to the displacement of St. Petersburg 500 km north to the latitude of Arkhangelck (the White Sea coast). As a whole, precipitation was lower in northern regions and slightly higher in the south.

Warming in all areas of our study began about the middle of the nineteenth century. It approached but did not reach the heights of the Little Climatic Optimum. In any case, the thermal conditions of the twentieth century are substantially the same as those in the time of the Vikings and differ from the seventeenth to nineteenth centuries. In our opinion, this last fact casts serious doubt on the direct use of upper C^{14} datings (usually about 400 to 300 years B.P.) for the estimation of present peat accumulation rates, because the nearest analog is the Little Climatic Optimum. We can, however, use the correlations shown in Figure 8 for some estimates. The data on peat accumulation rate during the Subatlantic as a whole are more numerous than for the Little Climatic Optimum.

Climate fluctuations during the Holocene show cyclic recurrence, but their quasiperiodic character makes forecasting difficult (Klimanov, 1994b). The long-term prediction of the natural fluctuations of climate is uncertain. Nevertheless, warming during this century may well be part of a natural cyclic change, and not only a response to anthropogenic CO_2 increases. This warming trend is within the range of previous historical examples of such changes. If it is a natural occurrence, cooling could follow in the twenty-first century.

According to the relationship between peat accumulation and thermal conditions obtained in our study, increased rates of peat accumulation will probably occur in Russia and adjacent countries. If the warming trend continues, then the

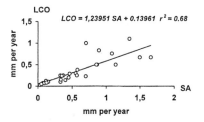

Figure 8 Relationship between apparent peat accumulation rate during Little Climatic Optimum and during the Subatlantic periods. SA — Subatlantic (2600 B.P. to present), LCO — Little Climatic Optimum (800 to 1100 A.D.).

present period of relatively low peat accumulation rates will also continue. Anthropogenic greenhouse effects would supplement this. The decrease of peat accumulation could enhance the greenhouse effect by increasing the microbial release of carbon in peatlands into atmospheric carbon dioxide and methane.

CONCLUSIONS

The variability in the rate of peat accumulation for the last 3000 years was studied in 28 profiles of peatlands from Russia and adjacent countries. Paleoclimatic curves were used to obtain the values of layer-by-layer peat accumulation rates. We believe these curves give the most substantiated and reliable time scale for stratigraphic separation of peatlands. The dates of warming and cooling peaks are well established by various types of geochronology. The peat layers studied are defined by natural climatic intervals.

Connections have been found between peat accumulation rate and thermal conditions during recent millennia: the rate increases in cold thermal phases and decreases in warm ones. These relationships are particularly striking in the Little Climatic Optimum (800 to 1100 A.D.) and the Little Ice Age (1550 to 1850 A.D.) periods. During these periods, the rate difference among the profiles studied averaged 350 percent.

Variations in the rate of accumulation are independent of the geographical location of a peatland, its type, or the average rate for the period under consideration. The mean peat accumulation rate for the Subatlantic period of the Holocene for the profiles studied differed by 15 times (from 0.07 to 1.1 mm/year).

The thermal conditions of the twentieth century are closer to those of the Little Climatic Optimum than the thermal conditions in the Little Ice Age. We believe that the use of radiocarbon dating for peat formed during the last 400 years is questionable for estimating present peat accumulation rate. A better estimate is obtained by using the results for the Little Climatic Optimum. It is possible to obtain an approximate estimation of peat accumulation using accumulation rates for the Little Climatic Optimum and for the Subatlantic period of the Holocene as a whole.

Warming during this century is within the range of previous warmings, and may well be part of a natural cyclic change. If this warming is a natural occurrence, cooling will follow in the twenty-first century. This will probably result in increased rates of peat accumulation in our study region. If warming lasts longer, then the present period of relatively low peat accumulation rates will also continue. Anthropogenic greenhouse effects would also favor low peat accumulation rates.

ACKNOWLEDGMENTS

The authors would like to acknowledge the contribution of Elon Sandy Verry (USDA Forest Service) and Richard Clymo (University of London School of

Biological Sciences) who helped a lot to revise the manuscript and make it more understandable to English readers.

REFERENCES

Anonymous, *Climate Variations during the Last 1000 Years,* Gidrometeoizdat, Leningrad, 1988, 408 pp. (in Russian).

Dinesman, L. G., Kiselyova, N. K., and Knyazev, A. V., *History of Steppe Ecosystems of Mongolian Peoples Republic*, Nauka Publ., Moscow, 1989, 215 pp. (in Russian).

El-Daoushy, F., Tolonen, E., and Rosenberg, R., Lead 210 and moss-increment dating of two Finnish sphagnum hummocks, *Nature,* 296, 429, 1982.

Franzen, L. G., Can Earth afford to lose the wetlands in the battle against the increasing greenhouse effect?, in Proc. 9th International Peat Congress, Vol. 1, 1992, p. 1.

Hotinsky, N. A., *Holocene of the North Eurasia*, Nauka Publ., Moscow, 1977, 198 pp. (in Russian).

Imbrie, J. and Imbrie, K. P., *Ice Age: Solving the Mystery*, Harvard University Press, Cambridge, MA, 1986, 224 pp.

Klimanov. V. A., A method of reconstruction of quantitative characteristics of past climate, *Moscow State University, Seria geograficheskaya, Vestnik,* 2, 1976, 92 pp. (in Russian).

Klimanov, V. A., Paleoclimatic reconstructions based on the information statistical method, in *Late Quaternary Environments of the Soviet Union*, Velichko, A. A., Ed., translated from Russian by Wright, H. E., Jr. and Barnosky, C. W., Eds., University of Minnesota Press, Minneapolis, MN, 1984, 297 pp.

Klimanov, V. A., Climate of the Little Climatic Optimum on the territory of Northern Eurasia, *Rep. Russ. Acad. Sci.,* 335 (2), 232, 1994a (in Russian).

Klimanov, V. A., Climate change pattern in North Eurasia in late Glacial and Holocene time, Bulliten Mockovskogo obthestva ispitatelej prirodi, Otdelenie geologii, 69 (1), 58, 1994b (in Russian).

Markhov, K. K., Lazukhov, G. I., and Nikolaev, V. A., *Quaternary Period*, Vol. 1, Moscow State Univ. Publ., Moscow, 1965, 372 pp. (in Russian).

Moore, P. D., Holocene paludification and hydrological changes as climate proxy data in Europe, in *Evaluation of Climate Proxy Data in Relation to the European Holocene*, Frenzel, B., Ed., Gustav Fisher Verlag, Stuttgart-Jena, 1991, p. 255.

Neishtadt, M. I., Mire formation processes in the Holocene, *Izvestija Acad. of Sci. of the USSR, Seria geograficheskaya*, 39 (1), 1985 (in Russian).

Ruuhijarvi, R., Finnish mire types and their regional distribution, in *Ecosystems of the World*, Vol. 4B, *Mires: Swamp, Bog, Fen and Moor,* Regional Studies, Gore, A. J. P., Ed., Elsevier, Amsterdam, 1983, p. 47.

Starkel, L., Ed., Evolution of the Vistula River Valley during the last 15,000 years, part III, *Geography Studies Special Issue* No. 5, Ossolineum, Wroclaw, The Publishing House of the Polish Academy of Science, 1990, 220 pp.

Tolonen, K., Natural history of raised bogs and forest vegetation in the Lammi area, southern Finland studied by stratigraphical methods, *Ann. Acad. Sci. Fenn. Ser. A*, III, 144, 46, 1987.

Velichko, A., Frenzel, B., and Pecsi, M., Eds., Atlas of Paleoclimates and Paleoenvironments of the Northern Hemisphere, late Pleistocene-Holocene, Budapest, 1991.

Vompersky, S. E., Role of mires in the cycle of carbon, in *Biogeocoenotical Peculiarities of Mires and Their Rational Use*, Nauka Publ., Moscow, 1994, p. 5 (in Russian).

Zelikson, E. M., Palynological investigation of a Holocene peat deposit from Spitsbergen, in *Palinology of Holocene,* Nauka Publ., Moscow, 1971, p. 199 (in Russian).

CHAPTER 23

The Effect of Peatland Forestry on Fluxes of Carbon Dioxide, Methane, and Nitrous Oxide

Hannu Nykänen, Jouko Silvola, Jukka Alm, and Pertti J. Martikainen

CONTENTS

INTRODUCTION

Mires are wet environments where a portion of the carbon and nutrients fixed in plant biomass accumulates as peat. Boreal and subarctic peatlands contain 455 Pg of sequestered carbon (Gorham, 1991), which is 30 percent of the total organic

1-56670-177-5/97/$0.00+$.50

carbon in the world's soil reservoir. Carbon dioxide (CO_2), methane (CH_4), and nitrous oxide (N_2O) are atmospheric gases that absorb electromagnetic energy from the solar radiation reflected from the earth's surface. An increase in the concentration of these gases changes global climate by warming the atmosphere (IPCC, 1990).

Virgin peatlands can be divided into two groups according to their present status. The origin of all mires is minerogenous. A mire may change to ombrotrophic when the *Sphagnum* peat thickens so that groundwater flow cannot feed the vegetation. As a result of nutrient supply, virgin minerotrophic mires (fens) are more fertile than virgin ombrotrophic mires (bogs). Groundwater level is an important control of the aerobic/anaerobic conditions in peat, and hence, the microbiological processes producing and consuming CO_2, CH_4, and N_2O.

Anthropogenic emissions are the most important sources for CO_2, whereas vegetation acts as a dynamic store of carbon. Because soil and vegetation processes are important in the fluxes of greenhouse gases, all changes in land use may affect their emissions. In Finland peatlands are used for forestry and agriculture, and peat is harvested for horticultural uses and fuel production. About 5.9 million ha are drained for forestry in Finland (Aarne, 1992), which is more than half of the original peatland area. Most studies of the effect of this land use change on greenhouse gas fluxes have dealt with changes in CO_2 flux (Silvola et al., 1985), but we also consider the dynamics of CH_4 and N_2O.

A FINNISH CASE STUDY

Study Sites

Study sites include the Lakkasuo mire complex located at central Finland (61°47′N and 24°18′E); and sites at Ilomantsi (62°46′N and 30°58′E) and Rääkkylä (62°14′N and 29°50′E) in eastern Finland (Figure 1). Mean annual temperature is 3°C, and annual precipitation is 650 mm, of which 240 mm is snow. Results from the Lakkasuo mire complex, with both drained and virgin fen and bog subsites and from a fen drained and afforested 50 years ago in Ilomantsi are presented here in more detail. In the Lakkasuo mire complex, a gradient of mire site types of different trophy occurred along a border ditch so that each virgin subsite had a corresponding subsite drained 30 years before (Laine et al., 1986).

Measurements

For the CO_2 measurements, a portable aluminum chamber (diameter 20 cm, height 5 cm) with foam plastic seals was put on the soil surface in fixed locations that were kept vegetationless by regular cutting. No collar was used for the chamber in peat. Plots (6 to 12) were prepared at each site or subsite, and the sites were visited once or twice a month. Air was pumped into and out of the chamber at a constant flow rate (76 L h⁻¹) and the difference in CO_2 concentration

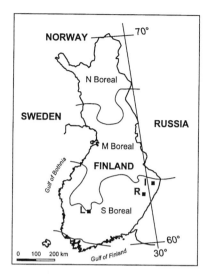

Figure 1   Finnish study site locations: Lakkasuo (L), Ilomantsi area (I), and Rääkkylä (R). Borders of southern (S), middle (M), and northern (N) boreal zones indicated.

between these input and output lines was measured with a portable infrared gas analyzer. During the measurements, the temperature of the peat profile (at 2, 5, and 10 cm) and the depth of water table were measured (Silvola and Alm, 1992).

CH$_4$ and N$_2$O fluxes were measured once or twice a month at both the Lakkasuo mire complex and the Ilomantsi sites. These measurements were conducted within about 5 m of the CO$_2$ measuring locations. A closed chamber method was used for the measurements of CH$_4$ and N$_2$O fluxes (Crill, 1991; Martikainen et al., 1995). The chamber covered an area of 0.36 m^2, and had a volume of 60 L. Chambers were located 20 to 30 m from the border ditch. In virgin areas of Lakkasuo, this distance was longer to eliminate the effect of ditches on the water table and gas emissions. The study sites were equipped with boardwalks to prevent peat disturbance during sampling. Chamber temperature, temperature in peat profiles (0 to 30 cm), and depth of water table near the chambers were recorded. Four gas samples of 50 mL were collected with plastic syringes during the 30-min measuring period, and gas concentrations were analyzed in the laboratory within 24 h. Gases were loaded to two gas chromatographs running in parallel by a 10-port valve having two 0.5 mL loops. CH$_4$ was analyzed by FID, which is in series with TCD, and N$_2$O was analyzed by ECD (Nykänen et al., 1995).

Results

Effect of Drainage on Water Table

Both at the virgin and drained sites, the water table generally followed the same climate-driven trend (changes in precipitation and evaporation), but was deeper at the drained sites (Figures 2 and 3). The study sites can be divided into

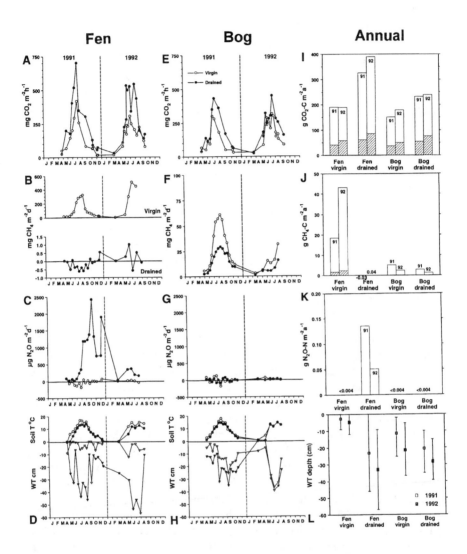

Figure 2 Fluxes of CO_2 (A, E), CH_4, and N_2O (C, G), soil temperature (-10 cm) and depth of groundwater table, WT (D, H) in the fen and bog sites, Lakkasuo mire complex. Open symbols denote virgin and solid symbols drained areas. Annual gas fluxes for the sites are given for 1991 and 1992 (I, J, K). Hatched parts of the bars denote winter fluxes for CO_2 and CH_4 (I, J). The mean WTs for the study years are shown with minimum and maximum depths. (Reprinted with permission from *Plant and Soil*, Kluwer Academic Publishers.)

two groups according to their water table: sites where the decrease in water table depth has been sufficient to enhance tree growth (Figure 2, fen; Figure 3, minerotrophic sites 1, 2, and 3) and those where tree growth was not affected (Figure 2, bog; Figure 3, ombrotrophic sites 4 and 5).

Maintenance of a low water table level at drained mires is primarily a result of ditching, but the fertility of the site is also associated with the change in water

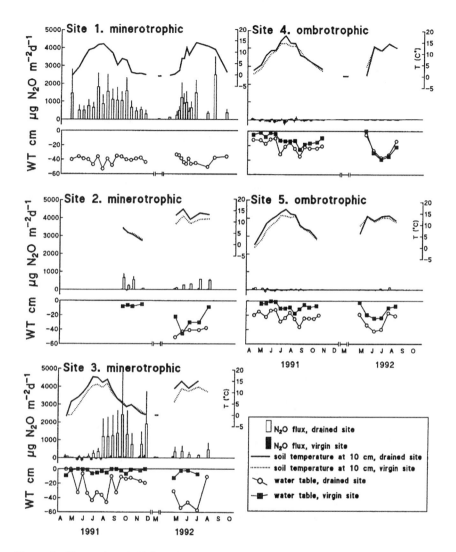

Figure 3 Fluctuations in N_2O emissions, water table, and soil temperatures in 1991 and 1992. Gas fluxes were measured by a static chamber method every second week from spring thaw to winter freeze in 1991. In 1992 fluxes were measured one to four times in a month. During winter 1992, Site 1 was measured twice, Sites 3 and 4 once. There were two to three chambers at each Site, mean fluxes and standard error for the measurements shown in the figure. (Reprinted with permission from *Nature*, Macmillan Magazines Ltd.)

table, as can be observed at Lakkasuo 30 years after drainage. There the water table is lower on fen sites than on bog sites, even when the original water table was lower on the bogs (Figure 2, D and H). Spacing and depth of the ditches were the same in all transects.

During the summer drought in 1992, the water table was lower at all ombrotrophic sites, but the reduced water supply was not sufficient to lower the water table at the minerotrophic fen site at Lakkasuo (Figure 2). In Ilomantsi (Martikainen et al., 1994), the water table also fell at the virgin minerotrophic site.

CO_2 Fluxes

Seasonal changes in soil respiration depended both on soil temperature and on the groundwater level. Flux rates also reflected the site productivity, so that the emissions were lowest at the poorest ombrotrophic site and increased with site productivity (Silvola et al., 1994). Even a temporary water table drawdown increased the flux rate. Efficient drainage clearly increased CO_2 emissions compared to those from virgin subsites of the same mire type. The predominant groundwater level, and thereby the drainage intensity, seemed to be related to CO_2 flux rates. Flux data were used to estimate the annual carbon release at some sites (Silvola et al., 1994). The lowest rates (50 to 150 g C m^{-2} yr^{-1}) were on poor virgin ombrotrophic sites and the highest rates (500 to 600 g C m^{-2} yr^{-1}) on effectively drained ombrotrophic dwarf shrub pine bog and minerotrophic treed fen sites.

CH_4 Fluxes

Peak CH_4 emissions from virgin fens, 600 mg m^{-2} d^{-1}, were higher than those from virgin bogs, 60 mg m^{-2} d^{-1} (Figure 2). Drainage decreased emissions at all sites. Some emissions occurred at the drained bog sites, but the annual emission of 20 to 40 g m^{-2} from a virgin fen could become net uptake after drainage (Figure 2J). At the drained bog site, the water table was 0 to 10 cm lower than at the virgin subsite (Figure 2H). The effect of drainage on CH_4 fluxes was most distinct in midsummer when the water table was lowest (Figure 2F). Annual emissions after drainage declined to half (2 g m^{-2}) of the original (5 g m^{-2}). The lowering of the water table induced by the exceptionally dry summer in 1992 reduced CH_4 emissions at the virgin bog site to the level found at the drained site in 1991 (Figure 2F), but at the fen site there was no clear decrease in water table and CH_4 emissions were higher than in the previous year (Figure 2B), possibly due to warming of the soil profile and earlier phenology of the vegetation.

N_2O Fluxes

The emissions of N_2O from virgin minerotrophic and ombrotrophic peatlands were low. Ditching induced emissions from minerotrophic sites (Figure 3). When the water table remained at 40 cm (Figure 3, site 1), annual emissions were stable, 0.14 g m^{-2} in 1991 and 0.13 g m^{-2} in 1992. At Lakkasuo (Figure 3, site 3), annual N_2O emissions varied from 0.14 g m^{-2} in 1991 to 0.051 g m^{-2} in 1992. Those emissions started at June and continued at high levels until winter. At Ahvensalo (Figure 3, site 1), emissions began earlier in spring and showed seasonal dynamics.

GENERAL DISCUSSION

In Scandinavia, forestry on natural peatlands is only possible in nutrient-rich, well-aerated, thin-peat spruce mires. On sites with a thick peat layer, the trees are sparse or nonexistent and the water table must be lowered by ditching to improve forest growth. The present area drained for forestry in Finland is greater than half of the original peatland area. By the end of 1990, total forest drainage was 5.9 million ha, with 1.5 million km of ditches (Aarne, 1992), or 17.6 percent of the area of the country. This change in land use has occurred during this century and is continuing to some extent. In previously drained areas, some ditch cleaning is also done. The adverse environmental effects of drainage on stream flow (Verry, 1988), quality of runoff water, including increases in carbon loss (Sallantaus, 1994), and in methylmercury load (Westling, 1991), and possible leaching of certain pathogenic microbes (Iivanainen et al., 1993) have been documented. However, studies on the impact of drainage on fluxes of greenhouse gases other than CO_2 (Silvola et al., 1985) are rare (Glenn et al., 1993; Roulet, 1994).

Mire Types and Draining Practice

Drained minerotrophic mires provide more fertile initial conditions for forest growth than do ombrotrophic mires. The developing tree stand increases evapotranspiration, and the resulting decrease in water table enhances aeration and mineralization of peat. On drained forested minerotrophic sites, the water table is lowered more than on drained ombrotrophic sites because of the higher evapotranspiration associated with better tree growth in the minerotrophic mires (Päivänen, 1973; Laine and Vanha-Majamaa, 1992). An exception to this generality are ombrotrophic dwarf-shrub pine bogs, where the original water table is low and decreases still after drainage (Silvola et al., 1994).

Tree roots can only grow in the upper 20 cm layer of drained peats (Paavilainen, 1967). Probable reasons for this are the physical peat compaction and increased bulk density below the rooting zone after drainage. Water level drawdown can stop the nutrient flow from groundwater to the rooting zone and change the site to ombrogenous. With time after drainage, the nutrient status may actually decrease in the rooting zone due to exploitation by growing tree roots. At the same time, however, peat compaction increases the volumetric nutrient content (Laiho and Laine, 1994). Poor tree growth can be expected on nutrient-poor sites. Efficient drainage may ultimately turn a former mire site to a heath forest on peat soil with typical upland ground vegetation (Reinikainen, 1994).

Mire types, as defined according to the Finnish peatland classification system (Ruuhijärvi, 1983), characterize the site nutrient and moisture regimes and can be used to predict forest development after drainage. The site class of a virgin peat site for forestry depends on both its climate and fertility (Heikurainen, 1973). Similarly, the fertility and moisture regime of peatland ecosystems also correlates with greenhouse gas fluxes.

Recommended drainage measures include trap ditches in mire margins to cut waterways leading to the mire, and ditches at spacing of 30 to 50 m, providing an average depth to water table from 30 cm to 60 cm during the growing period. Both ombrotrophic and minerotrophic peat sites have been drained for forestry in Finland, and at least 27 percent of such drainage has been estimated to be unprofitable (Eurola and Huttunen, 1990). These peat sites are not producing enough wood because of their high water table, poor nutrient status, or unsuitable climate.

The ability of the tree stand to maintain a low water table in ditched forested mires is demonstrated by the increase in water table level after tree harvesting (Heikurainen and Päivänen, 1970). The increase after harvesting is comparable to the changes occurring in harvested natural treed peatlands. In continuous peatland forestry, harvested sites have to be prepared after the first rotation with methods similar to those used in the drainage of virgin mires.

Greenhouse Gases from Peatlands

In their natural state, peatlands are waterlogged environments sequestering carbon as slowly decomposing organic matter, and are thus counteracting the increase in atmospheric CO_2 concentrations. CH_4 release from mires counteracts the decrease in radiative forcing caused by carbon accumulation in peat. Northern wetlands annually store 50 to 100 Tg carbon (Armentano and Menges, 1986) and emit 110 Tg CH_4 (Fung et al., 1991). CH_4 produced in anaerobic waterlogged peat is released to the atmosphere through diffusion, through plants, and by bubbling. CH_4 emissions from natural boreal mires contribute 19 to 27 percent of the total CH_4 emissions (Crill et al., 1993), and CH_4 contributes 15 percent to climatic warming (Rodhe, 1990). About half of virgin peat soil is carbon; 45 to 50 percent in poorly decomposed peat and 50 to 60 percent in well-humified peats (Karisto and Kaunisto, 1992; Hokkanen and Silvola, 1993). Only 2 to 20 percent of the annual primary production (Clymo, 1984) is stored as peat, and the remainder is released as CO_2, CH_4, and humic substances in water outflow.

Drainage increases the aerated peat volume, enhancing oxidative decomposition and the mineralization of peat. Biomass of ground vegetation decreases and that of trees increases after drainage. The changes in biomass production depend both on the site nutrient status (Keltikangas et al., 1986) and on drainage efficiency (Eurola and Huttunen, 1990). According to some preliminary results (Laine et al., 1994), there is a decrease in carbon storage in peat following drainage of nutrient-rich sites, but an increase due to lower peat decomposition rate and relatively greater fine root production of nutrient-poor sites (Finér and Laine, 1994).

The summertime CO_2 emissions from peat from this study (200 to 500 mg CO_2 m^{-2} yr^{-1}) are consistent with other results from northern virgin peatlands (Silvola and Heikkinen, 1979; Moore, 1989; Kim and Verma, 1992; Silvola et al., 1992). Estimates of annual CO_2 efflux from peat vary from 50 to 300 g C m^{-2} yr^{-1} (this study; Silvola et al., 1985; Moore, 1986, 1989), average carbon accumulation in peat is 15 to 30 g C m^{-2} yr^{-1} (Tolonen et al., 1992), and 5 to 10 g C m^{-2} yr^{-1} may be lost by leaching (Sallantaus, 1992). Both laboratory experiments and field studies have shown that lowering of the water table by 20 to 30 cm increases CO_2 flux from peat soil by 1.5- to 2.5-fold (Silvola et al., 1985; Moore and Knowles, 1989). Further, it seems that the changes in water table are more significant between 0 to 30 cm than between 30 to 60 cm (Silvola et al., 1985). The temperature-standardized CO_2 flux, converted to 12°C, indicates that the increase in soil respiration rate ceases when the water table reaches a depth that is characteristic of a given site type (Silvola et al., 1994). If the drainage is not efficient, the corresponding increase in CO_2 release is also small. On the average, lowering of the water table yields an additional carbon release of about 100 to 300 g C m^{-2} yr^{-1}. In addition to the increased oxidation of old peat carbon stores, especially in forested peatlands, a significant proportion of the increased respiration may be caused by increased allocation of carbon belowground (root respiration and root-derived microbial respiration; Silvola et al., 1992). This is also suggested by the fact that during the first decades after drainage, carbon loss from peat can be rather small (Laine et al., 1994) although examples of higher carbon losses have been published (Braekke, 1987). If the oxidation of upper peat layers gradually brings aerobic conditions to deeper peat layers, oxidation of these old peat layers will be slower due to more decay-resistant organic compounds (Hogg et al., 1992). A maximum estimate for oxidation of drying peat could be that from vegetationless peat harvesting fields, where the respiration rate at the average summertime temperature has been 150 to 200 mg CO_2 m^{-2} h^{-1} and the annual carbon loss ca. 250 g C m^{-2} yr^{-1} (Ahlholm and Silvola, 1990).

The most important factor moderating the emissions of CH_4 to the atmosphere is the depth of the water table (Roulet et al., 1992a,b; Moore and Roulet, 1993; Bubier et al., 1993; Dise et al., 1993). Water saturation of peat pore space restricts oxygen penetration into the soil, resulting in the formation of anaerobic conditions essential for methanogenic bacteria. In the aerated upper layer of the peat profile, CH_4 diffusing upward is partly or totally oxidized by methanotrophs. Therefore, the net flux of CH_4 from the peat surface depends not only on the amount of the CH_4 produced but also on its oxidation rate in the peat profile (Sundh et al., 1994). CH_4 oxidation in the upper aerobic soil profile may consume more CH_4 than is produced in the saturated peat column. This oxidation capacity is inversely proportional to nitrogen load (Crill et al., 1994).

Our experiments have shown that annual CH_4 emissions from drained and forested peatlands are 0 to 50 percent of emissions before drainage. After drainage, initially nutrient-rich fens with good tree growth may become sinks for CH_4. A drained fen soil can consume up to 3.3 mg CH_4 m^{-2} d^{-1} (Crill et al., 1994). In our drained bogs, the decrease in CH_4 emissions was clear, but CH_4 uptake never occurred.

The effect of evapotranspiration is one probable reason why drainage reduces CH_4 emissions more strongly from fens than from bogs (Martikainen et al., 1993, 1995). Because of a denser tree stand on more fertile sites, evapotranspiration is greater (Laine and Vanha-Majamaa, 1993). Several processes decrease CH_4 emissions after draining. First, there is active CH_4 oxidation in the uppermost aerobic peat profile (Crill et al., 1994; Lien et al., 1993; Sundh et al., 1994); the most active layer is above the mean water table (Sundh et al., 1994). Second, CH_4 production in the uppermost peat profile is reduced because of the presence of oxygen; the most active CH_4 producing layer in a natural peat column is a few centimeters below the mean water table (Sundh et al., 1994). Third, it has been shown that CH_4 emissions from wetlands are controlled by primary production (Whiting and Chanton, 1993). It is probable that a part of the fresh organic matter from photosynthesis (e.g., litter, root exudates) is oxidized in the aerobic soil surface of drained peatland and thus never reaches the deeper anaerobic methanogenic zones. Finally, nitrification and sulfide oxidation can generate high NO_3^- and SO_4^{2-} concentrations in the soil, leading to inhibition of methanogenesis (Amaral and Knowles, 1994).

N_2O is important as a greenhouse gas and also participates in reactions depleting stratospheric ozone (Cicerone, 1987), contributing about 5 percent to the greenhouse effect (Rodhe, 1990). In addition to carbon, nitrogen bound by vegetation is also stored as peat. This organic nitrogen is stabilized in peat by anoxia, but may be subject to microbial transformations when soil becomes aerated.

N_2O is produced mainly in nitrification and denitrification. Decay of organic matter generates ammonium. In nitrification, ammonium is first oxidized to nitrite by ammonium-oxidizing bacteria, and nitrite is further oxidized to nitrate by nitrite-oxidizing bacteria. Ammonium oxidizers produce N_2O even in well-aerated conditions, but absence of oxygen favors N_2O production (Goodroad and Keeney, 1984). In an environment lacking oxygen, nitrate and nitrite are reduced to N_2O by denitrifying bacteria. Because denitrification requires the presence of nitrate and/or nitrite, it is closely associated with nitrification. Addition of nitrate to soil directly stimulates N_2O production. N_2O is an intermediate product in denitrification, and can further be reduced to N_2. In acidic and low temperature environments, the reduction of N_2O to N_2 is prevented (Ivarson, 1977). In these environments N_2O, not the radiatively inert N_2, can be the main product of denitrification.

Emissions of N_2O from virgin study sites were low (Martikainen et al., 1993). Decreases in the water table depth induced by draining of nutrient-poor bog sites did not enhance their N_2O emissions. *In situ* chamber measurements or laboratory experiments with intact peat cores have measured emissions from virgin mires below 0.025 grams N_2O–N per square meter per year (Urban et al., 1988; Freeman et al., 1993; Martikainen et al., 1993). In our fertile drained sites, annual emissions varied from 0.05 to 0.14 grams N_2O–N m^{-2} yr^{-1} (Martikainen et al., 1993). A reason for the low emission of N_2O from virgin mires may be their low nitrification activity (Ivarson, 1977; Rosswall and Granhall, 1980; Rangeley and Knowles, 1988; Williams and Wheatley, 1988; Koerselman et al., 1989), and this may also be true for drained bogs. Drained fen peat profiles, however, have the capacity for nitrification (Lång et al., 1994), providing nitrate for denitrification. Ditching and drying of boreal fens may increase their N_2O emissions. Annual emission rates similar to those from drained fens (0.08 g N_2O–N m^{-2} yr^{-1}) have been reported for a Welsh peatland after lowering of the water table in laboratory experiments (Freeman et al., 1993). N_2O emissions from drained boreal fens are lower than emissions from drained agricultural organic soils, but they are 10 to

100 times higher than the emissions from coniferous forest soils (see Martikainen et al., 1993).

The Climatic Impact of Drainage

The proportion of ditch area in Finnish drainages is 2 to 5 percent. Our unpublished results suggest that CH_4 release from ditches corresponds to that from natural mires, so emissions from ditches may increase greenhouse gas fluxes from drained areas. If the original CH_4 fluxes from virgin peatlands are low, the moderate fluxes from ditches can increase the CH_4 emissions of the ecosystem (Roulet, 1994).

The climatic impact of gases released from mires to the atmosphere after drainage can be roughly evaluated using the Global Warming Potential (GWP). The concept of GWP has been developed for policymakers as a measure of the warming effect of a greenhouse gas relative to CO_2 (IPCC, 1990). Updated GWP coefficients as equivalent CO_2 emissions are 24.5 and 320 for CH_4 and N_2O, respectively, corresponding to a period of 100 years after disturbance (IPCC, 1994). There are, however, severe problems in applying GWP-figures. A good example is the comparison of short-lived CH_4 (atmospheric lifetime 10 years) (IPCC, 1990) to carbon stores released to the atmosphere due to increased oxidation from accumulated peat deposits that are thousands of years old. After drainage, the old carbon store (<10,000 years) is gradually changed to carbon that is only temporarily bound to tree biomass (<100 years) or understory vegetation. After harvesting, sites must be redrained and nutrient amendments added in order to retain the tree stand in a state of carbon binding. During this process, the ancient carbon store decreases. GWP thus gives only an apparent measure of the greenhouse gas exchange. The amount of carbon input in drained peatlands, especially as below-ground biomass and root litter, is also difficult to estimate. The use of the GWP concept with current data can only assess changes in CH_4 and N_2O fluxes from drained peatlands. At drained fens, decreases in CH_4 fluxes are partly compensated by increases in N_2O fluxes. It is possible that during a relatively short period (100 years) in the efficiently drained afforested fen, the outputs of greenhouse gases are balanced by the growth of the tree stand. In a longer perspective, taking into account the harvesting cycles and wood product life cycles, the cumulative carbon losses from oxidizing peat deposits probably become larger than the cumulative inputs, leading to a climatic warming effect from drained nutrient-rich fens.

REFERENCES

Aarne, M., Ed., Yearbook of forest statistics 1990–91, Maa-ja metsätalous 3, *Folia For.,* 790, 88, 1992.

Ahlholm, U. and Silvola, J., CO_2 release from peat-harvested peatlands and stockpiles, Peat 90, 2: 1–12. International Conference on Peat Production and Use, June 11–15 1990, Jyväskylä, Finland, The Association of Finnish Peat Industries, 1990.

Amaral, J. A. and Knowles, R., Methane metabolism in a temperate swamp, *Appl. Environ. Microbiol.,* 60, 3945, 1994.

Armentano, T. V. and Menges, E. S., Patterns of change in the carbon balance of organic soil-wetlands of the temperate zone, *J. Ecol.,* 74, 755, 1986.

Braekke, F. H., Nutrient relationships in forest stands: effects of drainage and fertilization on surface peat layers, *For. Ecol., Manage.,* 21, 269, 1987.

Bubier, J., Moore, T. R., and Roulet, N. T., Methane emissions from wetlands in the mid-boreal region of Northern Ontario, Canada, *Ecology,* 74, 2240, 1993.

Cicerone, R. J., Changes in stratospheric ozone, *Science,* 237, 35, 1987.

Clymo, R. S., The limits to peat bog growth, *Phil. Trans. R. Soc. Lond.,* B303, 605, 1994.

Crill, P. M., Seasonal patterns of methane uptake and carbon dioxide release by a temperate woodland soil, *Global Biogeochem. Cycles,* 5, 319, 1991.

Crill, P. M., Martikainen, P. J., Nykänen, H., and Silvola, J., Temperature and N fertilization effects on methane oxidation in a drained peatland soil, *Soil Biol. Biochem.,* 26, 1331, 1994.

Crill, P. M., Bartlett, K., and Roulet, N., Methane fluxes from boreal peatlands, *Suo,* 43, 173, 1992 (1993).

Dise, N. B., Gorham, E., and Verry, E. S., Environmental factors controlling methane emissions from peatlands in northern Minnesota, *J. Geophys. Res.,* 98, 10583, 1993.

Eurola, S. and Huttunen, A., Suoekosysteemin toiminnallinen ryhmitys (Summary: The functional grouping of mire ecosystems and their response to drainage), *Suo,* 41, 15, 1990.

Finér, L., Biomass and nutrient cycle in fertilized and unfertilized pine, mixed birch and pine and spruce stands on a drained mire, *Acta For. Fenn.,* 208, 1, 1989.

Finér, L. and Laine, J., Fine root production and decomposition on drained peatlands, *The Finnish Research Programme on Climate Change. Second Progress Report,* Kanninen, M. and Heikinheimo, P., Eds., Painatuskeskus Oy, Helsinki, 1994, p. 267.

Freeman, C., Lock, M. A., and Reynolds, B., Fluxes of CO_2, CH_4, and N_2O from a Welsh peatland following simulation of water table draw-down: potential feedback to climatic change, *Biogeochemistry,* 19, 51, 1993.

Fung, I., John, J., Lerner, J., Matthews, E., Prather, M., Steele, L. P., and Fraser, P. J., Three-dimensional model synthesis of the global methane cycle, *J. Geophys. Res.,* 96, 13033, 1991.

Glenn, S., Heyes, A., and Moore, T., Carbon dioxide and methane emissions from drained peatland soils, southern Quebec, *Global Biogeochem. Cycles,* 7, 247, 1993.

Goodroad, L. L. and Keeney, D. R., Nitrous oxide emissions from soil during thawing, *Can. J. Soil Sci.,* 64, 187, 1984.

Gorham, E., Northern peatlands: role in the carbon cycle and probable response to climatic warming, *Ecol. Appl.,* 1, 182, 1991.

Heikurainen, L., Soiden metsänkasvatuskelpoisuuden laskentamenetelmä, *Acta For. Fenn.,* 131, 1, 1973.

Heikurainen, L. and Päivänen, J., Effect of thinning, clearcutting, and fertilization on the hydrology of peatland drained for forestry, *Acta For. Fenn.,* 104, 1, 1970.

Hogg, E. H., Lieffers, V. J., and Wein, R. W., Potential carbon losses from peat profiles: effects of temperature, drought cycles and fire, *Ecol. Appl.,* 2(3), 298, 1992.

Hokkanen, T. J. and Silvola, J., Respiration of cultivated histosols in field and laboratory measurements and the relationships between respiration and soil properties, in *Biogeochemistry of Global Change: Radiatively Active Trace Gases,* Oremland, R. S., Ed., London, Routledge, Chapman and Hall, 1993, p. 386.

Iivanainen, E. K., Martikainen, P. J., Väänänen, P. K., and Katila, M.-L., Environmental factors affecting the occurrence of mycobacteria in brook waters, *Appl. Environ. Microbiol.,* 59, 398, 1993.

IPCC, Scientific Assessment of Climate Change, WMO/UNEP Intergovernmental Panel on Climate Change, Geneva, 1990, 347 pp.

IPCC, Radiative Forcing of Climate Change, Summary for Policy Makers, WMO/UNEP, The 1994 report of the Scientific Assessment Working Group of IPCC, 1994.

Ivarson, K. C., Changes in decomposition rate, microbial population and carbohydrate content of an acid peat bog after liming and reclamation, *Can. J. Soil Sci.,* 57, 129, 1977.

Karisto, M. and Kaunisto, S., Effect of drainage intensity and fertilization on peat carbon compounds and their mineralization, *The Finnish Research Programme on Climate Change. Progress Report,* Kanninen, M. and Anttila, P., Eds., Painatuskeskus Oy, Helsinki, 1992, p. 216.

Keltikangas, M., Laine, J., Puttonen, P., and Seppälä, K., Peatlands drained for forestry during 1930–1978: results from field surveys of drained areas (in Finnish with English summary), *Acta For. Fenn.,* 193, 1, 1986.

Kim, J. and Verma, S., Soil surface CO_2 flux in a Minnesota peatland, *Biogeochemistry,* 18, 37, 1992.

Koerselman, W., De Cluwe, H., and Kieskamp, W. M., Denitrification and dinitrogen fixation in two quaking fens in the Vechtplassen area, The Netherlands, *Biogeochemistry,* 8, 153, 1989.

Laiho, R. and Laine, J., Nitrogen and phosphorus stores in peatlands drainerd for forestry in Finland, *Scand. J. For. Res.,* 9, 25, 1994.

Laine, J., Minkkinen, K., Puhalainen, A., and Jauhiainen, S., Effect of forest drainage on the carbon balance of peatland ecosystem, *The Finnish Research Programme on Climate Change. Second Progress Report,* Kanninen, M. and Heikinheimo, P., Eds., Painatuskeskus Oy, Helsinki, 1994, p. 303.

Laine, J., Päivänen, J., Schneider, H., and Vasander, H., Site Types at Lakkasuo Mire Complex, Field Guide — Publication from the Department of Peatland Forestry, University of Helsinki, 1986.

Laine, J. and Vanha-Majamaa, I., Vegetation ecology along a trophic gradient on drained pine mires in southern Finland, *Ann. Bot. Fenn.,* 29, 213, 1992.

Lång K., Lehtonen M., and Martikainen, P. J., Nitrification potentials at different pH in peat samples from various layers of a drained mire, *Geomicrobiol. J.,* 11, 141, 1994.

Lien, T., Martikainen, P., Nykänen, H., and Bakken, L., Regulating factors for methane fluxes in peat, *Suo,* 43, 231, 1992 (1993).

Martikainen, P. J., Nykänen, H., Crill, P., and Silvola, J., Effect of a lowered water table on nitrous oxide fluxes from northern peatlands, *Nature,* 366, 51, 1993.

Martikainen, P. J., Nykänen, H., Crill, P., and Silvola, J., Emissions of methane and nitrogen oxides from peatland ecosystems, *The Finnish Research Programme on Climate Change. Second Progress Report,* Kanninen, M. and Heikinheimo, P., Eds., Painatuskeskus Oy, Helsinki, 1994, p. 279.

Martikainen, P. J., Nykänen, H., Alm, J., and Silvola, J., Changes in fluxes of carbon dioxide, methane and nitrous oxide due to forest drainage of mire sites of different trophy, *Plant Soil,* 168–169, 571, 1995

Moore, T. R., Carbon dioxide evolution from subarctic peatlands in eastern Canada, *Arctic Alpine Res.,* 18 (2), 189, 1986.

Moore, T. R., Plant production, decomposition, and carbon efflux in a subarctic patterned fen, *Arctic Alpine Res.,* 21 (2), 156, 1989.

Moore, T. R. and Knowles, R., The influence of water table levels on methane and carbon dioxide emissions from peatland soils, *Can. J. Soil Sci.,* 69, 33, 1989.

Moore, T. R. and Roulet, N. T., Methane flux: water table relations in northern peatlands, *Geophys. Res. Lett.,* 20, 587, 1993.

Nykänen, H., Alm, J., Lang, K., Silvola, J., and Martikainen, P. J., Emission of CH_4, N_2O and CO_2 from a virgin fen and a fen drained for grassland in Finland, *J. Biogeogr.,* 22, 351, 1995.

Paavilainen, E., Männyn juuriston suhteista turpeen ilmatilaan (Relationship between the root system of Scots pine and the air content of peat), *Commun. Inst. For. Fenn.,* 63, 1, 1967.

Päivänen, J., Hydraulic conductivity and water retention in peat soils, *Acta For. Fenn.,* 129, 1, 1973.

Rangeley, A. and Knowles, R., Nitrogen transformation in a Scottish peat soil under laboratory conditions, *Soil Biol. Biochem.,* 20, 385, 1988.

Reinikainen, A., Effect of drainage and fertilization on plant biomass and primary production in mire ecosystem, *Suo,* 32, 110, 1981.

Reinikainen, A., The similarity of vegetation on old drained peatlands and upland forest site types. Abstract in English, in Workshop of Forest Site Classification, Vantaa, Finland, October 27, 1994, *Finn. For. Res. Inst. Res. Pap.,* 531, 11, 1994.

Rodhe, H., A comparison of the contribution of various gases to the greenhouse effect, *Science,* 248, 1217, 1990.

Rosswall, T. and Granhall, U., Nitrogen cycling in a subarctic ombrotrophic mire, *Ecol. Bull.* (Stockholm), 30, 209, 1980.

Roulet, N. T., The effect of forestry drainage practices on the emission of methane from northern peatlands. Technical Program, International Symposium on the Ecology and Management of Northern Forested Wetlands, Traverse City, MI, August 29–31, 1994.

Roulet, N. T., Ash, R., and Moore, T. R., Low boreal wetlands as a source of atmospheric methane, *J. Geophys. Res.,* 97, 3739, 1992a.

Roulet, N., Moore, T., Bubier, J., and Lafleur, P., Northern fens: methane flux and climatic change, *Tellus,* 44B, 100, 1992b.

Ruuhijärvi, R., The Finnish mire types and their distribution, in *Ecosystems of the World,* Vol. 4B, *Mires: Swamp, Bog, Fen and Moor,* Regional studies, Gore, A. J. P., Ed., Elsevier, Amsterdam, 1983, p. 47.

Sallantaus, T., Leaching and the material balance of peatlands — preliminary results, *Suo,* 43, 253, 1992.

Sallantaus, T., Response of leaching from mire ecosystems to changing climate, *The Finnish Research Programme on Climate Change. Second Progress Report,* Kanninen, M. and Heikinheimo, P., Eds., Painatuskeskus Oy, Helsinki, 1994, 291.

Silvola, J. and Alm, J., Dynamics of greenhouse gases in virgin and managed peatlands, *The Finnish Research Programme on Climate Change. Progress Report,* Kanninen, M. and Anttila, P., Eds., Painatuskeskus Oy, Helsinki, 1992, p. 193.

Silvola, J., Alm, J., and Ahlholm, U., The effect of plant roots on CO_2 release from peat soils, *Suo,* 43, 259, 1992.

Silvola, J., Alm, J., Ahlholm, U., Nykänen, H., and Martikainen, P. J., CO_2 fluxes in peatlands under varying temperature and moisture conditions, *The Finnish Research Programme on Climate Change. Second Progress Report,* Kanninen, M. and Heikinheimo, P., Eds., Painatuskeskus Oy, Helsinki, 1994, p. 273.

Silvola, J. and Heikkinen, S., CO_2 exchange in the Empetrum nigrum-Sphagnum fuscum community, *Oecologia,* 37, 273, 1979.

Silvola, J., Välijoki, J., and Aaltonen, H., Effect of draining and fertilization on soil respiration at three ameliorated peatland sites, *Acta For. Fenn.,* 191, 1, 1985.

Sundh, I., Nilsson, M., Granberg, G., and Svensson, B. H., Depth distribution of microbial production and oxidation of methane in northern boreal peatlands, *Microb. Ecol.,* 27, 253, 1994.

Tolonen, K., Vasander, H., Damman, A. W. H., and Clymo, R. S., Preliminary estimate of long-term carbon accumulation and loss in 25 boreal peatlands, *Suo,* 43, 277, 1992.

Urban, N. R., Eisenreich, S. J., and Bayley, S. E., The relative importance of denitrification and nitrate assimilation in midcontinental bogs, *Limnol. Oceanogr.,* 33, 1611, 1988.

Verry, E. S., Brooks, K. N., and Barten, P. K., Streamflow response from an ombrotrophic mire, in Proc. Symp. on the Hydrology of Peatlands in Temperate and Cold Regions, 6–8 June 1988, Joensuu, Finland, Publ. 1 of the Academy of Finland, Helsinki, 1988, p. 52.

Westling, O., Mercury in runoff from drained and undrained peatlands in Sweden, *Water, Air, Soil Pollut.,* 56, 419, 1991.

Whiting, G. J. and Chanton, J. P., Primary production control of methane emission from wetlands, *Nature,* 364, 794, 1993.

Williams, B. L. and Wheatley, R. E., Nitrogen mineralization and water-table height in oligotrophic deep peat, *Biol. Fertil. Soils,* 6, 141, 1988.

CHAPTER 24

Nutrient Removals Associated with Harvesting Peatland Black Spruce Forest

Y. Teng, N. W. Foster, P. Hazlett, and I. K. Morrison

CONTENTS

INTRODUCTION

The harvesting impacts issue has been debated vigorously since it was first raised 40 years ago (Rennie, 1957). Long-term impacts of current forestry practices on forest production, nutrient cycling, and ecosystem dynamics are still not well understood today. Concerns have been expressed that nutrient removal associated with clearcut harvesting operations alone, or in combination with other processes associated with site disturbance, may deplete nutrient reserves in forests (e.g., Federer et al., 1989) and reduce the long-term sustainability of forests (Powers and Van Cleve, 1991). Good forestry practices should ensure environmental protection of nontimber forest benefits (e.g., water yield and quality, wildlife habitat) yet not erode the capability of a site to sustain wood production.

In general, biomass reserves and harvesting impacts in boreal black spruce (*Picea mariana* [Mill.] B.S.P.) forests have been well documented (Bonan, 1990; Foster and Morrison, 1987, 1988; Gordon, 1981, 1983; Moore and Verspoor, 1973; Timmer et al., 1983), though not with the combination of stand age, climate, and soil conditions in our current study. In this paper, we explore the impacts of alternate industrial forestry operations in managing nutrients in a boreal black spruce peatland. The vegetation structure associated with second-growth (resulting from a 1928 cut) vs. old-growth stands provided historical evidence of forest harvesting impacts on a boreal black spruce peatland ecosystem. Based on this information, a relatively new approach to forest harvesting and regeneration that could minimize the disturbance of forest nutrient reserves and nutrient cycles typically associated with clearcutting is examined. The objective of the study is to determine potential nutrient removals from sites associated with specific harvesting prescriptions when applied to uneven-aged stands in a black spruce swamp in the Lake Abitibi Model Forest.

STUDY AREA

The study sites (lat. 49°0'N; long. 81°30'W) are located beside Wade Lake, 30 km east of Cochrane, Ontario, in the Northern Clay Section of the boreal forest (Rowe, 1972). Mean total precipitation (Cochrane) is 885 mm, the average January temperature is −18°C and the average July temperature is 17°C (Anonymous, 1982a,b). The sites are occupied by uneven-aged black spruce forests (Groot and Horton, 1994), dominated by a Ledum site type (OG 11) (Jones et al., 1983), with 35 to 175 cm of peat accumulated over lacustrine clay (Berry and Jeglum, 1991). The mean height of dominant and codominant trees was 11.5 and 13.4 m in the second- and old-growth forest, respectively. The mean DBH and basal area of all trees (>1.3 m in height) were 7.6 cm, and 29.6 m^2 ha^{-1} in the second-growth stand, and 8.4 cm and 23.6 m^2 ha^{-1} in the old-growth stand. The age of trees of the old-growth forest ranged from 40 to 280 years; and the second-growth (horse-logged) forest was largely composed of trees 65 years or older (Sundstrom, 1992).

MATERIALS AND METHODS

Field Survey and Laboratory Analysis

The field investigation was conducted during the 1993 season with supplementary work done in 1994. Twelve 10 × 50 m experimental plots in the second-growth and fifteen in the old-growth stands were established. Live trees with height >1.3 m (overstory) and dead stems (standing and fallen) with DBH >5 cm in each plot were measured for their DBH, and their number was counted and recorded by DBH classes of 5.0 cm intervals. Heights of seedlings (trees <1.3 m in height) in 75 and 100 quadrats measuring 2×2 m were recorded in the second- and old-growth stands, respectively, and their numbers were counted by height classes of 20.0 cm intervals. In the late summer of 1993, seven understory and ten mature black spruce trees of various sizes from each of the two stands were destructively sampled and separated into foliage, fruit, live branch, dead branch, stem wood, and stem bark. The heights of the trees were measured. The samples were oven dried at 70°C to a constant weight to derive biomass components for each sampled tree. Subsamples of the components from ten trees were then ground for chemical analysis. Analysis for phosphorus (P), potassium (K), calcium (Ca), and magnesium (Mg) was by a Thermo Jarrel-Ash ICAP 1100 Spectrometer after wet ashing with nitric-perchloric acids. Nitrogen was determined using a Tecator 1030 Analyzer after semimicro-Kjedahl digestion (Mader and Hoyle, 1964).

Data Analysis and Model Development

Student t-test was performed to compare the difference in number of trees of each DBH class between the old- and second-growth stand. Difference in seedling number in each height class between the two stands was also tested by the same method. Data from all sampled trees were used to construct regression models taking the form of:

$$\log W = a_0 + a_1(\log D) + a_2(\log H) \tag{1}$$

where W was total aboveground biomass or its components, D was measured DBH, H was height of the tree, and a_0, a_1, and a_2 were regression coefficients estimated using REG program in SAS (SAS Institute, 1985). A quadratic regression model was built for relating tree height (H) to DBH (D):

$$H = a_0 + a_1 D + a_2 D^2 \tag{2}$$

where the regression coefficients a_0, a_1, and a_2 were estimated also using REG program in SAS.

Equation 2 was used to compute height for each tree, and then total aboveground biomass of each tree and its components (fruit, live branch, dead branch, stem wood, and stem bark) were calculated using Equation 1 for each experimental plot. The biomass per hectare by components were computed based on the number of trees in each diameter class, and were converted to carbon (C) content using a conversion factor of 0.5. Nutrient content of the components was calculated by multiplying their biomass and nutrient concentration. Nutrient removals, estimated from the biomass and nutrient component data, were examined for significance of difference among different harvesting methods by Student t-Test.

RESULTS AND DISCUSSION

Aboveground Biomass Reserve and Distribution

The amount of aboveground biomass removed from the site by timber harvesting, and the associated nutrient drain, are dependent on the silvicultural system used. Selection of silvicultural system is guided by, in part, the existing forest-stand structure and the management objective for the residual or newly established forest. Analysis of the distribution of the stand biomass in the second-growth forest at Wade Lake indicated that 25 percent of the total biomass was in trees <10 cm in DBH (Figure 1a), while 75 percent was in trees >10 cm in DBH (Figure 1b). The biomass groupings were selected on the assumption that trees of DBH <10 cm would survive careful logging and could form a residual forest. Biomass of seedlings (<1.30 m in height) and shrubs was omitted because they were relatively small (Gordon, 1983). In general, aboveground biomass (Figure 1c) was relatively low (102 Mg ha^{-1}). Documented data, which was based largely on examination of upland forests, suggested an average black spruce biomass of 147 Mg ha^{-1}, ranging from 50 to 325 Mg ha^{-1} (Foster and Morrison, 1987). Furthermore, the crown (cones, foliage and branches) to stem (stem bark and stem wood) ratio (0.52) at Wade Lake was much higher than that for boreal upland black spruce stands (cf. 0.21, Foster and Morrison, 1987). This higher ratio may be explained by a combination of lower productivity, younger tree ages, and the uneven-aged structure in the peatland spruce stand, which we examined and would significantly contribute to the nutrient distribution pattern of the system.

Nutrient Reserves and Distribution

Although relatively small (13 percent) in terms of biomass, foliage contained the largest pool of N (46 percent), P (36 percent), and K (37 percent) in the forest vegetation component of the second-growth stand (Table 1). In general, the canopy represented a larger reservoir than the stems for all nutrients except for Ca. With only 35 percent of the aboveground tree biomass, the crown contained an overwhelmingly large part of N (82 percent) and P (65 percent), while the other three macronutrients were almost equally distributed between the crown and the stem. This dramatic difference in chemical composition was best depicted by C:N ratio of the components (Table 1): it was 21 times higher for the stem wood (1207) than for the foliage (56), and was nine-fold as high in the stem as a whole as in the crown. The ratio was higher in merchantable trees (>10 cm in DBH) than that for submerchantable ones (<10 cm in DBH) (Table 1).

Nutrient Removals by Harvesting

Conventionally, only merchantable stems (>10 cm in DBH) were removed from the sites. Most of the trees <10 cm in DBH as well as the crowns of the removed stems were retained on the sites during clearcut logging, although they would be cut or run over to facilitate access to merchantable trees. The removal

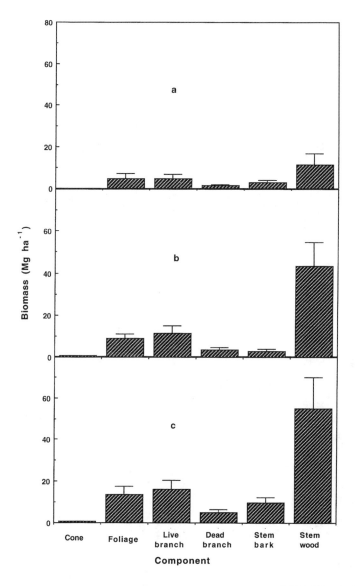

Figure 1   Biomass distribution in aboveground tree components of (a) DBH <10 cm, (b) DBH >10 cm, and (c) the stand in Wade Lake peatland black spruce second-growth forest. Top bars indicate standard deviations (n = 12).

of N, P, K, Ca, and Mg from the site by this conventional operation would be 32, 6, 36, 152, and 11 kg ha^{-1}, respectively (Table 2). Due to the high proportion of crown biomass (crown to stem ratio 0.52) in the stand (Figure 1), a more severe nutrient loss would be caused by full-tree logging, even if only trees >10 cm in DBH were harvested. This additional 52 percent of biomass (e.g., the crown) could be harvested only at the cost of removing much more N (418 percent), P (150 percent), K (97 percent), Ca (88 percent) and Mg (109 percent) from the

Table 1 Element Distribution in Aboveground Tree Components in the Second-Growth Stand in Wade Lake Peatland Black Spruce Forest

	N (kg ha⁻¹)	P (kg ha⁻¹)	K (kg ha⁻¹)	Ca (kg ha⁻¹)	Mg (kg ha⁻¹)	C (kg ha⁻¹)	C:N (g g⁻¹)
DBH <10 cm							
Cone	0 (<1)	0 (0.0)	0.0 (0.0)	0 (0)	0.0 (0.0)	24 (7)	126 (11)
Foliage	41 (11)	2.9 (1.1)	12.8 (3.2)	36 (8)	3.8 (0.9)	1987 (623)	49 (7)
Live branches	17 (6)	1.1 (0.4)	2.4 (0.7)	19 (7)	2.1 (0.7)	2121 (605)	118 (13)
Dead branches	5 (1)	0.5 (0.1)	2.0 (0.4)	6 (2)	1.6 (0.3)	510 (165)	122 (20)
Stem bark	6 (2)	0.8 (0.3)	4.0 (1.2)	27 (9)	1.3 (0.4)	1209 (353)	201 (33)
Stem wood	5 (1)	1.0 (0.2)	6.8 (1.5)	23 (6)	2.1 (0.4)	5884 (1534)	1176 (83)
DBH >10 cm							
Cone	2 (<1)	0.2 (0.0)	0.3 (0.1)	0 (0)	0.2 (0.0)	265 (93)	127 (12)
Foliage	76 (21)	5.4 (1.5)	23.5 (6.1)	66 (15)	7.1 (2.0)	4568 (1452)	60 (7)
Live branches	42 (15)	2.6 (1.0)	5.9 (1.9)	49 (15)	2.8 (1.1)	5670 (1954)	135 (14)
Dead branches	15 (4)	1.3 (0.5)	5.9 (2.0)	18 (5)	1.8 (0.5)	1898 (635)	126 (9)
Stem bark	14 (4)	1.9 (0.5)	10.0 (3.5)	68 (20)	3.2 (1.0)	3357 (1124)	241 (41)
Stem wood	18 (3)	3.9 (0.6)	25.6 (4.2)	84 (13)	7.8 (1.2)	21814 (7565)	1212 (77)
Whole stand							
Cone	2 (<1)	0.2 (0.0)	0.3 (0.1)	0 (0)	0.2 (0.0)	289 (96)	127 (13)
Foliage	117 (27)	8.3 (1.9)	36.2 (9.1)	102 (25)	8.9 (2.0)	6555 (2110)	56 (7)
Live branches	59 (17)	3.7 (0.8)	8.3 (2.5)	68 (19)	4.9 (1.1)	7791 (2374)	132 (18)
Dead branches	20 (5)	1.8 (0.7)	7.9 (1.9)	24 (6)	3.4 (0.8)	2408 (812)	120 (11)
Stem bark	20 (4)	2.7 (0.5)	14.0 (2.9)	95 (19)	4.5 (0.9)	4566 (1545)	231 (41)
Stem wood	23 (7)	4.9 (1.4)	32.4 (9.8)	107 (23)	9.9 (2.9)	27298 (938)	1207 (65)

Note: Values in parentheses are standard deviations (n = 12).

Table 2   Predicted Aboveground Biomass and Nutrient Removals and Retentions by
Conventional and Full-Tree Harvesting[a] Operations in the Second-Growth
Stand in Wade Lake Peatland Black Spruce Forest

	Biomass (mg ha⁻¹)	N (kg ha⁻¹)	P (kg ha⁻¹)	K (kg ha⁻¹)	Ca (kg ha⁻¹)	Mg (kg ha⁻¹)
Removal						
Conventional	50 a[b]	32 a	6 a	36 a	152 a	11 a
Full-tree	76 b	166 b	15 b	71 b	285 b	23 b
Retention						
Conventional	52 b	208 b	16 b	63 b	244 b	21 b
Full-tree	26 a	74 a	7 a	28 a	111 a	9 a

[a] Conventional harvesting refers to removal of stems of trees >10 cm in DBH, retaining the crowns and trees <10 cm on the site; full-tree harvesting refers to removal of all aboveground portions of trees >10 cm in DBH, leaving only trees <10 cm on the site.
[b] Column values followed by different letters indicate significant difference ($p < 0.05$) between harvesting method for removal or retention, judged by SNK test.

site. As Foster and Morrison (1987) and Timmer et al. (1983) pointed out, removing such large quantities of bases, particularly Ca (285 kg ha⁻¹), from the ecosystem may cause nutrient depletion and also have an acidifying effect on the soil. Since foliage and branches of black spruce have had no significant industrial use so far, the crown biomass is of little merchantable value. However, if left on the site as slash, a substantially greater amount of N (181 percent), P (129 percent), K (125 percent), Ca (120 percent), and Mg (133 percent) would be retained (Table 2) on the site. This would be a significant nutrient source for the succeeding rotation and represent significant contribution to the sustainability of stand production (Timmer et al., 1983), especially considering the low nutrient availability of many peat deposits (e.g., Wright, 1959) and the slow mineralization of nutrients from peat organic matter (Clymo, 1965; Flanagan and Van Cleve, 1983).

Nutrient Reserves of Undersized Trees and Advantage of Careful Logging

In addition to removals in the form of harvested products, nutrients released from decomposition of the fresh litter are especially liable to be removed from the system by leaching and runoff during the first few years after harvesting (Gessel and Cole, 1965; Hornbeck and Kropelin, 1982) because of the severe disturbance to the forest floor, reduced vegetation coverage, and the associated reduction in nutrient absorption. Nutrient release from decomposition of fresh litter is much faster than from decomposing peats, even in boreal forests (Gordon and Van Cleve, 1983). Elevated release of inorganic nutrients, follows immediately after clearcut harvesting (Gordon and Van Cleve, 1983) or litter fall (Johnson et al., 1969). Excessive nutrients availability on cutover, especially of N and K (Likens et al., 1977), could stimulate competition by other vegetation and ultimately extend the length of the coniferous rotation. Thus, the main advantage of tree-length harvesting, that is, leaving greater nutrient reserves for crop trees, would be compromised.

Table 3   Differences in Overstory DBH Class Distribution (stem ha^{-1}) Between Old-Growth Plots (n = 15) and Second-Growth Plots (n = 12) in Wade Lake Peatland Black Spruce Forest

DBH class (cm)	Old growth	Second growth
Live (height >1.30 m)		
< 5.0	1235 a[a]	2584 b
5.0–9.9	727 a	1608 b
10.0–14.9	717 a	1028 b
15.0–19.9	551 a	388 b
20.0–24.9	77 a	50 b
25.0–29.9	3 a	8 a
Live total	3310 a	5666 b
Dead (standing and fallen)		
5.0–14.9	484 a	127 b
>15.0	158 a	41 b
Dead total	642 a	168 b

[a] Means within a row followed by the same letter are not significantly different at 0.05 level of Student t-Test.

However, black spruce on peatland is well known for its vigorous asexual propagation by layering (Stanek, 1961, 1968). Usually there are abundant small trees, saplings, and seedlings of varying ages in a mature stand, as we observed at Wade Lake (Table 3 and 4). Hence, careful logging, by which undersized trees (DBH <10 cm) are not harvested or destroyed and, in fact, are preserved, would provide a possible solution to the new stand establishment problem. The absorption and assimilation of nutrient by these residual trees would form a strong sink to minimize nutrient output from the system. The biomass (Figure 1) of trees of DBH <10 cm was only 25 percent of the total aboveground biomass, but their number (Table 3) was threefold (in the second-growth stand) as many as that of the trees of DBH >10 cm. Seedlings (height <1.30 m) were even more numerous (Table 4). For mature and overmature trees (DBH >10 cm), nutrient metabolism involves relocation or reutilization in the plant, in addition to absorption from the soil (Cole, 1981). Younger trees, on the other hand, are more vigorous in terms of nutrient absorption (Miller, 1984), so that careful harvesting of mature trees probably would not initially result in significant reduction in nutrient absorption by the stands. Harvesting would also open up the canopy and thus provide more resources for younger trees and advanced regeneration, and immediately accelerate their nutrient absorption. This is because their root systems are already well established in the soil. Further increases in tree growth would probably result from suppressing the growth of competing vegetation due to a more competitive nutrient absorption capability of the trees. Released trees could benefit from

Table 4 Differences in Prerelease Advanced
Regeneration Pattern (Number of
Seedlings of Height <1.30 m) Between
Old-Growth Plots (n = 100) and Second-
Growth Plots (n = 75) in Wade Lake
Peatland Black Spruce Forest

Height class (cm)	Old growth	Second growth
20.0–39.9	12,575 a[a]	10,200 b
40.0–59.9	4,808 a	4,925 a
60.0–79.9	1,167 a	1,605 a
80.0–99.9	525 a	750 a
>100.0	358 a	202 a
Subtotal	19,433 a	18,862 a
<20.0	10,892 a	6,325 b
Total	30,325 a	25,007 b

[a] Means within a row followed by the same letter are
not significantly different at 0.05 level of Student t-
Test.

increased nutrient availability if crowns of the harvested trees were retained on site by careful logging combined with stem-only logging.

Harvesting and Forest Sustainability

A more profound advantage of careful logging is the mitigation of the impacts of harvesting on the sustainability of the forests. Table 3 shows the stand structure of two adjacent peatland black spruce stands in our current investigation. The old-growth stand had never been cut, while the second-growth stand was horse-logged (merchantable trees were felled and logs were skidded to roadside by horses) in 1928. Horse-logging could be considered a form of "careful logging." In the second-growth stand, many small trees were left at time of harvesting, and the stand was fully stocked by natural regeneration (Sundstrom, 1992). In the old-growth stand, we observed a large number of trees of >15 cm in DBH dead (standing or fallen) (Table 3), probably reflecting the overmaturity of the stand. In the second-growth stand, significantly fewer standing trees of >15 cm in DBH existed than in the old-growth stand, presumably due to the harvesting conducted 60 years ago. However, the second-growth stand had significantly more trees of <15 cm, especially those <10 cm in DBH, than the old-growth stand. This might indicate that horse-logging had made more resources available for small trees to grow by opening up the stand, so that more trees of these classes survived. If careful logging were conducted now, fewer trees would be harvested, but many more could be harvested from the second-growth than from the old-growth stand after trees of <10 cm in DBH matured. Furthermore, if only half of the biomass (e.g., the stems >10 cm in DBH) were removed by careful logging, loss of N, P, K, Ca, and Mg (Table 2) would be reduced by 650, 272, 178, 160, and 186 percent, respectively.

Comparison of the distribution of seedling height classes between the two sites (Table 4) indicates that the regeneration patterns were probably historically similar, since their populations of seedlings >20 cm in height were not significantly different. However, seedlings of <20 cm in height were more numerous in the old-growth than in second-growth stand. This, as well as the observation that many trees were dead and fallen, might indicate that the old-growth stand was overmature and naturally breaking up (Robichaud and Methven, 1993). Sundstrom (1992) revealed that this stand was much older (some trees were 280 years in 1985) than the second-growth stand (about 60 to 70 years). Both the previously horse-logged and the undisturbed natural forest, hence, would be suitable candidates for careful logging on even shorter rotations than in clearcut management stands. Careful logging would probably not only avoid severe interruption of production but also would minimize those adverse effects of clearcut logging, such as acceleration of nutrient export, rising water tables, expansion of swamp, and retardation of reforestation of black spruce. In the long term, careful logging may increase forest production by shortening rotations and promoting smoother renewal of the stands, thereby contributing to the rapid restoration of nutrient cycling between trees and soil and to sustainable forest production.

CONCLUSION

Maintaining sustained wood production without deteriorating the peatland forest ecosystem is the ultimate goal of the Lake Abitibi Model Forest research program. This study, as one part of the program, found that the peatland black spruce forest at Wade Lake was vulnerable to bioelement depletion under a full-tree clearcut cropping system. Nutrient losses could be significantly reduced with a conventional harvesting system by which only merchantable stems are extracted from the stands. Careful logging would avoid severe interruption of forest production and minimize postharvest off-site nutrient losses. We are presently investigating, after careful logging, how the smaller trees will be preserved and how released stems take advantage of the canopy openings and the nutrient release from the slash decomposition, thus increasing forest production in the long term by shortening rotations and promoting smoother renewal of the stands.

ACKNOWLEDGMENTS

We thank the following individuals who contributed to the study: R. Boudah, A. Creighton, C. Faulkner, F. Gerbone, K. Jones, K. Kallo, P. Kibbee, and L. Hawdon in sample collection and processing; L. Irwin and J. Ramakers in chemical analysis; A. Creighton and T. Hardiman in data compilation; and S. Dominy and A. Groot in manuscript reviewing. The staff of Abitibi-Price Inc., Iroquois Falls Division assisted in site selection and protected the study area. The financial support from the Lake Abitibi Model Forest is gratefully recognized.

REFERENCES

Anonymous, Canadian Climate Normals (1951–80). Vol. 2. Temperature, Department of the Environment, Atmospher. Environ. Serv., Downsview, Ontario, 1982a, 186 pp.

Anonymous, Canadian Climate Normals (1951–80). Vol. 3. Precipitation, Department of the Environment, Atmospher. Environ. Serv., Downsview, Ontario, 1982b, 602 pp.

Berry, G. J. and Jeglum, J. K., Hydrology of drained and undrained black spruce peatlands: Groundwater table profiles and fluctuations, COFRDA Report 3307, Forestry Canada, Ontario Region, Sault Ste. Marie, Ontario, 1991, 31 pp.

Bonan, G. B., Carbon and nitrogen cycling in North American boreal forests, II. Biogeographic patterns, *Can. J. For. Res.*, 20, 1077, 1990.

Clymo, R. S., Experiments on breakdown of sphagnum in two bogs, *J. Ecol.*, 53, 747, 1965.

Cole, D. W., Nitrogen uptake and translocation by forest ecosystems, in Terrestrial Nitrogen Cycles, Clark, F.E., and Rosswall, T., Eds, *Ecol. Bull.*, 33, 219, 1981.

Federer, C. A., Hornbeck, J. W., Tritton, L. M., Martin, C. W., Pierce, R. S., and Smith, C. T., Long-term depletion of calcium and other nutrients in eastern US forests, *Environ. Manage.*, 13, 593, 1989.

Flanagan, P. W. and Van Cleve, K., Nutrient cycling in relation to decomposition and organic-matter quality in taiga ecosystem, *Can. J. For. Res.*, 13, 795, 1983.

Foster, N. W. and Morrison, I. K., Alternate strip clearcutting in upland black spruce, IV. Projected nutrient removals associated with harvesting, *For. Chron.*, 63, 451, 1987.

Foster, N. W. and Morrison, I. K., Effects of site preparation and full tree logging on nutrient cycling, in Compiler of Aspects of Site Preparation Biology and Practice, Proc. of Workshop Held 27–28, Sept. 1988, Fort Frances, Ontario, Ontario Ministry of Natural Resources Northwestern Ontario For. Tech. Devel. Unit, Thunder Bay, Ont., 1988, 28.

Gessel, S. P. and Cole, D. W., Influence of removal of forest cover on movement of water and associated elements through soil, *J. Am. Water Works Assoc.* 57, 1301, 1965.

Gordon, A. G., Impacts of harvesting on nutrient cycling in the boreal mixedwood forest, Proc. of Boreal Mixedwood Symposium Held at 16–18, Sept. 1980, Thunder Bay, Ont., COJFRC Symposium Proc. O-P-9, Whitney, R. D. and McClain, K. M., Eds., Sault Ste. Marie, Ontario, 1981, p. 121.

Gordon, A. G., Nutrient cycling dynamics in differing spruce and mixedwood ecosystems in Ontario and the effects of nutrient removals through harvesting, in Resources and Dynamics of the Boreal Zone, Proc. of a Conference Held at Thunder Bay, Ont., August, 1982, Wein, R. W., Wein, R. R., and Methven, I. R., Eds., Assoc. Can. Univ. North. Stud. 1983, p. 97.

Gordon, A. G. and Van Cleve, K., Seasonal patterns of nitrogen mineralization following harvesting in the white spruce forests of interior Alaska, in Resources and Dynamics of the Boreal Zone, Proc. of a Conference Held at Thunder Bay, Ont., August 1982, Wein, R. W., Wein, R. R., and Methven, I. R., Eds., Assoc. Can. Univ. North. Stud. 1983, p. 119.

Groot, A. and Horton, B., Age and size structure of natural and second-growth peatland *Picea mariana* stands, *Can. J. For.*, 24, 225, 1994.

Hornbeck, J. W. and Kropelin, W., Nutrient removal and leaching from a whole-tree harvest of northern hardwoods, *J. Environ. Qual.*, 11, 309, 1982.

Johnson, N. M., Likens, G. E., Bormann, F. H., Fisher, D. W., and Pierce, R. S., A working model for the variation in streamwater chemistry at the Hubbard Brook Experimental Forest, New Hampshire, *Water Resources Res.*, 5, 1353, 1969.

Jones, R. K., Pierpoint, G., Wickware, G. M., Jeglum, J. K., Arnup, R. W., and Bowles, J. M., Field Guide to Forest Ecosystem Classification for the Clay Belt, Site Region 3E, Ontario Ministry of Natural Resources, Maple, Ontario, 1983, 122 pp.

Likens, G. E., Bormann, F. H., Pierce, R. S., Eaton, J. S., and Johnson, N. M., *Biogeochemistry of a Forested Ecosystem*, Springer-Verlag, New York, 1977, 146 pp.

Mader, D. L. and Hoyle, M. C., An inexpensive portable unit for perchloric acid digests and semimicro-Kjedahl determinations, *Soil Sci.*, 98, 295, 1964.

Miller, H. G., Dynamics of nutrient cycling in plantation ecosystems, in *Nutrition of Plantation Forests*, Bowen, G. D. and Nambiar, E. K. S., Eds., Academic Press, London, 1984, pp. 53–78.

Moore, T. R. and Verspoor, E., Aboveground biomass of black spruce stands in subarctic Quebec, *Can. J. For. Res.*, 3, 596, 1973.

Powers, R. F. and Van Cleve, K., Long-term ecological research in temperate and boreal forest ecosystems, *Agron. J.*, 83, 11, 1991.

Rennie, P. J., The uptake of nutrients by mature timber forest and its importance to timber production in Britain, *Quart. J. For.*, 51, 101, 1957.

Robichaud, E. and Methven, I., The effect of site quality on the timing of stand breakup, tree longevity, and the maximum attainable height of black spruce, *Can. J. For. Res.*, 23, 1514, 1993.

Rowe, J. S., Forest Regions of Canada, Canadian Publication No. 1300, Forest Service, Ottawa, Ontario, 1972, 172 pp.

SAS Institute, SAS User's Guide: Statistics, Version 5, SAS Institute, Cary, NC, 1985.

Stanek, W., Natural layering of black spruce in northern Ontario, *For. Chron.*, 37, 245, 1961.

Stanek, W., Development of black spruce layers in Quebec and Ontario, *For. Chron.*, 44, 25, 1968.

Sundstrom, E., Five-year growth response in drained and fertilized black spruce peatland, I. Permanent growth plot analysis, Forestry Canada Investigative Report O-X-417, Forestry Canada, Ontario Region, Sault Ste. Marie, 1992, 19 pp.

Timmer, V. R., Savinsky, H. M., and Marek, G. T., Impact of intensive harvesting on nutrient budgets of boreal forest stands, in Resources and Dynamics of the Boreal Zone, Proc. of a Conference Held at Thunder Bay, Ontario, August, 1982, Wein, R. W., Wein, R. R., and Methven, I. R., Eds., Assoc. Can. Univ. North. Stud. 1983, p. 131.

Wright, T. W., Use of fertilizers in the afforestation of deep peat, *J. Sci. Food Agric.*, 10, 645, 1959.

CHAPTER 25

The Effects and Fate of Inorganic Nitrogen Inputs to Oligotrophic Peat Soils

Berwyn L. Williams and Deborah J. Silcock

CONTENTS

INTRODUCTION

In ombrotrophic peatland ecosystems, precipitation is the main source of nutrients for the vegetation and their contents of essential nutrients, such as phosphorus (P) and of nitrogen (N), are low compared with other soils. These

1-56670-177-5/97/$0.00+$.50
© 1997 by CRC Press, Inc.

bogs represent net accumulations of organic matter and N over millennia, relying mainly on inputs of N from atmospheric deposition (Damman, 1988). However, the current trend of increasing N inputs from atmospheric sources threatens the nature of the vegetation and the carbon balance in these ecosystems (Francez, 1991). Damman (1988) calculated that less than half of the total N input to a peat bog could be accounted for and suggested that N losses occurred because the immobilization was limited by a low activity of microorganisms. Verhoeven et al. (1990) reported little effect of the greater atmospheric N input in the Netherlands on the decomposition of cellulose in ombrotrophic bogs measured using cotton strips as substrates compared with values obtained in other regions. Microbial activity in these situations was considered to be limited by the low availability of inorganic nutrients, in particular P. A budget for N in Thoreau's Bog, MA showed that runoff was the major pathway for N loss (Hemond, 1983), whereas this process was neglible for a subarctic ombrotrophic mire (Rosswall and Granhall, 1980). Evidently, many factors determine the fate of N that reaches the bog surface. In this paper we consider the information currently available on the fate of N added to peat soils and present some experimental results of the effects that N additions have on the size and activity of the microbial biomass.

NITROGEN INPUTS

Atmospheric Inputs

Atmospheric inputs of N are associated with wet and dry deposition and inputs in fogs and mist (Morris, 1991). Inputs from cloud droplets (occult deposition) and dry deposition interact with the vegetation, which varies in its catching efficiency. Dry deposition also includes gaseous ammonia in the atmosphere, which contributes to the N input in some areas (Pearson and Stewart, 1993). Average values for wet deposition vary geographically and range in Europe from approximately 2 kg ha^{-1} yr^{-1} in northern Scandinavia to 80 kg ha^{-1} yr^{-1} in the Netherlands (Department of the Environment, 1994). In North America, values for inputs of N to wetlands in wet deposition of up to 12 kg N ha^{-1} yr^{-1} have been reported (Morris, 1991). Where deposition values include an analysis of soluble organic N, approximately a third to one-half of the N is in this form. The proportion of ammonium in the inorganic component varies from 30 to >90 percent, depending upon the source of the atmospheric N. Most values reported refer to inorganic N inputs, and little is known about the organic component except that after passage through the vegetation canopy, the organic component can increase significantly (Qualls et al., 1991).

Biological N$_2$ Fixation

Estimates of biological fixation range from 2 kg N ha^{-1} yr^{-1} in the arctic tundra (Rosswall and Granhall, 1980) to 10 kg N ha^{-1} yr^{-1} in Thoreau's Bog (Hemond,

1983). At both of these sites, biological fixation and atmospheric deposition contributed similar amounts of N, although the former is greater in more nutrient-rich sites (Rosswall and Granhall, 1980).

PLANT UPTAKE

In nutrient-poor ecosystems during the growing season, plant uptake probably accounts for the major proportion of inorganic N deposited in atmospheric inputs. Annual uptake of inorganic N by vascular plants in oligotrophic peatland ecosystems ranges from 7 kg N ha^{-1} yr^{-1} in tundra (Rosswall and Granhall, 1980) to 30 kg N ha^{-1} yr^{-1} in more temperate conditions (Hemond, 1983). Coniferous trees and shrubs growing in acid soils utilize mainly ammonium, but assimilation of available inorganic N by the soil microbial biomass can compete effectively in some situations, such that fertilizer-N recovery by trees can be as low as 10 percent (Hulm and Killham, 1990). Studies have shown that *Sphagnum* moss species are equipped with nitrate reductase enzymes and have adapted to ombrotrophic conditions by using nitrate in wet deposition (Woodin et al., 1985). The rates of nitrate assimilation by bryophytes in midcontinental bogs of North America exceeded the rates of denitrification (Urban et al., 1988) confirming that in this case nitrate is removed mainly by the vegetation. However, in areas with high inputs of N and acidic pollution, bog mosses have been eliminated and marked reductions in growth have been observed when mosses have been transferred from low to high input sites (Baxter et al., 1992). Very high fertilizer N inputs, >200 kg N ha^{-1}, can influence the competitive ability of *Molinia caerulea* and *Erica tetralix* in wet heathlands and lead to changes in the species composition of the vegetation, particularly if phosphorus is also available (Aerts and Berendse, 1988).

Bog mosses from an unpolluted site responded to inorganic N additions by a transient and very high production of amino acids in the plant tissues during a period of three weeks (Baxter et al., 1992). Vegetation receiving N in excess may release organic N into canopy throughfall and into the litter and soil horizons. In the absence of plant uptake, atmospheric N is more likely to penetrate the vegetation canopy and reach the litter and peat horizons, provided hydrological conditions do not favor losses in surface runoff. It follows that there ought to be a seasonal pattern in the degree of penetration of inorganic N into the litter layers and underlying peat where the fate of the inorganic N will be determined by several processes.

The major pathway for N into the peat horizons is as organic N in plant litter, and on ombrotrophic bogs this input ranges from 6 kg N ha^{-1} yr^{-1} in arctic conditions (Rosswall and Granhall, 1980) to 38 kg N ha^{-1} yr^{-1} in Massachusetts (Hemond, 1983). This litter passes into the acrotelm, the aerobic zone of the peat profile where decomposition is active (Ingram, 1978).

RUNOFF AND LEACHING

Investigations on the inorganic N concentrations of waters draining upland parts of catchments containing natural vegetation on peaty soils generally show that inputs exceed outputs and that upland areas behave as sinks for inorganic N (Edwards et al., 1985). In another study in northeastern Scotland, nitrate in stream water was associated with acid peaty soils that were not actively nitrifying, and biological activity was not sufficient to prevent losses in runoff (Black et al., 1993). The reasons for this paradox are not clear, and more information about the characteristics of the vegetation, soils, and land use in the different catchments is required. Losses from ombrotrophic bogs appear to be neglible (Rosswall and Granhall, 1980) although Damman (1988) calculated that only half of the total N input was immobilized in the peat and therefore losses did occur. Runoff waters draining the surface 5 cm of reseeded blanket bog contained less inorganic N than the incoming rain, 0.2 compared with 4.9 kg N ha^{-1} yr^{-1}, respectively, and on plots fertilized with 150 kg N ha^{-1} yr^{-1} the amount of inorganic N in runoff increased to 2.5 kg N ha^{-1} yr^{-1} (Williams and Young, 1994). Perturbations, such as ploughing up grassland or harvesting tree crops, can change this and result in increased N losses (Stevens and Wannop, 1987). Analyses of the drainage waters from clear-felled forest and from improved blanket bog have shown that the concentrations of dissolved organic N can exceed those of either ammonium or nitrate (Stevens and Wannop, 1987; Williams and Young, 1994). The significance of this soluble organic N for N losses and for transformations downstream is not clear and requires further investigation.

NITRIFICATION AND DENITRIFICATION

Autotrophic nitrifying bacteria have not been cultured from ombrotrophic bogs (Rosswall and Granhall, 1980), and active nitrate production has not been found in peats having low ash contents (Williams, 1984). The main source of nitrate to ombrotrophic bogs is atmospheric deposition, and any nitrate is utilized mainly by the vegetation. In well-aerated upland soils, nitrate levels in surface runoff have been known to increase during winter as plant uptake declines (Edwards et al., 1985), but in a limed blanket bog, no such increase was obtained (Williams and Young, 1994). This occurred at a site where denitrification rates were high during spring and autumn and suggested that nitrate reduction was active even during winter when soil temperatures had fallen below 5°C. In acid peats, denitrification is not active unless pH has been raised to the region of pH 6 by liming (Klemmedtsson et al., 1977), and losses of N as nitrate in runoff may occur on virgin bogs during the winter.

IMMOBILIZATION

Assimilation of inorganic N by microorganisms is also generally regarded as an important pathway for N to be immobilized in peats and other soils. The dynamics of this process are influenced by the carbon (C) to N ratio of the organic matter undergoing decomposition; this ratio is invariably high in plant litter from peatland vegetation, particularly if *Sphagnum* mosses are present (Williams, 1983). Damman (1988) has suggested that limitations on the size and activity of the microbial biomass in peats by deficiencies of inorganic elements, particularly P, could also restrict the degree of N immobilization resulting in N losses. Certainly, peatland that has been limed and fertilized for agriculture or forestry is characterized by increased net immobilization of N, leading to poor responses by grass to applied N (Williams and Wheatley, 1992) and reductions in rates of tree growth (Dickson, 1972). The degree of immobilization of N could, therefore, depend on the amounts of P, K, and other inorganic elements in the peat, which in turn are determined by the topography of the peatland and the composition of the underlying rocks in the area (Pyatt et al., 1979).

Microbial Biomass

Clymo (1992) has described four zones in the surface 50 cm of a raised mire distinguished by their physical structure and bulk density and implies that the microbial activities in these zones are also different. The size of the microbial biomass can be measured as microbial C using the Substrate Induced Respiration (SIR) method (Anderson and Domsch, 1978). In cores from a raised mire, Moidach More near Grantown on Spey in the northeast of Scotland, microbial biomass C increased in the 10- to 20-cm layer corresponding to the moss litter horizon (Figure 1). This corresponded to Clymo's second zone beneath the living moss, where aerobic decomposition is the dominant process. Beneath this layer there is a transitional zone in which conditions become increasingly anoxic. Microbial N, measured by fumigation with chloroform followed by extraction with 0.5 M K_2SO_4, showed a peak at the same depth as microbial C, indicating that this horizon could possibly be the site of greatest immobilizing activity and decomposition.

Activity *in situ* may be a better indicator of the boundary between the acrotelm and the more inert anaerobic catotelm. Variations in ammonium contents are subject to the effects of immobilization and ammonification and are not correlated with microbial biomass (Figure 2). Williams and Wheatley (1988) reported sharp declines in the ammonium contents and in the numbers of ammonifying bacteria with depth in the profile of a blanket bog peat. Lowering the water table in the surrounding ditches for a period of almost 20 years had the effect of increasing the ammonifying activity and numbers of bacteria in the surface horizons. The contents of other nutrients such as P and K are also greater in the surface layer of the peat profile, and deficiences of these elements may limit increases in microbial activity at lower depths. There is as yet insufficient data on the size and activity of the microbial biomass in different peats to prove the hypothesis of Damman (1988), that microbial activity and hence immobilization of N are limited in some bogs by low concentrations of inorganic nutrients. Additions of inorganic N to reseeded blanket bog significantly ($P < 0.05$) increased microbial biomass P (Table 1), indicating that the impact of added inorganic N may well depend on the P status of the peat.

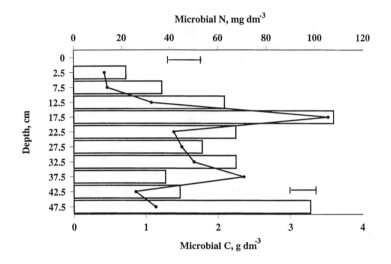

Figure 1   Amounts (g dm⁻³ fresh peat) of microbial C (line) and N (bar) in the profile, 0- to 50-cm depth from the surface of *Sphagnum magellanicum* moss in a raised bog at Moidach More, near Grantown on Spey, Scotland.

Values for microbial C in peats range from 0.2 to 2.3 g C per kilogram dry matter and are comparable with values from the organic horizons of forest soils when expressed on a dry weight basis (Williams and Sparling, 1988). Values for microbial C in the surface 15 cm of a blanket bog increased significantly on lowering the depth of the water table in surrounding ditches from 0 to 50 cm. In peat beneath lodgepole pine (*Pinus contorta*), microbial biomass C was less than in adjacent unplanted blanket peat, indicating that the physical and chemical

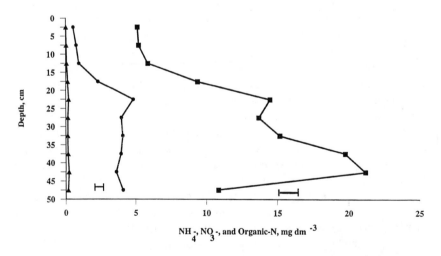

Figure 2   Amounts (mg dm⁻³ fresh peat) of NH₄ (●), NO₃ (▲), and organic-N (■) in the profile, 0- to 50-cm depth from the surface of *Sphagnum magellanicum* moss in a raised bog at Moidach More, near Grantown on Spey, Scotland.

**Table 1   Amounts (mg P dm⁻³) of P Extracted with
10 mM CaCl₂ from Peat Cores 1 d After
Treatment in situ with 112.5 kg N per Hectare.
Microbial-P Flush Measured After Fumigation
Overnight with Chloroform**

	Control	NH₄NO₃	sed[a]
PO₄–P	1.8	1.6	1.47
Microbial-P flush	17.1	26.3[b]	3.73

[a] sed = Standard error of difference.
[b] Significantly (p <0.05) greater than the control.

changes that result from afforestation, such as increased acidity (Williams et al., 1978), also reduce microbial activity and biomass. Incubation studies of peats under aerobic conditions showed that net mineralization of N was greater in samples from forested than unplanted peatland probably because of lower rates of immobilization (Williams et al., 1979). Hence peatland that has been drained and afforested could be expected to react with inorganic N quite differently from unplanted virgin bogs. Comparisons of microbial biomass N values for samples from seminatural and reseeded blanket bog indicated no changes as a result of liming and additions of N fertilizers, suggesting that N does not accumulate in the biomass (Williams and Wheatley, 1989). Microbial phosphorus, however, did show increases following fertilization of peat with superphosphate, and in highly organic soils can account for 25 percent of the total P (Williams and Sparling, 1984).

Microbial Activity

Rates of assimilation of N into the microbial biomass can be measured either by following the labeling of microbial N or by computing the rate of immobilization from the rate of dilution of the ^{15}N-labeled inorganic N pool (Davidson et al., 1991). Given a choice of ammonium and nitrate, soil microorganisms have been shown to prefer ammonium-N, although nitrate is utilized if present as the sole source and at low levels of inorganic N (Jansson et al., 1955). In a limed peat treated with ^{15}N-labeled ammonium or nitrate, the retention of ^{15}N in the microbial biomass 30 d after addition was significantly greater for ammonium than nitrate (Table 2). However, competition for nitrate by denitrifying microorganisms could have been an important factor in this limed peat.

The seasonality of the processes of mineralization-immobilization will determine the fate of added N at different times of the year. Zak et al. (1990) suggested that high uptake rates of N by the ephemeral understory vegetation in a hardwood forest during spring was a mechanism for N conservation in the ecosystem. However, their measurements showed that immobilization of labeled ^{15}N was almost ten times greater in the soil microbial biomass than in the understory vegetation. On blanket bog, peat improved for grazing by liming, net immobilization of inorganic-N tended to predominate during April and May (Williams and Wheatley, 1992). This pattern contrasted with results obtained beneath Sitka

Table 2 Amounts (mg core⁻¹) and ¹⁵N Abundance (atom %) of Extractable NH_4– and NO_3–N in Peat Cores 30 d After Treatment *in situ* with $^{15}NH_4$ or $^{15}NO_3$ (5 atom %) at 112.5 kg N per Hectare Applied as NH_4NO_3

	Control	$^{15}NH_4NO_3$	$NH_4{}^{15}NO_3$	sed[a]
NH_4–N	0.61	0.82	0.53	0.276
$^{15}NH_4$ atom %	0.367	0.719	0.412	0.048
NO_3–N	0.06	0.06	0.05	0.004
$^{15}NO_3$ atom %	0.367	0.381	0.380	0.004

[a] sed = Standard error of difference.

spruce (*Picea sitchensis* (Bong.) Carr.) planted in pure stands and in mixture with Scots pine (*Pinus sylvestris* L.) on a heathland soil, where net mineralization of N in the forest floor was greatest during spring (Williams, 1992). The high activity of N immobilization in the blanket bog during spring suggested that an easily decomposable carbon substrate was available, although its not clear whether this originated from the vegetation or the soil microbial biomass.

Rates of decomposition of organic matter, as measured by CO_2 production, appear to be reduced by additions of inorganic N to the forest floor (Baath et al., 1981), and this apparent negative effect on the breakdown of organic matter is not uncommon in material with C/N ratios >30. Additions of ¹⁵N-labeled ammonium sulfate of up to 87 kg N ha⁻¹ yr⁻¹ did not increase rates of decomposition of *Calluna vulgaris* (L.) Hull and *Molinia caerulea* (L.) leaves as measured by weight loss, but there was a temporary increase in the decomposition of the leaves of *Deschampia flexuosa* (L.) (van Vuuren and van der Eerden, 1992). Of the three litters used in this study, *Molinia* retained all N, whereas *Calluna* and *Deschampsia* released mainly unlabeled N. Several mechanisms have been suggested to explain the effects of added N on decomposition rates, ranging from chemical effects to competition between decomposer organisms and the repression of enzymes by ammonium (Fog, 1988). Some of these effects will be better explained from an improved understanding of the effects of inorganic N on the size and activity of the microbial biomass associated with decomposing litters.

Soluble Organic N

Dissolved organic N has received relatively little attention in soil N investigations, and its origin is not clear. In forest soils, throughfall can be enriched or depleted in soluble organic matter (Qualls et al., 1991), and other vegetation canopies behave in a similar way (Seastedt, 1985). Throughfall can also become enriched in soluble organic matter passing through the litter layer on the forest floor (Brown and Iles, 1991). Death and decay of microorganisms may also contribute to the soluble organic fraction. The mobility of this pool and its mineralization could have important implications for the eutrophication of surface waters. Studies on the composition of runoff waters draining a reseeded blanket bog showed that the organic component far exceeded the inorganic N fraction (Williams and Edwards, 1993).

It has been reported that *Sphagnum* mosses respond to additions of inorganic N by the production of amino acids (Baxter et al., 1992). Dadd et al. (1953) detected the presence of amino acids in waters extruded from *Sphagnum* mosses, and there have been concerted attempts to identify and determine the origin of free amino acids in soils (Ivarson and Sowden, 1966). The size of the soluble organic N pool in peats is much greater than the inorganic N, which consists almost entirely of ammonium (Figure 2); this raises questions about the susceptibility of the dissolved organic-N to ammonification. Isotope tracer studies in the forest floor of paper birch showed that ^{15}N label added as inorganic N quickly became incorporated into the water soluble organic N fraction (Van Cleve and White, 1980). This suggested that the fraction is relatively active, though the mechanism and roles of the soil microbial biomass and the vegetation in these processes are not clear. Depending on the molecular size and chemical nature of the N, the soluble organic N pool may simply be a product of rapid assimilation of inorganic N into the microbial biomass and a form of temporary immobilization. On the other hand, it could also contain recalcitrant organic molecules formed by chemical reactions (Fog, 1988).

ABIOTIC INTERACTIONS

Added ammonium N can enter into chemical exchange with native soil N, though some of the chemical reactions between organic matter and ammonium occur at high pH. The Maillard reaction between carbohydrates and ammonium or amino acids is a reaction that can occur in peats at ambient temperatures (Benzing-Purdie et al., 1986). Other chemical reactions include condensation reactions between proteins and phenols (Handley, 1961). These reactions tend to be slow relative to biological processes, but may be more significant when biological activity is low. Fertilizer urea N has been shown to interact chemically more strongly than ammonium when applied to organic soils. Urea applied to litter and humus horizons causes dissolution of organic matter and leaching (Ogner, 1972). In peats, the distribution of particle size fractions is changed (Benzing-Purdie et al., 1992). In the latter study, NMR spectroscopy showed that some of the complexes formed in a raised bog peat were derivatives of urea formed by direct reactions between organic matter and urea before its hydrolysis. There is no quantitative information on the proportion of urea N combined in this way or on the potential availability of the products for mineralization and plant uptake.

LOCATION OF IMMOBILIZED N

Peats are composed of plant residues, microorganisms, and fragments covering a range of sizes from >1 to <5 μm, and their total N content increases as the particles get smaller and more decomposed (Williams, 1983). Strickland et al. (1992) reported that after 60 d, ^{15}N-labeled ammonium added to mineral soils was located as organic N in the heavy fraction, indicating that the organo-mineral component of the soil was active in immobilizing N. Some of this N was released as dissolved organic N after incubation and sonication, and only a relatively small proportion was released from the killed biomass after fumigation with chloroform. Incubation of samples from a cutover raised bog with ^{15}N-labeled urea and ammonium sulfate showed that immobilized N is located in different particle size fractions of peat, depending on the form of the added N (Benzing-Purdie et al., 1992). In the case of ammonium sulfate, incorporated ^{15}N was found mainly in the fine fraction (<53 μm), where the microbial material would also be expected to reside, while a large proportion of exchangeable ^{15}N-labeled ammonium occurred in the water-soluble fraction. Urea N was found associated with larger particles, which may have been the result of the direct interaction between urea and plant debris as described above. In general, urea N is retained by acid organic soils for longer periods than ammonium or nitrate salts, and probably accounts for the long-term fertilizer effects of urea compared with inorganic forms (Johnson, 1979).

CONCLUSIONS

Peatland ecosystems behave as sinks for inorganic N inputs, and provided these amounts are small, the N is assimilated by the vegetation. At high inorganic N input rates from atmospheric deposition, the composition of bog vegetation may change, and sensitive species, such as *Sphagnum* mosses, may be damaged. Mosses have been shown to respond to added inorganic N by producing amino acids. Experiments with undisturbed peat cores demonstrated the marked removal of inorganic N and the predominance of extractable soluble organic N over inorganic forms in the profile of a raised bog. The mobility and turnover of this organic N could have important consequences for potential losses of N in drainage waters from peatlands. The effects of added inorganic N on rates of decomposition of organic matter are not clear, and it is uncertain if N alone can cause increased decomposition rates. Added inorganic N can increase microbial biomass P, and the impact of N may depend on the P status of the peat. Phosphorus and the concentrations of base cations may well have a controlling influence on the processes of assimilation and retention of inorganic N in peatland ecosystems.

ACKNOWLEDGMENTS

This work was financed by the Scottish Office of Agriculture and Fisheries Department and the European Commission. The authors thank Miriam Young for chemical analyses, Braemoray Estates, Scottish Natural Heritage, and Fountain Forestry International for access to experimental sites.

REFERENCES

Aerts, R. and Berendse, F., The effect of increased nutrient availability on vegetation dynamics in wet heathlands, *Vegetatio,* 76, 63, 1988.

Anderson, J. P. E. and Domsch, K. H., A physiological method for the quantitative measurement of microbial biomass in soils, *Soil Biol. Biochem.,* 1, 215, 1978.

Baath, E., Lundgren, B., and Soderström, B., Effects of nitrogen fertilization on the activity and biomass of fungi and bacteria in a podzolic soil, *Zentralbl. Bakteriol. Mikrobiol. Hygiene Originale,* 1 Abt., C 2, 90, 1981.

Baxter, R., Emes, M. J., and Lee, J. A., Effects of an experimentally applied increase in ammonium on growth and amino-acid metabolism of *Sphagnum cuspidatum* Ehrh. ex. Hoffm. from differently polluted areas, *New Phytol.,* 120, 265, 1992.

Benzing-Purdie, L. M., Cheshire, M. V., Williams, B. L., Ratcliffe, C. I., Ripmeester, J. A., and Goodman, B. A., Interactions between peat and sodium acetate, ammonium sulphate, urea or wheat straw during incubation studied by ^{13}C and ^{15}N NMR spectroscopy, *J. Soil Sci.,* 43, 113, 1992.

Benzing-Purdie, L. M., Cheshire, M. V., Williams, B. L., Sparling, G. P., Ratcliffe, C. I., and Ripmeester, J. A., Fate of [^{15}N] glycine in peat as determined by ^{13}C and ^{15}N CP-MAS NMR spectroscopy, *J. Agric. Food Chem.,* 34, 170, 1986.

Black, K. E., Lowe, A. H., Billett, M. F., and Cresser, M. S., Observations on the changes in nitrate concentrations along streams in seven upland moorland catchments in northeast Scotland, *Water Res.,* 27, 1195, 1993.

Brown, A. H. F. and Iles, M. A., Water chemistry profiles under four tree species at Gisburn N. W. England, *Forestry,* 64, 169, 1991.

Clymo, R. S., Productivity and decomposition of peatland ecosystems, in Proc. Int. Symp. Peatland Ecosystems and Man: An Impact Assessment, Bragg, O. M., Hulme, P. D., Ingram, H. A. P., and Robertson, R. A., Eds., University of Dundee, 1992, p. 3.

Dadd, C. C., Fowden, L., and Pearsall, W. H., An investigation of the free amino acids in organic soil types using paper partition chromatography, *J. Soil Sci.,* 4, 69, 1953.

Damman, A. W. H., Regulation of nitrogen removal and retention in *Sphagnum* bogs and other peatlands, *Oikos,* 51, 291, 1988.

Davidson, E. A., Hart, S. C., Shanks, C. A., and Firestone, M. K., Measuring gross nitrogen mineralization, immobilization, and nitrification by ^{15}N isotopic pool dilution in intact soil cores, *J. Soil Sci.,* 42, 335, 1991.

Department of the Environment, Impacts of Nitrogen Deposition in Terrestrial Ecosytems, Report of the United Kingdom Review Group on Impacts of Atmospheric Nitrogen, 1994.

Dickson, D. A., Effects of limestone, phosphate and potash on the early growth and nutrient uptake of Sitka spruce (*Picea sitchensis* (Bong.) Carr.) planted on deep peat in Northern Ireland, *Proc. 4th Int. Peat Congr.,* Finland, III, 1972, p. 479.

Edwards, A. C., Creasy, J., and Cresser, M. S., Factors influencing nitrogen inputs and outputs in two Scottish upland catchments, *Soil Use Manage.,* 1, 83, 1985.

Fog, K., The effect of added nitrogen on the rate of decomposition of organic matter, *Biol. Rev.,* 63, 433, 1988.

Francez, A.-J., Production Primaire et Accumulation de Matière Organique dans les Tourbières à Spaignes des Monts du Forez (Puy-de-Dôme). Influence des Activités Humaines sur leur Fonctionnement et leur Évolution, These de Doctorat de l'Université de Paris, 1991.

Handley, W. R. C., Further evidence for the importance of residual leaf protein complexes in litter decomposition and the supply of nitrogen for plant growth, *Plant Soil,* 15, 37, 1961.

Hemond, H. F., The nitrogen budget of Thoreau's bog, *Ecology,* 64, 99, 1983.

Hulm, S. C. and Killham, K., Response over two growing seasons of a Sitka spruce stand to ^{15}N-urea fertilizer, *Plant Soil,* 124, 65, 1990.

Ingram, H. A. P., Soil layers in mires: function and terminology, *J. Soil Sci.,* 29, 224, 1978.

Ivarson, K. C. and Sowden, F. J., Effect of freezing on the free amino acids in soil, *Can. J. Soil Sci.,* 46, 115, 1966.

Jansson, S. L., Hallam, M. J., and Stojanovic, B. F., Preferential utilization of ammonium over nitrate by micro-organisms in the decomposition of oat straw, *Plant Soil,* 6, 382, 1955.

Johnson, D. W., Some nitrogen fractions in two forest soils and their changes in response to urea fertilization, *Northwest Sci.,* 53, 22, 1979.

Klemmedtsson, L., Svensson, B. H., Lindberg, T., and Rosswall, T., The use of acetylene in quantifying denitrification in soils, *Swed. J. Agric. Res.,* 1, 179, 1977.

Morris, J. T., Effects of nitrogen loading on wetland ecosystems with particular reference to atmospheric deposition, *Ann. Rev. Ecol. Sys.,* 22, 257, 1991.

Ogner, G., Leaching of organic matter from a forest soil after fertilization with urea, *Meddeleser fra det Norske Skogsforsoksuesen,* 30, 425, 1972.

Pearson, J. and Stewart, G. R., The deposition of atmospheric ammonia and its effects on plants, *New Phytol.,* 125, 283, 1993.

Pyatt, D. G., Craven, M. M., and Williams, B. L., Peatland classification for forestry in Great Britain, Proc. Int. Symp. Classification of Peat and Peatlands, Hyytiälä, Finland, 1979, p. 351.

Qualls, R. G., Haines, B. L., and Swank, W. T., Fluxes of dissolved organic nutrients and humic substances in a deciduous forest, *Ecology,* 72, 254, 1991.

Rosswall, T. and Granhall, U., Nitrogen cycling in a subarctic ombrotrophic mire, in Ecology of a Subarctic Mire, Sonesson, M., Ed., *Ecol. Bull.,* Stockholm, 30, 209, 1980.

Seastedt, T. R., Canopy interception of nitrogen in bulk precipitation by annually burned and unburned tallgrass prairie, *Oecologia,* 66, 88, 1985.

Stevens, P. A. and Wannop, C. P., Dissolved organic nitrogen and nitrate in an acid forest soil, *Plant Soil,* 102, 137, 1987.

Strickland, T. C., Sollins, P., Rudd, N., and Schimel, D. S., Rapid stabilization of ^{15}N in forest and range soils, *Soil Biol. Biochem.,* 24, 849, 1992.

Urban, N. R., Eisenreich, S. J., and Bayley, S. E., The relative importance of denitrification and nitrate assimilation in midcontinental bogs, *Limnol. Oceanogr.,* 33, 1611, 1988.

Van Cleve, K. and White, R., Forest-floor nitrogen dynamics in a 60-year-old paper birch ecosystem in interior Alaska, *Plant Soil,* 54, 359, 1980.

Verhoeven, J. T. A., Maltby, E., and Schmitz, M. B., Nitrogen and phosphorus mineralization in fens and bogs, *J. Ecol.,* 78, 713, 1990.

van Vuuren, M. M. I. and van der Eerden, L. J., Effects of three rates of atmospheric nitrogen deposition enriched with ^{15}N on litter decomposition in a heathland, *Soil Biol. Biochem.,* 24, 527, 1992.

Williams, B. L., The nitrogen content of particle size fractions separated from peat and its rate of mineralization during incubation, *J. Soil Sci.,* 34, 113, 1983.

Williams, B. L., The influence of peatland type and the chemical characteristics of peat on the content of readily mineralized nitrogen, Proc. 7th Int. Peat Congr. Dublin, 4, 410, 1984.

Williams, B. L., Nitrogen dynamics in humus and soil beneath Sitka spruce (*Picea sitchensis* (Bong.) Carr.) planted in pure stands and in mixture with Scots pine (*Pinus sylvestris* L.), *Plant Soil,* 144, 77, 1992.

Williams, B. L., Cooper, J. M., and Pyatt, D. G., Effects of afforestation with *Pinus contorta* on nutrient content, acidity and exchangeable cations in peat, *Forestry,* 51, 29, 1978.

Williams, B. L., Cooper, J. M., and Pyatt, D. G., Some effects of afforestation with lodgepole pine on rates of nitrogen mineralization in peat, *Forestry,* 52, 151, 1979.

Williams, B. L. and Edwards, A. C., Processes influencing dissolved organic nitrogen, phosphorus and sulphur in soils, *Chem. Ecol.,* 8, 203, 1993.

Williams, B. L. and Sparling, G. P., Extractable nitrogen and phosphorus in relation to microbial biomass in acid organic soils, *Plant Soil,* 76, 139, 1984.

Williams, B. L. and Sparling, G. P., Microbial biomass carbon and readily mineralized nitrogen in peat and forest humus, *Soil Biol. Biochem.,* 20, 579, 1988.

Williams, B. L. and Wheatley, R. E., Nitrogen mineralization and water-table height in oligotrophic deep peat, *Biol. Fertil. Soils,* 6, 141, 1988.

Williams, B. L. and Wheatley, R. E., Nitrogen transformations in poorly drained reseeded blanket peat under different management treatments, *Int. Peat J.,* 3, 96, 1989.

Williams, B. L. and Wheatley, R. E., Mineral nitrogen dynamics in poorly drained blanket peat, *Biol. Fertil. Soils,* 13, 96, 1992.

Williams, B. L. and Young, M. E., Nutrient fluxes on reseeded blanket bog, limed and fertilized with urea, phosphorus and potassium, *Soil Use Manage.,* in press.

Woodin, S., Press, M. C., and Lee, J. A., Nitrate reductase activity in *Sphagnum fuscum* in relation to wet deposition of nitrate from the atmosphere, *New Phytol.,* 99, 381, 1985.

Zak, D. R., Groffman, P. M., Pregitzer, K. S., Christensen, S., and Tiedje, J. M., The vernal dam: plant-microbe competition for nitrogen in northern hardwood forests, *Ecology,* 71, 651, 1990.

Wetland Management

CHAPTER 26

Forest Management Impacts on Undrained Peatlands in North America

D. F. Grigal and K. N. Brooks

CONTENTS

INTRODUCTION

In this chapter, the effects of timber harvesting, site preparation, and associated activities on forested peatland ecosystems in North America are discussed.

1-56670-177-5/97/$0.00+$.50
© 1997 by CRC Press, Inc.

The impacts on soils and hydrological processes are emphasized, and these effects are related to the sustainability of productivity.

In contrast to most discussions of forest management of peatlands, this chapter will consider only undrained systems. In nearly all forested peatlands in Europe, forest management entails drainage. In fact, Heikurainen (1964) noted that drainability is one of the major criteria used to classify peatlands for forestry purposes. In North America, because of both economic considerations and more recent environmental concerns, peatland drainage has rarely been practiced. As a consequence, this chapter departs from much of the literature discussing peatland forestry and speculates in instances where little has been published.

FOREST MANAGEMENT PRACTICES

Forest management on peatlands in eastern North America is nearly synonymous with management of black spruce (*Picea mariana*). Although other species such as tamarack (*Larix laricina*) and eastern white-cedar (*Thuja occidentalis*) also occur on peatlands and wet mineral soils, black spruce is dominant. Productivity of peatland forests is often low, however. For example, in Ontario only about half of the estimated area of black spruce-dominated peatland forest (8 of 15.7 million ha) is considered to be productive (capable of producing a merchantable yield of wood) (Haavisto et al., 1988). In the Great Lakes states of Minnesota, Wisconsin, and Michigan, similarly defined black spruce timberland occupies 900,000 ha (Raile and Smith, 1983; Spencer et al., 1988; Miles and Chen, 1992).

Because black spruce is the focus of most forest management on peatlands of North America, this chapter emphasizes its management. Clearcutting is the recommended method of harvesting black spruce on peatlands in the Great Lakes states (Johnston, 1977). Forest management in Ontario similarly involves clearcutting, often followed by site preparation, regeneration practices, and sometimes tending practices (Jeglum and Kennington, 1993). Recent silvicultural approaches include taking advantage of advanced reproduction to regenerate the site.

In Ontario, stripcutting of black spruce has been practiced since the early 1900s (Jeglum and Kennington, 1993). With this practice, uncut strips are left as a seed source for natural regeneration. These strips are then harvested after the cut areas begin to regenerate. In some cases these seed-tree strips may be left uncut, but at a cost of lost harvest volume. An alternative to the strip-cut method is small patch-cuts or clearcuts, leaving a "checkerboard" pattern. Patches must be kept small enough so that they can benefit from seed dispersal from nearby stands.

When site preparation is deemed necessary, either mechanical treatments or prescribed burning can be used (Jeglum and Kennington, 1993). If the newly harvested and regenerating site has intense competition from species such as speckled alder (*Alnus rugosa*), trembling aspen (*Populus tremuloides*), willow (*Salix* spp.), or sedges and grasses, either chemical or manual weeding can be used. Because North American peatlands are not usually drained, competition from woody species normally found on uplands, such as aspen, is reduced.

EFFECTS ON SOILS

The effects of forest management activities on soils will be discussed with respect to biogeochemical and physical properties. Soil biogeochemical properties of interest are those related to plant nutrients and their availability. Because of the low nutrient status of many peatlands and the reduced effectiveness of natural processes of nutrient replacement as compared to uplands, nutrient removal by harvesting may reduce the ability of a site to sustain productivity. The physical properties of soils are largely affected by the methods used to extract forest products, including road construction, skid trails, and logging decks. The effects of heavy equipment are usually concentrated in specific areas and can alter properties so that soils are no longer productive. In addition, such alteration can influence hydrological properties and processes of the soils and thus of the entire peatland system for long periods of time.

Biogeochemical Properties

Product Removal

Removal of forest products removes nutrients from forest systems; nutrient elements are an integral part of the biomass that is harvested. Three critical questions must be addressed when evaluating the impact of nutrient removal on continued site productivity — the initial nutrient capital of the site, the amount removed by harvest, and the natural rate of replacement. Even approximate answers to these questions can provide insight into the potential impacts of harvest on black spruce forests.

Initial Nutrient Capital

Peatlands vary widely in nutrient status, from highly minerotrophic fens to ombrotrophic bogs. However, a review of the literature for forested peatlands (Damman, 1978; Paavilainen, 1980; Waughman, 1980; Grigal and Nord, 1983; Waughman and Bellamy, 1984; Grigal, 1991) indicates a relatively narrow range of elemental concentration and content of peat within the acrotelm, the surface layer of biological activity above the water table (Ingram, 1978). Nitrogen (N) and calcium (Ca) average near 10 mg g^{-1}; phosphorus (P) and potassium (K) average near 1 mg g^{-1}; and although magnesium (Mg) is more variable, it averages near 3 mg g^{-1}.

Both Ca and Mg concentrations are especially influenced by the degree of minerotrophy of the peatland via inputs from surface flow and groundwater. The proximity of a specific peatland to the sea and the resulting atmospheric deposition of sea salts through precipitation can also affect Mg concentrations. Although inputs of N, P, and K are derived from groundwater and precipitation, their concentrations in peatlands are more uniform. Nitrogen, and to a lesser degree Ca, are integral components of the organic materials in peatlands, and the organic matter to N ratio is relatively constant among peatlands. Phosphorus is present in low concentrations in both groundwater and precipitation, and organic soil materials do not strongly retain P unless they have high contents of aluminum and iron-bearing minerals (Richardson, 1985). Potassium concentrations in groundwater are low, and K is neither retained structurally in the organic molecules of peat nor is well retained by ion exchange reactions compared to the divalent cations Ca and Mg (Russell, 1973).

Assuming an acrotelm thickness of 25 cm and a bulk density of ca. 0.1 g cm^{-3}, the total storage of nutrients in the rooting zone of peatlands ranges from 2500 kg ha^{-1} for N and Ca, to 750 kg ha^{-1} for Mg, to 250 kg ha^{-1} for P and K. The low concentrations and quantities of P and K represent potential deficiencies for forest production, and hence fertilization is a required part of intensive management (Paavilainen, 1980). The need for N fertilization in peatlands is usually related to N availability and not to its total amount.

Nutrient Removal in Harvest

The amount of nutrients removed by harvest depends on both the total amount of product and on nutrient concentrations in that product. Yield tables (Plonski, 1956; Perala, 1971) quantify the obvious; gross stem volume of spruce stands increases with stand age and site quality. To determine nutrient loss, however, other tree components must be quantified. Stand basal area is a measure that allows conversion between yield tables and nutrient status. Based on Plonski's (1956) yield tables, volume of black spruce increases as

$$\ln (VL) = 2.34 + 0.093 \ BA, \ r^2 = 0.98, \ s_{y.x} = 0.15 \qquad (1)$$

where ln (VL) is the natural logarithm of gross total volume in $m^3 \ ha^{-1}$, and BA is stand basal area in $m^2 \ ha^{-1}$. Because this equation is based on smoothed data from the yield tables, the fit is very good.

Data from studies of biomass and nutrient status of black spruce stands (Table 1) can be fit to a similar relationship:

$$\ln (BIOM) = 2.76 + 0.062 \ BA, \ r^2 = 0.80, \ s_{y.x} = 0.40 \qquad (2)$$

where ln (BIOM) is the natural log of total aboveground biomass and is expressed in Mg ha^{-1}. These studies (Table 1) also provide data to determine the mass and nutrient content of foliage, crown, and woody bole as a function of stand basal area (Table 2). Although the studies span a wide range of locations and site quality in North America, including both uplands and peatlands, the data have surprisingly low variability (Figure 1). Although stem volume and biomass increase rapidly with basal area, crown and foliage mass increase more slowly. This reflects competition for space in canopies of closed stands. Removal of nutrients, measured as mass per unit area, increases with basal area (and volume and biomass) of the harvested stand. As basal area increases, however, removal will be proportionally less for nutrients with greater proportions in the crown.

Table 1   Sources of Data for Estimation of Biomass and Nutrient Content of Black
Spruce Stands in North America

Source	Location	Age (year)	Basal area (m² ha⁻¹)	Aboveground biomass (Mg ha⁻¹) Total	Aboveground biomass (Mg ha⁻¹) Crown
Van Cleve (1981)	AK	51	8	17	9
Grigal (1991)	MN	75	10	31	7
Van Cleve (1981)	AK	55	11	24	10
Foster and Morrison (1987)	ON	110	23	142	35
Van Cleve et al. (1983)	AK	95	25	50	14
Van Cleve (1981)	AK	130	27	113	27
Grigal (1991)	MN	110	27	101	18
Freedman et al. (1982)	NS	75	28	113	39
Timmer et al. (1983)	ON	126	32	138	23
Gordon (1983)	ON	48	33	141	36
Gordon (1983)	ON	108	40	161	40
Weetman and Webber (1972)	NB	65	42	107	19
Timmer et al. (1983)	ON	126	49	325	39

To quantify these relationships, tree-length harvest (removing only woody boles and bark) of a black spruce stand with basal area of 30 m²/ha removes approximately 75 Mg ha⁻¹, or about 170 m³ ha⁻¹. The harvest will also remove about half of the nutrients in the aboveground biomass. Although this removal is important, removal with respect to the total nutrient capital of the site is probably more important. This tree-length harvest would remove 13 kg ha⁻¹ P, or 5 percent of the nutrient capital of the acrotelm; 60 kg ha⁻¹ K, 25 percent; 139 kg ha⁻¹ N, 6 percent; 238 kg ha⁻¹ Ca, 10 percent; and 20 kg ha⁻¹ Mg, 3 percent. These proportions from this simple computation indicate that the most serious threat of nutrient depletion appears to be associated with K loss.

Although tree-length harvest is common, full-tree harvest (where the entire aboveground portion is removed) of black spruce is also common; this exacerbates the removal of nutrients. Even if the tree crown is removed at the landing, nutrients therein are effectively removed from the growing site. With full-tree harvest of a 30 m² ha⁻¹ stand, the following nutrient capital of the acrotelm is removed: 27

Table 2   Aboveground Biomass (Mg ha⁻¹) and Nutrient Content
(kg ha⁻¹) of Black Spruce Stands as a Function of Stand
Basal Area (m² ha⁻¹)

Dependent variable	Total aboveground b0	Total aboveground b1	Total aboveground r^2	Total aboveground $S_{y,x}$	Bole and bark b0	Bole and bark b1	Bole and bark r^2	Bole and bark $S_{y,x}$
Biomass	2.78	0.061	0.80	0.40	2.22	0.070	0.80	0.46
Nitrogen	3.87	0.057	0.68	0.50	2.86	0.069	0.60	0.74
Phosphorus	1.54	0.058	0.78	0.40	0.52	0.069	0.67	0.63
Potassium	3.05	0.055	0.67	0.50	1.97	0.071	0.67	0.65
Calcium	3.96	0.065	0.65	0.62	3.46	0.067	0.63	0.67
Magnesium	2.07	0.051	0.69	0.44	1.14	0.062	0.67	0.57

Note: Data from studies of 13 stands reported in the literature; stand details in Table 1. Equations of the form $\ln_e(Y) = b0 + b1 \ast$ basal area. Standard error of the estimate expressed in logarithmic units.

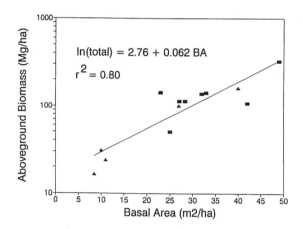

Figure 1   Relationship between aboveground biomass and basal area of black spruce stands sampled across North America (see Table 1). Stands located on peatlands indicated by filled triangles, upland stands by squares.

kg ha^{-1} P, 11 percent; 110 kg ha^{-1} K, 45 percent; 265 kg ha^{-1} N, 11 percent; 367 kg ha^{-1} Ca, 15 percent; and 37 kg ha^{-1} Mg, 5 percent. Thus, when full-tree harvest is practiced, proportional depletion of K appears to be even more serious.

There are other issues that must be considered when advantages and disadvantages of tree-length vs. full-tree harvest are considered. For example, full-tree harvesting has been advocated as a method to dispose of slash and aid spruce reproduction, and it may lead to greater yield of wood fiber per unit harvested area because of efficiencies at the landing (Johnston, 1975).

Mechanisms of Nutrient Restoration

There are three natural processes that potentially restore nutrients that are lost through harvest: mineral weathering, atmospheric deposition, and inputs via water flow. Because peatland soils are organic, mineral weathering is not important. Atmospheric deposition can, however, restore lost nutrients. Grigal and Bates (1992) established relationships between ionic concentrations in precipitation and latitude and longitude for the Great Lakes states and surrounding area. Based on these relationships, annual precipitation, and a continent-wide relationship that dry deposition is approximately equal to wet deposition (Johnson and Lindberg, 1992), P lost via tree-length harvest of a 30 m^2 ha^{-1} stand would be replaced in about 25 years, while losses in full-tree harvest would require 50 years. K would be restored in about 96 to 180 years; N in 10 to 18 years; Ca in 100 to 150 years; and Mg in 200 to 350 years. These data indicate that K, Ca, and Mg are the nutrients with the lowest rates of natural replacement via atmospheric deposition.

By definition fens are minerotrophic, and significant base cations can be supplied to these peatlands by groundwater and lesser amounts by surface flow. Ca and Mg are especially important constituents of groundwater flow. Although groundwater chemistry varies widely, groundwater from 15 fen peatlands in northern Minnesota averaged 1.2 Mg L^{-1} K, 0.08 Mg L^{-1} P, 1.8 Mg L^{-1} N, 7.9 Mg L^{-1} Mg, and 27 Mg L^{-1} Ca (Clausen and Brooks, 1983). Total quantities of these ions retained by peatlands are difficult to estimate. Verry and Timmons (1982) measured retention of ions for a small peatland in northern Minnesota. This site was weakly minerotrophic, low in nutrient status, and thus could be expected to provide an upper limit on ion retention. Of the ions delivered by surface flow and interflow (excluding precipitation inputs), the peatland retained 10 percent of the K, 20 percent of the P and Mg, 40 percent of the Ca, and all of the N. At the concentrations cited above and with runoff of 30 cm per year, P lost by tree-length harvest would be replaced by water flow in about 250 years, K in 170 years, N in 25 years, Ca in 7 years, and Mg in about 4 years. Phosphorus and K are likely to be the nutrients with lowest rates of natural replacement via water flow.

Nearly all soil profiles in peatlands show a sharp decrease in K concentration with depth (Damman, 1986). Based on ratios of K to Rb in bog systems, the K that is present is concentrated at the surface by efficient biocycling by living vegetation (Buttleman and Grigal, 1985). This pattern is a consequence of the low rate of natural K replacement in peatlands and implies a high potential for K deficiencies to occur with intensive harvest. Low levels of K and P are common in peatlands throughout northern areas, and are often mitigated by fertilization in Fennoscandinavia (Paavilainen and Päivänen, 1988).

Modification of Organic Matter Decomposition

A second major mechanism by which forest harvesting can alter the nutrient status of a site is by enhanced elemental transport in streamflow following the harvest. Increased export of nutrients after harvesting has been documented for a variety of upland sites (Hornbeck et al., 1986; Swank, 1988). One important cause of this loss is increased water yield, which increases transport of nutrient ions from the system even if their concentrations in waters do not increase. Nutrient concentrations usually also increase in uplands due to reductions in uptake because of the removal of the overstory and because of increases in rates of nutrient release from decomposing organic debris, including slash and forest floor materials (Hornbeck et al., 1986). Peatland systems are not completely analogous to uplands; the same mechanisms are operative, albeit in different relative importance. For example, increased water yield and associated nutrient loss may or may not occur in peatlands. Effects of harvesting undrained peatlands on water yield vary with the characteristics of the individual peatlands and are discussed later in this chapter.

Effects of harvest on rates of nutrient uptake and organic matter decomposition are not well documented for undrained peatlands. Increased soil temperature following harvest on peatlands leads to an increased potential for decomposition (Trettin and Jurgensen, 1992) and is likely to increase nutrient concentrations in drainage waters. Nicolson (1988) examined effects of black spruce harvest in northern Ontario. He studied small basins primarily containing shallow soils over bedrock, with approximately 20 percent of the basins composed of peatlands and wet mineral soils. In these systems, water yield increased with harvest from 33 to 47 percent of annual precipitation over the 7 years of measurement. More important for implications to peatland harvest, however, were the documented increases in concentrations of nutrient elements; Ca and Mg concentration increased by 1.5 times and K by 2 times compared to uncut controls. Neither total N nor total P increased in concentration. These changes imply increased decomposition and cation release but retention of N and P by decomposers. Increases in water yield led to increased loss of all nutrients, but a greater proportional loss of the cations.

Knighton and Stiegler (1980) studied P dynamics following spruce harvest and slash burning on both a bog and a fen in northern Minnesota. In both cases, concentrations of P in streamflow increased, leading to losses over a 6-year postharvest period of about 0.5 kg ha^{-1} from the bog and about 3.5 kg ha^{-1} from the fen. Compared to controls, increased losses of P from the bog occurred even before the slash was burned. Phosphorus concentrations in waters from the fen were highest in warm-wet years, implying increased microbial decomposition. Laboratory studies indicated release of P following both incubation and simple air-drying of the peat. Fluctuating water levels and associated cycles of wetting and drying following harvest may enhance P loss. The differences in behavior of P in the systems studied by Nicolson (1988) and Knighton and Stiegler (1980) are likely due to the important role of mineral material in P retention (Richardson, 1985).

Physical Properties

Rutting/Compaction

Forest harvest can also affect the physical properties of peatland soils. Soil compaction and rutting can reduce infiltration, redirect flow, and alter pathways by which water moves through and from a peatland, thus affecting the character of the peatland itself. Jeglum et al. (1983) consider physical damage to the site to be one of the two most severe problems in peatland forestry in Ontario (the second being inadequate regeneration). Unfortunately, little quantitative information exists on effects of harvest on physical properties and thence on hydrological processes. Anecdotal data and extrapolations from specific events must be used to assess potential changes.

In undrained and undisturbed peatlands, saturated hydraulic conductivity diminishes with peat decomposition and with depth. In northern Minnesota, saturated hydraulic conductivities of peat characteristic of the upper 20 cm of peatlands range from about 40 to over 270 m d^{-1}, whereas conductivities of deeper peats diminish from 7 to less than 0.5 m d^{-1} (Gafni and Brooks, 1990). As a result, nearly all horizontal water flow in peatlands occurs in the acrotelm, the surface zone of macropores and high hydraulic conductivity (Ingram, 1978).

Because of these macropores, the acrotelm is susceptible to disturbance by harvesting (Berry and Poskitt, 1972). This disturbance affects the saturated hydraulic conductivity and in turn affects the hydrology of the entire peatland. Peatland hydrology can be characterized by the occurrence of a shallow water table and active movement of water through the acrotelm. Disruption of flow across the acrotelm can significantly affect tree growth. For example, crossing peatlands with roads or pipelines, without adequate provision for cross-drainage, frequently leads to flooding on the upslope side and subtle drainage on the downslope side of such crossings (Stoeckeler, 1967; Boelter and Close, 1974). This changes the hydrological character of the peatland and removes significant areas from tree growth.

Summer harvest on peatlands can markedly affect the soil surface. For example, summer harvest of peatlands in Ontario produced rutting on about 15 percent of the surface; nearly all were deep ruts when a conventional rubber-tired skidder (tires 60 cm wide) was used, but were shallow ruts with a wide-tire skidder (127 cm) (Groot, 1987). Winter harvest with a narrow-tired skidder disturbed about 10 percent of the surface with shallow ruts. Shallow ruts, defined as having evidence of compaction, scraping, or shallow mixing, have a minor impact on system hydrology. Conversely, water-filled and heavily disturbed deep ruts may significantly affect surface water flow (Groot, 1987). A survey of over 5000 ha in Ontario that had been harvested primarily in summer with conventional narrow-tired skidders found that over 40 percent of the area was silviculturally untreatable because of deep rutting (Groot, 1987). This statistic relates strictly to operability and does not directly address the long-term hydrological consequences of the site disturbance.

There are many economic circumstances that necessitate summer harvest, including peatlands that comprise a large proportion of a geographic area, and the necessity for year-round operations (Barrett, 1966). Such operations have the potential to severely damage approximately one-third of the peatlands by deep-rutting (Sleep, 1979). Although this proportion is greater than that reported experimentally by Groot (1987), the difference may be related to secondary disturbance of the hydrological system.

Winter harvest is an obvious remedy to these dramatic site disturbances. For example, because of environmental considerations, right-of-way construction for an electric transmission line across a large peatland in Minnesota was carried out exclusively in winter. Evidence of obvious disturbance was minimal, with no apparent vehicle tracks or other soil displacement (Grigal, 1983). Subtle increases in acrotelm bulk density (about 5 percent) were measured in one vegetation type, a treed fen, but not in four other types that were monitored (Grigal, 1983). Based on established relationships (Boelter, 1969), hydraulic conductivity was estimated to remain high, at 3.6 m d^{-1}, even where the maximum changes in bulk density had occurred. One caveat with winter harvest is the freezing depth into the peat. Early winter operations or heavy snow cover can minimize the depth of freezing and keep the soil in a fragile state.

EFFECTS ON HYDROLOGICAL PROCESSES

Much of the research on the hydrology of forested peatlands has dealt with effects of drainage and the response of drained forest lands to various management schemes. This discussion considers the hydrology of undrained peatlands and what is known about forest management effects on water quantity and quality, using the limited research and empirical data that are available. Forested peatlands in North America occur where there is an excess of water, usually where annual precipitation exceeds potential evapotranspiration and where water flow velocities are extremely low. Consequently, the hydrology of peatlands is governed by shallow groundwater, organic soils, flat topography, and vegetation that makes these ecosystems unique (Boelter and Verry, 1977; Brooks, 1992). The depth to the water table, whether it is a perched groundwater or regional groundwater system, determines evapotranspiration (for a given input of energy) and the runoff response of peatlands at any point in time. The source of water, whether regional groundwater or precipitation, influences the type and density of vegetation, as well as the quality of water that is moving through or being discharged from a peatland (Glaser et al., 1981; Clausen and Brooks, 1983).

Northern forested peatlands generally yield the greatest amount of runoff during the snowmelt and early spring period when water tables are normally near or above the soil surface (Boelter and Verry, 1977). Later in the growing season, evapotranspiration losses become dominant, water tables recede, and streamflow from and groundwater flow through peatlands diminish. By late summer, water tables on ombrotrophic bogs can drop to the point that surface discharge can become negligible. Streamflow from minerotrophic fens, on the other hand, is less variable and maintains greater discharge throughout the summer (Boelter and Verry, 1977).

Forest management affects the hydrology of undrained peatlands in terms of specific hydrological processes and the overall hydrologic response of peatland watersheds. It also affects water yield, pattern of flow, and water quality, while processes such as evapotranspiration, snowmelt, runoff, and groundwater flow are also changed by forest management.

Water Yield

Clearcutting of forest overstory is generally considered to be one of the most drastic forest management practices affecting water yield. In most instances, clearing results in an increase in water yield due to the reduction of evapotranspiration (Whitehead and Robinson, 1993). However, strip cutting and clearcutting black spruce in an undrained bog in northern Minnesota had no effect on annual water yield (Verry, 1980). In this case, clearcutting produced higher water yield in the wet springtime, but these increases were offset by lower yields during dry periods. Removal of the overstory reduced interception losses in the rainy season, but these effects were apparently offset by higher evapotranspiration rates by residual vegetation during the summer. Therefore, there was no effect of harvesting on annual water yield.

In drained peatlands, water tables have been reported to rise following clearcutting as a result of reduced transpiration and interception (Heikurainen, 1967), a phenomenon referred to as "watering up." Whether water tables rise and the extent of the rise apparently depend on the depth of the water table prior to harvesting, the type of vegetation, and amount of biomass removed (Heikurainen, 1967; Heikurainen and Päivänen, 1970; Päivänen, 1982). If water tables are shallow prior to harvest, there tends to be little effect of harvesting on residual water tables. This response seems to be consistent with that reported for undrained peatlands in Minnesota, although some evidence of watering-up has been reported for undrained peatlands in Ontario (Haavisto et al., 1986). In any case, because depth of water table governs both evapotranspiration and streamflow generation from peatlands, the water yield response to forest harvesting is directly linked to water table behavior.

The effects of summer harvesting on water yield may differ if surface soils become rutted and compacted, as described earlier. If saturated hydraulic conductivities decrease sufficiently, more surface runoff could occur during spring and early summer, with the possibility of a higher annual water yield. However, the literature does not suggest such an effect for undrained peatlands.

Streamflow Pattern

Streamflow pattern refers to changing levels of streamflow over time, but greatest interest is on the extreme high and low flow durations and amounts. Examples of questions in this area would be, "Will forest harvesting increase peak flows with a potential for increased flooding?" or "Will harvesting reduce streamflows during low-flow periods?" Some interesting effects of forest management have emerged with respect to these questions.

In the study by Verry (1980), stripcutting and clearcutting of black spruce did not increase annual streamflow, but changed streamflow pattern. When forest overstory was removed, interception losses were reduced. This caused water tables to rise 10 cm over that predicted from the behavior of a control watershed, resulting in higher discharges during spring and early summer when rainfall amounts are the greatest. However, during mid to late summer, water tables fell 19 cm below predictions; this reduced discharges during the dry season. Reduced water tables and streamflow during the summer were caused by higher evapotranspiration and were considered to be the result of

- higher available energy and wind velocities in the clearcut area at the peat soil surface,
- a nearly fivefold increase in grass and sedge biomass, which, coupled with the fact that these plants are physiologically more active than are black spruce, resulted in higher transpiration rates.

As Verry (1980) points out, this response differs from that expected had the area been drained. In the latter case, as mentioned earlier, water tables rise and streamflow increases (Heikurainen, 1967). In the undrained case, grasses and sedges exposed to higher energy exchanges transpired more water than the spruce overstory that had previously occupied the site. Plant physiology, and specifically stomatal response of plants, thus influenced overall hydrological response.

The effects of winter harvesting on soil surface compaction are expected to be negligible, as described earlier, and would therefore not usually affect the runoff response (and thus the flooding response) of peatlands to rainfall events. However, summer harvesting has the potential to alter the thickness of the acrotelm and its characteristics via rutting and enhanced organic matter decomposition. This disturbance can alter the water table–streamflow relationship of the peatland. More surface flow could result, although the absence of steep hydraulic gradients and the lack of an efficient conveyance system (ditches) will prevent these undrained peatlands from responding quickly to rainfall. The effects of summer harvesting over vast areas could have the potential to increase localized runoff and flooding.

Snowmelt runoff from forested areas can differ from cleared areas in the flat to rolling topography that is typical of the Great Lakes region of the United States. Whether drained or undrained, the snowpack on cleared areas will melt earlier in the season and at a faster rate than that within forested areas. With these differences in snowmelt, a mixture of cleared and forested areas within a given watershed exhibits a desynchronized snowmelt runoff response. This has implications in terms of peak discharges and the potential for flooding from snowmelt. When between 20 and 50 percent of a watershed is clearcut, peak discharges from snowmelt can actually be reduced compared to those from fully forested areas; however, when greater than 60 percent of a watershed area is clearcut, such peak flows and thus the potential for flooding can increase dramatically (Verry et al., 1983). If strip-cutting practices are followed for black spruce, the desynchronization of melt should diminish snowmelt peaks, or at the minimum, not cause them to increase. As larger percentages of a watershed become cleared, the snowmelt peak discharges could be expected to remain at higher levels than before cutting for extended periods of time because of the slow growth rate of black spruce on undrained peatlands. From a hydrological perspective, therefore, regenerating black spruce stands of ages up to 30 to 50 years, depending on site quality, could be considered to be equivalent to clearcut conditions. A watershed perspective is needed to address such issues, because the percentage of a watershed in various cover conditions governs streamflow response. The influence of forest management on seasonal groundwater yield would be expected to parallel that of surface water but could be lagged in time, depending on the surface–groundwater linkage.

The above effects would be important in designing culverts for roads in forested and cleared areas. Because of their lack of topographic relief and poor conveyance systems, undrained peatlands naturally attenuate floods from rainfall and snowmelt events (Verry et al., 1983). Therefore, unless vast areas are clearcut, forest harvesting would not be expected to cause increased flooding.

Water Quality

Water quality characteristics of peatlands in northern Minnesota have been described by Glaser et al. (1981) and Clausen and Brooks (1983). In general, the export of nutrients in streamflow is higher from fens than bogs. Water quality characteristics of pH, specific conductance, alkalinity, and Ca and Mg concentrations can all be used to differentiate fens from bogs. When considering water quality changes due to forest management, it is important to recognize that significant natural differences can be expected. To date, few reports have been published concerning nutrient export via streamflow or groundwater associated with forest management on undrained peatlands. A general review of effects of forest management on water quality in various wetland types has been published (Shepard, 1994). However, most of this review was concerned with southern forests and drained systems.

The delivery of sediment to water bodies as a result of logging and associated activities is of concern in many forest management schemes. In undrained peatlands, mineral sediments are unlikely to be a problem, although organic soil particles may be transported from logging sites and can be considered as a nonpoint source pollutant. Paired watershed experiments in northern Minnesota have not reported increases in sedimentation following clearcutting of either the upland or peatland portions of the watersheds (Verry, 1986). The export of nutrients from peatlands following forest harvest is not likely to be of serious concern in terms of adversely affecting quality of receiving waters. The long residence time of water within the shallow water table of undrained peatlands and the opportunities for exchange with peat soils suggest that water quality changes due to harvesting would be minimal. However, as discussed earlier, the implications of even small losses of nutrients on site productivity can be important.

We have previously discussed the effects of site preparation on water quality following clearcutting of a minerotrophic fen and ombrotrophic bog in northern Minnesota (Knighton and Stiegler, 1980). Phosphorus export in streamflow was studied because of the implications for downstream receiving waters; small increases in P loading in Minnesota lakes can lead to eutrophication. The clearcutting and slash burning increased P concentrations from both peatlands. The increases in concentrations were greatest the first year following burning of slash. For the bog, P export was 0.9 kg ha^{-1} the first year after harvest, compared to 0.3 kg ha^{-1} before harvest; after 2 years, levels of P export returned to preharvest levels. For the fen, P export was 2.2 kg ha^{-1} the first year following harvest, in contrast to 0.7 kg ha^{-1} before harvest, and remained higher than preharvest levels for 6 years.

EFFECTS ON RIPARIAN SYSTEMS

The current concern for wetlands and riparian systems has become a dominant factor affecting the way in which natural resources are managed; this represents a challenge to forest managers. This is also the case for black spruce management on undrained peatlands, which represent some of the largest contiguous wetlands in North America. Management activities that are carried out on black spruce peatlands, and particularly road construction and skid trail and loading deck requirements for logging, will affect these wetlands. These effects, in large part, can be minimized with proper planning to minimize the area that is disturbed. Winter harvesting is perhaps the best method to minimize effects on soil and water resources.

The occurrence of streams and lakes within peatlands presents special management considerations, just as they do in mineral soil systems and upland forests. In most undrained peatlands, there is not a dense drainage system, and in fact often there is not a well-defined channel or watershed boundary. Streams flowing through such peatlands and lakes occurring within watershed boundaries should receive special attention by managers. Jeglum and Kennington (1993) recommend a "terrain-adapted" strip-cut approach for black spruce management that allows for irregular strips to be left along streams, lakes, wetlands, and other landscape features as desired by managers. Protection of such riparian zones along streams and lakes can benefit fish and wildlife and aquatic ecosystems in general. Furthermore, such practices are being required as parts of Best Management Practices adopted by many states, aimed at reducing loading of sediment, nutrient, and other pollutants into water bodies.

CONCLUSIONS — EFFECTS ON SUSTAINED FOREST PRODUCTIVITY

By many measures, forested peatlands are fragile systems. They have the capacity to produce considerable wood fiber, even without intensive management activities such as drainage and fertilization. However, that capacity can be significantly reduced by disruption of both the hydrological and the nutrient cycles of these landscapes. Disturbance of flow paths by road building, rutting, and other activities can remove significant area from the productive land base. Full-tree harvesting has the potential to significantly reduce the K capital of the system, and natural rates of replacement are inadequate to replenish that capital within a reasonable rotation length. Because of its low capital and amounts of removal by harvesting, P is also of concern. Continued full-tree harvest may have serious repercussions on sustained productivity of peatlands.

In contrast to drained peatlands, the effects of forest management activities on water yield, stormflow, nutrients, and other material exported via water flow are dampened in undrained systems. The very act of drainage increases aerobic microbial activity and provides a more efficient conveyance system to transport materials from the peatland. The low hydraulic gradients and lack of an efficient drainage system that characterize most undrained peatlands help explain their overall hydrological behavior. As a result, undrained peatlands that are harvested in winter are less disturbed, and harvesting can be expected to affect soil and water to a lesser degree than would similar activities on drained peatlands.

Awareness of the unique nature of these systems, and measures such as winter harvesting and bole-only removal, can assure that these peatlands can continue to supply forest products well into the future.

REFERENCES

Barrett, P. N., Rubber-tired skidders on unfrozen swamp, *Pulp Pap. Mag. Can.,* April, WR222, 1966.

Berry, P. L. and Poskitt, T. J., The consolidation of peat, *Geotechnique,* 22, 27, 1972.

Boelter, D. H., Physical properties of peats as related to degree of decomposition, *Soil Sci. Soc. Am. Proc.,* 33, 606, 1969.

Boelter, D. H. and Close, G. E., Pipelines in forested wetlands, *J. For.,* 72, 561, 1974.

Boelter, D. H. and Verry, E. S., Peatland and Water, U.S. Department of Agriculture, Forest Service, Gen. Tech. Report NC-31, 1977.

Brooks, K. N., Surface hydrology, in *The Patterned Peatlands of Minnesota,* Wright, H. E., Coffin, B. A., and Aaseng, N. E., Eds., University of Minnesota Press, Minneapolis, MN, 1992, chap. 10.

Buttleman, C. G. and Grigal, D. F., Use of the Rb/K ratio to evaluate potassium nutrition of peatlands, *Oikos,* 44, 253, 1985.

Clausen, J. C. and Brooks, K. N., Quality of runoff from Minnesota peatlands: I. A characterization, *Water Res. Bull.,* 19, 763, 1983.

Damman, A. W. H., Distribution and movement of elements in ombrotrophic peat bogs, *Oikos,* 30, 480, 1978.

Damman, A. W., Hydrology, development, and biogeochemistry of ombrogenous peat bogs with special reference to nutrient relocation in a western Newfoundland bog, *Can. J. Bot.,* 64, 384, 1986.

Foster, N. W. and Morrison, I. K., Alternate strip clearcutting in upland black spruce. IV. Projected nutrient removals associated with harvesting, *For. Chron.,* 63, 451, 1987.

Freedman, B., Duinker, P. N., Barclay, H., Morash, R., and Prager, U., Forest Biomass and Nutrient Studies in Central Nova Scotia, Canadian Forestry Service Information Report M-X-134, 1982.

Gafni, A. and Brooks, K. N., Hydraulic characteristics of four peatlands in northern Minnesota, *Can. J. Soil Sci.,* 70, 239, 1990.

Glaser, P. H., Wheeler, G. A., Gorham, E., and Wright, H. E., Jr., The patterned mires of Red Lake peatlands, northern Minnesota, *J. Ecol.,* 69, 575, 1981.

Gordon, A. G., Nutrient cycling dynamics in differing spruce and mixed wood ecosystems in Ontario and the effects of nutrient removals through harvesting, in Resources and Dynamics of the Boreal Zone, Wein, R.W., Riewe, R.R., and Methven, I.R., Eds., Assoc. Can. Univ. North. Stud., 1983, pp. 97–118.

Grigal, D. F., Impact of right-of-way construction on organic soil bulk density in the Red Lake Peatland, *Can. J. Soil Sci.,* 63, 557, 1983.

Grigal, D. F., Elemental dynamics in forested bogs in northern Minnesota, *Can. J. Bot.,* 69, 539, 1991.

Grigal, D. F. and Bates, P. C., Forest Soils. A Technical Paper for a Generic Environmental Impact Statement on Timber Harvesting and Forest Management in Minnesota, Jaakko Poyry Consulting, Tarrytown, NY, 1992.

Grigal, D. F. and Nord, W. S., Inventory of Heavy Metals in Minnesota Peatlands, Report to Minnesota Department of Natural Resources, Division of Minerals, St. Paul, MN, 1983.

Groot, A., Silvicultural Consequences of Forest Harvesting on Peatlands: Site Damage and Slash Conditions, Canada Forest Service, Information Report O-X-384, 1987.

Haavisto, V. F., Groot, A., and Jeglum, J. K., Climate, water, and drainage in peatlands, in Climate applications in forest renewal and forest production, Proc. Forest Climate '86, Geneva Park, Orilla, Ontario, 17–20 Nov. 1986, MacIver, D. C., Street, R. B., and Auclair, A. N., Eds., Canada Forestry Service, Sault Ste. Marie, Ontario, 1986, pp. 150–155.

Haavisto, V. F., Jeglum, J. K., and Groot, A., Management practices — black spruce ecosystem, in Proc. Ecology and Management of Wetlands, Charleston, SC, 1986, Timber Press, Portland, OR, 1988, chap. 21.

Heikurainen, L., Improvement of forest growth in poorly drained soil, *Int. Rev. For. Res.*, 1, 39, 1964.

Heikurainen, L., Hakkuun vaaikutus ojjitettujen soiden vesitalouteen (Summary: on the influence of cutting on the water economy of drained peatlands), *Acta For. Fenn.*, 82, 1, 1967.

Heikurainen, L. and Päivänen, J., The effect of thinning, clearing and fertilisation on the hydrology of peatland drained for forestry, *Acta For. Fenn.*, 104, 1, 1970.

Hornbeck, J. W., Martin, C. W., Pierce, R. S., Bormann, F. H., Likens, G. E., and Eaton, J. S., Clearcutting northern hardwoods: effects on hydrologic and nutrient ion budgets, *For. Sci.*, 32, 667, 1986.

Ingram, H. A. P., Soil layers in mires: function and terminology, *J. Soil Sci.*, 29, 224, 1978.

Jeglum, J. K., Haavisto, V. F., and Groot, A., Peatland forestry in Ontario: an overview, in Symposium '82, Symposium on Peat and Peatlands, Shippegan, N. B., 12–15 Sept. 1982, Sheppard, J. D., Musial, J., and Tibbetts, T. E., Eds., Canadian National Committee, International Peat Society, Halifax, N.S., 1983, p. 127.

Jeglum, J. K. and Kennington, D. J., Strip Clearcutting in Black Spruce: a Guide for the Practicing Forester, Forestry Canada, Ontario Region, Great Lakes Forestry Center, Sault Ste. Marie, 1993.

Johnson, D. W. and Lindberg, S. E., Eds., *Atmospheric Deposition and Nutrient Cycling in Forest Ecosystems*, New York, Springer-Verlag, 1992.

Johnston, W. F., Full-Tree Skidding Black Spruce: Another Way to Favor Reproduction, USDA Forest Service, Research Note NC-188, 1975.

Johnston, W. F., Manager's Handbook for Black Spruce in the North Central States, U.S. Department of Agriculture, Forestry Service, General Technical Report NC-31, North Cent. For. Exp. Stn., St. Paul, MN, 1977.

Knighton, M. D. and Stiegler, J. H., Phosphorus release following clearcutting of a black spruce fen and a black spruce bog, in Proc. 6th Int. Peat Congr., Duluth, MN, 1980, p. 577.

Miles, P. D. and Chen, C. M., Minnesota Forest Statistics, 1990, USDA Forest Service Resource Bulletin NC-141, 1992, 130.

Nicolson, J. A., Alternate strip clearcutting in upland black spruce. V. The impact of harvesting on the quality of water flowing from small basins in shallow-soil boreal ecosystems, *For. Chron.*, 64, 52, 1988.

Paavilainen, E., Effect of fertilization on plant biomass and nutrient cycle on a drained dwarf shrub pine swamp, *Commun. Inst. For. Fenn.*, 98 (5), 1, 1980.

Paavilainen, E. and Päivänen, J., Use and management of forested hydric soils, in *Proc. Ecology and Management of Wetlands*, Charleston, S.C., 1986, Timber Press, Portland, OR, 1988, chap. 22.

Päivänen, J., The effect of cutting and fertilisation on the hydrology of an old forest drainage area, *Folia For.* (Helsinki), 51, 1, 1982.

Perala, D. A., Growth and Yield of Black Spruce on Organic Soils in Minnesota, USDA Forest Service Research Paper NC-56, 1971.

Plonski, W. L., Normal Yield Tables for Black Spruce, Jack Pine, Aspen and White Birch in Northern Ontario, Ontario Department of Lands and Forests, Division of Timber Management, Report 24, 1956.

Raile, G. K. and Smith, W. B., Michigan's Forest Statistics, 1980, USDA Forest Service Resource Bulletin NC-67, 1983.

Richardson, C. J., Mechanisms controlling phosphorus retention capacity in freshwater wetlands, *Science,* 228, 1424, 1985.

Russell, E. W., *Soil Conditions and Plant Growth,* 10th ed., Longman, New York, 1973, 849 pp.

Shepard, J. P., Effects of forest management on surface water quality in wetland forests, *Wetlands,* 14, 18, 1994.

Sleep, V., Logging on claybelt sites, in Proc. Second Ontario Conf. Forest Regeneration, 6–8 March 1979, Kapuskasing, Ontario, Ontario Ministry of Natural Resources, Toronto, Ontario, 1979.

Spencer, J. S., Jr., Smith, W. B., Hahn, J. T., and Raile, G. K., Wisconsin's Fourth Forest Inventory, 1983, USDA Forest Service Resource Bulletin NC-107, 1988, 158 pp.

Stoeckeler, J. H., Wetland Road Crossings: Drainage Problems and Timber Damage, USDA Forest Service, Research Note NC-27, 1967, 4 pp.

Swank, W. T., Stream chemistry responses to disturbance, in *Forest Hydrology and Ecology at Coweeta,* Swank, W. T. and Crossley, D. A., Jr., Eds., Springer-Verlag, New York, 1988, chap. 25.

Timmer, V. R., Savinsky, H. M., and Marek, G. T., Impact of intensive harvesting on nutrient budgets of boreal forest stands, in Resources and Dynamics of the Boreal Zone, Wein, R. W., Riewe, R. R., and Methven, I. R., Eds., Assoc. Can. Univ. North Stud., 1983, p. 131.

Trettin, C. C. and Jurgensen, M. F., Organic matter decomposition response following disturbance in a forested wetland in northern Michigan, in *Proc. 9th Int. Peat Congr.,* Uppsala, Sweden, International Peat Society, Helsinki, Finland, 2 (3), 392, 1992.

Van Cleve, K., Data for black spruce, paper birch, in Elemental cycling in forest ecosystems, Cole, D. W. and Rapp, M., Eds., 341–409, in *Dynamic Properties of Forest Ecosystems,* Reichle, D. E., Ed., Cambridge University Press, Cambridge, U.K., 1981, pp. 376–379.

Van Cleve, K., Oliver, L., Schlenter, R., Viereck, L. A., and Dyrness, C. T., Productivity and nutrient cycling in taiga forest ecosystems, *Can. J. For. Res.,* 13, 747, 1983.

Verry, E. S., Water table and streamflow changes after stripcutting and clearcutting an undrained black spruce bog, in Proc. 6th Int. Peat Congr., Duluth, MN, 1980, p. 493.

Verry, E. S. Forest harvesting and water: the Lake States experience, *Water Resources Bull.,* 22, 1039, 1986.

Verry, E. S., Lewis, J. R., and Brooks, K. N., Aspen clearcutting increases snowmelt and stormflow peaks in north central Minnesota, *Water Resources Bull.,* 19, 59, 1983.

Verry, E. S. and Timmons, D. R., Waterborne nutrient flow through an upland-peatland watershed in Minnesota, *Ecology,* 63, 1456, 1982.

Waughman, G. J., Chemical aspects of the ecology of some South German peatlands, *J. Ecol.,* 68, 1025, 1980.

Waughman, G. J. and Bellamy, D. J., The distribution of major elements between some ecosystem components in different peatland zones, Proc. 7th Int. Peat Congr., Dublin, Ireland, 2, 32, 1984.

Weetman, G. F. and Webber, B., The influence of wood harvesting on the nutrient status of two spruce stands, *Can. J. For. Res.*, 2, 351, 1972.

Whitehead, P. G. and Robinson, M., Experimental basin studies — an international and historical perspective of forest impacts, *J. Hydrol.*, 145, 217, 1993.

CHAPTER 27

Peatland Forestry in Finland: Problems and Possibilities from the Nutritional Point of View

Seppo Kaunisto

CONTENTS

INTRODUCTION

About 5.7 million ha (containing 0.7 million ha of drained, paludified forests) or more than 50 percent of Finnish wetlands have been drained for forestry, and about 0.7 million additional ha for farming. Drainage of virgin, tree-covered peatlands has usually resulted in satisfactory wood production, but poor results have also been experienced in large areas, estimated to be about 10 to 17 percent of the drained area (Keltikangas et al., 1986; Eurola et al., 1988). A portion of farmland on peat soils has been abandoned from farming and afforested, and plans are to afforest other areas. The results have mainly been unsatisfactory (Hynönen and Saksa, 1991; Hytönen, 1991; Valtanen, 1991; Rossi et al., 1993).

About 622,000 ha of peatlands are considered suitable for peat harvesting, and about 120,000 ha have been reserved for this purpose; peat harvesting is presently conducted on 47,000 to 48,000 ha. The peat cutaway areas currently cover 2500 to 3000 ha. This area will increase by about 2000 ha annually in the near future and is estimated to reach 15,000 to 16,000 ha by the year 2000. Research on the afforestation of peat cutaway areas has been conducted in Finland for 40 years and more intensively since the early 1980s. Thus far, the studies have shown poor, good, and excellent tree growth (Mikola, 1975; Kaunisto, 1979, 1986, 1987a; Ferm and Kaunisto, 1983; Kaunisto and Viinamäki, 1991).

An important reason for the failure or success of wood production on the sites mentioned above seems to be nutrition. Tree nutrition on wetlands depends on the quantity, availability, and balance of nutrients. Drainage is merely a technical means to improve the availability and uptake of nutrients. The quantity of nutrients depends on such factors as the quality of the subsoil, thickness of the peat layer, topography, and the quality and source of water penetrating through the peat layers. The availability of nutrients depends on the climatic conditions, microbial activity, ratios between carbon and organically bound nutrients, and chemical fixation.

NUTRITIONAL BASES IN FINNISH WETLAND FORESTS

Nitrogen

A prerequisite for growing forests on wetlands is satisfactory nitrogen mineralization. This depends on both the temperature sum and the carbon to nitrogen (C:N) ratio and thus on the total nitrogen content of the peat. The mean annual cumulative temperature (degree-days, dd, with the threshold value of 5°C) on the southern coast of Finland is 1350 dd and in the northernmost part 400 to 600 dd. The total nitrogen concentration in surface peat may vary from 0.4 to more than 3.0 percent and the C:N ratio from less than 15 to more than 100. Thus, the length of the growing season varies considerably in the different parts of the country, and the conditions for nitrogen mineralization vary both among locations and among sites.

The nitrogen nutrition of trees can be improved by fertilization. However, at least in young pine trees, mere nitrogen fertilization may cause needle and bud damage due to an imbalance between nitrogen and phosphorus, even on quite nitrogen-poor sites (Paavilainen, 1976; Kaunisto and Paavilainen, 1977). On the other hand, the effect of nitrogen + phosphorus + potassium (NPK) fertilization on sites with poor nitrogen mineralization is inadequate and only of short duration, lasting less than 10 years (Kaunisto, 1977; Paavilainen, 1977b; Moilanen and Issakainen, 1990; Moilanen, 1993). The short duration of the effect of nitrogen fertilization on nitrogen-poor sites is due to nitrogen immobilization by microbes, as demonstrated in liming experiments (Meshechok, 1971; Gardiner, 1975; Kaunisto and Norlamo, 1976). Wood production on sites where natural nitrogen mineralization is insufficient is thus unprofitable (Rantala and Moilanen, 1993).

In some experiments, height growth in 7- and 14-year old PK-fertilized Scots pine (*Pinus sylvestris*) plantations increased with the peat nitrogen concentration up to about 1.6 to 1.7 percent. Above that concentration, growth remained the same or even declined (Kaunisto, 1982, 1987b). In these areas the average temperature sum varied between 1050 and 1100 dd. NPK fertilization compared with PK increased height growth below a peat nitrogen concentration of about 1.4 to 1.6 percent, and decreased it above that concentration (Kaunisto, 1982). On the other hand, NPK fertilization compared with PK increased the volume growth of older, natural stands at peat nitrogen concentrations of 2.1 percent (Moilanen, 1993) and 2.8 percent (Paavilainen, 1990) in areas where the average temperature sum varied between 900 and 1000 dd. It seems that higher peat nitrogen concentrations are required for satisfactory nitrogen mineralization in the northern part of Finland compared to the south.

The results reported by Kaunisto (1982, 1987b) imply that the reasons for the harmful effects of nitrogen fertilization are growth disturbances due to boron deficiency and frost damage of buds and needles due to an imbalance between nitrogen and other nutrients. Damage was explained best by needle N:P and N:K ratios. Ferm et al. (1990) have also found decreased frost resistance of Scots pines at elevated N concentrations or N:P and N:K ratios. In a report by Moilanen (1993), the N:B ratio explained 55 percent of the variation in leader dieback of Scots pine. Excessive nitrogen causes accumulation of arginine in needles, which implies problems in protein synthesis (Ferm et al., 1990; Pietiläinen, 1994). The arginine content in needles was negatively correlated with boron, phosphorus, and especially with potassium concentrations in needles (Pietiläinen, 1994).

Mineral Nutrients

Peat soils in Finland generally contain low amounts of mineral nutrients (Table 1). While the N:P:K ratios are 100:10 to 13:32–39 in pine stands (Paavilainen, 1980; Finer, 1989), they may be 100:3:1 in the corresponding drained surface peats in old drainage areas (70 to 80 years) (Kaunisto and Paavilainen, 1988). N to P ratios in 0 to 20 cm surface peat were 100:5 to 6 in somewhat younger (31 to 55 years) meso-oligotrophic drainage areas and 100:6 to 7 in oligo-ombrotrophic areas (41 to 55 years) (Laiho and Laine, 1994). Nearly 95 percent of the phosphorus in peat may be in an organic form (Kaila, 1956), from which it can only be released by microbial activity. Alexander (1979) stated that organically bound nitrogen and phosphorus are released by microbes in the same ratio as they exist in organic matter. Thus, there generally is a shortage of available phosphorus in Finnish peat soils compared with nitrogen and with the requirements of trees. This has been demonstrated by several fertilization experiments

Table 1   Nutrient Amounts (kg ha⁻¹) in the 0 to 20 cm
           Surface Peat Layer in Some Virgin Peatlands
           and Old (80 to 90 years) Drainage Areas

	Site type[a]	Amount, kg ha⁻¹		
		N	P	K
Drained	Rhk	5500	225	64
	Mk	6300	210	58
	Pk	4300	180	46
	RhSR	7000	240	37
	VSR	5600	250	60
	IR	3000	90	34
Virgin	RhSR	2800	140	110
	VSR	1200	60	90
	IR	1000	40	70

[a] Rhk = herbrich spruce swamp; Mk = *Vaccinium myrtillus*
spruce swamp; Pk = *Vaccinium vitis idaea* spruce swamp;
RhSR = herbrich pine fen; VSR = tall-sedge pine fen; IR
= dwarf shrub pine bog.

From Westman, C. J., *Acta For. Fenn.*, 172, 1981; and from
Kaunisto, S. and Paavilainen, E., *Commun. Inst. For. Fenn.*,
145, 1988. With permission.

(e.g., Huikari and Paarlahti, 1966; Mannerkoski and Seppälä, 1970; Kaunisto, 1972, 1992; Paavilainen, 1977b; Kaunisto and Paavilainen, 1977). The response of trees to fertilization depends on the peat nitrogen content (Paavilainen, 1977b, 1979; Kaunisto, 1982, 1987; Moilanen, 1993). On nitrogen-rich sites, P fertilization may increase tree growth for more than 20 years if potassium nutrition is adequate (Kaunisto, 1989).

There are generally limited amounts of potassium in peat compared to the requirements of trees (Kaunisto and Paavilainen, 1988). Potassium, however, is either in soil solution or on exchange sites of humus colloids and so is fully available to trees; it is also highly leachable (Ahti, 1983; Malcolm and Cuttle, 1983). Potassium deficiency often occurs on originally wet, deep peat sites such as open fens, sparsely stocked mixed open and pine fens (Kaunisto and Tukeva, 1984; Kaunisto, 1989, 1992), or sparsely stocked mixed open and spruce swamps.

The potassium nutrition of trees can be improved by fertilization. The effect, however, is of shorter duration than that of P fertilization: only 10 to 20 years on nitrogen-rich sites (Kaunisto, 1989, 1992). The reasons seem to be greater use by trees (Paavilainen, 1980; Finer, 1989) and higher leaching losses compared to phosphorus (Ahti, 1983; Kaunisto, 1992). Potassium in fertilizers is in the form of KCl, which is completely water soluble, whereas phosphorus additions have been in water insoluble forms, first as rock phosphate and today as apatite. A slowly soluble potassium mineral, flogobite, has been used quite successfully in fertilization experiments (Kaunisto et al., 1993). Its low potassium content (total K about 8.5 percent, HCl soluble about 4.5 percent) is a problem that may lead to high application rates. An application of 700 kg ha⁻¹, however, has given a satisfactory result for 14 years (Kaunisto et al., 1993).

Boron is the most problematic micronutrient in Finnish peatland forests. Its shortage causes growth disturbances that are quite common on deep-peat sites with high nitrogen and phosphorus nutrition and especially on open mires (Veijalainen, 1983; Reinikainen and Veijalainen, 1983). It is as easily leached from organic matter as potassium, but it may form poorly soluble compounds with calcium.

PROBLEM AREAS

Drainage of Too Poor Sites

In the late 1950s and early 1960s forest cutting surpassed forest growth in Finland. A general concern about sustainable wood production for industry initiated several programs that aimed at improving forest growth. Some of the programs were focused on improving forest growth on peatlands, mainly by drainage and fertilization. Forest improvement measures on peatlands increased rapidly and reached their peak in the late 1960s and early 1970s. During the peak years, 294,000 ha were drained (in 1969) and 116,000 ha were fertilized (in 1974).

The extensive improvement activity also led to the drainage of too poor mires, partly because drainage was supported by the state, partly because it was so inexpensive that private forest owners used their own financing, and partly because fertilization was inexpensive and there was inadequate knowledge about tree nutrition on peatlands.

Because of unsuitable climatic and/or nutritional conditions, about 0.5 to 0.8 million ha of peatlands drained for forestry have not been able to support satisfactory tree growth (Keltikangas et al., 1986; Eurola et al., 1988). While drainage of a certain mire site type in southern Finland may produce a high growth increment, effects of drainage may be negligible on a similar site in the north (Keltikangas et al., 1986). Areas of misdrainage suffer primarily from too low mineralization of organically bound nutrients such as nitrogen and phosphorus, either due to low temperature or to a high C:N ratio or to both.

Most of the profitable originally tree-covered drainage sites can be separated from the unprofitable ones on the basis of site types and climatic regions (Heikurainen, 1973). Changes in the ground vegetation and nutrient status in old drainage areas, however, have created some difficulties in determining problem sites. It is therefore important to also use other measures (peat nitrogen content, C:N ratio, degree of humification) for evaluating mineralization of peat nitrogen in different climatic regions and in stands at different stages of development. A sufficient knowledge of nitrogen status is also important because, as stated earlier, most peatland forests benefit from phosphorus and potassium application if nitrogen mineralization is sufficient.

Open (treeless or almost treeless) mires form a specific group among the problem sites. In most cases drained open mires represent site types where nitrogen mineralization should be satisfactory (in southern Finland peat nitrogen concentration >1.3 to 1.4 percent, Kaunisto, 1982, 1987), but about one-fourth of the drained open mires represent nitrogen-poor site types (Keltikangas et al., 1986). In addition to the problems of mineralization of organically bound nutrients, the basic nutritional problem on open mires is the shortage of mineral nutrients such as phosphorus, potassium, and boron, and the imbalance between nitrogen and these nutrients. Phosphorus is required at afforestation (Huikari and Paarlahti, 1966; Mannerkoski and Seppälä, 1970; Kaunisto, 1972; Laine and Mannerkoski, 1980). On nitrogen-poor sites, seedlings may grow satisfactorily for over 10 years with only phosphorus or NP fertilization (Laine and Mannerkoski, 1980). As the stand ages, boron and potassium are needed, and potassium fertilization has to be repeated several times (Kaunisto and Tukeva, 1984).

About 0.4 million ha of open mires have been drained (Keltikangas et al., 1986) and about 250,000 ha of them afforested. The drainage of open mires in Finland has been discontinued because of nutritional problems. Very high yields, however, have been achieved on afforested nitrogen-rich open mires with wood ash fertilization (Silfverberg and Huikari, 1985; Silfverberg and Hotanen, 1989), the effects of which may last for 30 to 40 years.

The poorest group of drained peatlands, according to Eurola and Huttunen (1990), cover 31 percent of the drained area but account for only 9 percent of the yield. However, it is necessary to remember that stand volume in these areas was quite small at the time of drainage, and the recovery was slow. Most of the peatlands that were originally judged to be suitable for forestry have been drained. Meanwhile the prerequisites for forest drainage have been made stricter, especially in northern Finland, in order to avoid drainage of too poor sites. New drainage in Finland has ceased in practice, and drainage activities are focused on maintenance of old drainage areas. No maintenance is supported by the government in areas with the temperature sum lower than 750 dd, or the growing stock less than 25 m^3 ha^{-1}.

Drainage of Sites with Unbalanced Nutrient Regime

Originally sparsely stocked, wet tall-sedge and herbrich pine and spruce mires have been considered suitable for drainage in most parts of the country (Keltikangas et al., 1986). Although the growth rate has generally been high, some studies show that they later may develop a potassium deficiency that causes a growth reduction (Kaunisto, 1989, 1992) and may cause severe deterioration of stands. Old drainage areas of these site types are especially subject to a potassium shortage (Kaunisto and Paavilainen, 1988; Kaunisto, 1989). This shortage could result from potassium losses in tree harvesting, because no decrease in potassium in peat has been reported in peatland forests that have not been harvested (Laiho and Laine, 1992). Whole-tree harvesting is especially harmful to potassium stores, because two-thirds of the potassium in the aboveground biomass of peatland forests is in branches and needles (Paavilainen, 1980; Finér, 1989).

Abandoned Agricultural Peatland Fields

Since the 1960s, about 170,000 ha of Finnish farmland have been afforested. Because of agricultural overproduction, about one-fourth of the remaining 2.2 million ha will be retired from farming. Plans are to afforest most of this land, a large proportion of which is peatland fields. They are considered to be low-yielding compared with mineral-soil fields, are usually rather small, are often located in remote areas, and are difficult to farm with modern methods. The exact land area is not known.

Stand establishment by sowing or natural seeding does not usually succeed on peatland fields. Only downy birch (*Betula pubescens*) and Scots pine can be used as the first tree generation for afforestation of peatland fields. Norway spruce (*Picea abies*) can very rarely be used. Thorough site preparation and chemical weed control are needed at afforestation. However, chemical weed control is not as effective on peatlands as on mineral soils because of the high organic matter content of the substratum.

In some experiments, the early development of Scots pine plantations on peatland fields has been quite successful (Paavilainen, 1970, 1977a). However, in many surveys covering operational sites, afforestation has been unsatisfactory and is often a complete failure (Hynönen and Saksa, 1991; Hytönen, 1991; Rossi et al., 1993; Valtanen, 1991). One of the reasons for the different outcome has inevitably been competition by ground vegetation, especially root competition (Ferm et al., 1994). The experimental sites were weeded regularly for several years, unlike the operational sites. Plantations on peatland fields compared to those on mineral-soil fields also seem to have more moose and vole damage (because of the remote location and excessive grass vegetation, respectively), problems with poor drainage, and fungus diseases, especially *Ascocalyx abietina*.

The initial growth of trees has not generally been hindered by poor nutrition on peatland fields (Paavilainen, 1970, 1977a). However, unbalanced nutrition seems to be quite common later (Kaunisto, 1991; Hynönen, 1992; Hytönen and Ekola, 1993). The nutrient regime in peatland fields has been affected by farming and is quite different from normal peatlands. Fields are usually quite rich in nitrogen (total N of organic matter >2.0 percent). Mineral soil from adjacent upland forests were quite often used for reclaiming peat soil for farming, resulting in fairly high amounts of phosphorus and potassium in the plow layer. The amounts of these nutrients are related to soil density, i.e., with the amounts of mineral soil that were used (Kaunisto, 1991; Hytönen and Ekola, 1993). Boron contents, however, are generally low (Kaunisto, 1991; Hytönen and Ekola, 1993). Liming that was also used in farming enhances nitrogen mineralization, but it may also chemically fix boron.

Imbalances between boron and other nutrients, especially nitrogen, have caused severe growth disturbances (Raitio and Rantala, 1977; Veijalainen, 1983; Hytönen and Ekola, 1993). High aluminum and iron contents in some fields may cause phosphorus fixation. Deep-peat fields without any mineral soil application may suffer from a shortage of other nutrients, especially boron and potassium (Hytönen and Ekola, 1993). These deficiencies can be corrected by fertilization with commercial fertilizers, but an imbalance between nutrients may continue. Good results have been achieved by using wood ash (Ferm et al., 1992). An excess of nitrogen causes problems in wood quality, and pine, especially, develops thick branches on peatland fields (Kaunisto and Tukeva, 1986).

It is difficult to visually estimate the nutritional suitability of an abandoned peatland field for growing forests, because its chemical properties are often quite complicated. A fairly good assessment can be achieved by determining bulk density and content of organic matter and by analyzing the amounts of critical nutrients per unit volume *in situ* (Kaunisto, 1991), both within and below the till layer. Foliar analysis can be used quite successfully to determine the nutrient status of trees on planted sites (Veijalainen, 1983; Ferm et al., 1992; Hytönen and Ekola, 1993). It seems, however, that one should be very careful when making a decision for afforestation of an abandoned peatland field because of the many nutritional and other problems.

Peat Cutaway Areas

Peat harvesting is practiced in deep peat areas. The surface peat, at least, has developed under ombrotrophic conditions. Surface peat is used for garden peat and deeper, more humified layers for fuel. To be a good material for fuel, peat should have a low ash content. The nutrient regime on these sites depends on the thickness of the remaining peat layer and on the quality of the subsoil.

After harvesting, a peat layer of varying thickness (from 0 to more than 1 m) remains on the site. Variation is caused by the stoniness or surface formation of the mineral subsoil or by the harvesting techniques. The peat is rich in nitrogen but contains only small amounts of any mineral nutrients and is quite acid, often below pH 4.0 and seldom above pH 5.0 (in H_2O). Subsoil varies from coarse sand to heavy clays. In the long run, the quality of the subsoil and the depth of the remaining peat layer determine the fertility of the site. On sites with sandy subsoils, young trees may suffer from severe potassium and phosphorus deficiency even if in contact with mineral soil, but on silty subsoils trees may grow as well as with PK-fertilized treatments (Kaunisto, 1987a). A peat layer that is 10 cm in thickness may prevent roots of small seedlings from obtaining mineral nutrients from the subsoil. On average, the root penetration of 20-year-old pine stands was less than 20 cm (Kaunisto and Viinamäki, 1991) and even roots of 30-year-old birch trees were unable to penetrate a 40-cm thick peat layer (Kaunisto and Saarinen, 1989). Tree growth correlated negatively with the thickness of the peat layer if that layer was more than 20 cm thick (Kaunisto and Saarinen, 1989; Kaunisto and Viinamäki, 1991). An earlier result of positive correlation between tree growth and peat thickness (Kaunisto, 1977) was attributed to the positive correlation between peat phosphorus and peat thickness.

Variation in peat thickness and subsoil quality causes random variation in the mineral nutrient regime of the substrate.

Trees can be supplied with mineral nutrients either by leaving a thin peat layer, by lifting subsoil in connection with drainage, or by fertilization (Mikola, 1975; Kaunisto, 1979, 1985, 1986, 1987a). Ditches are usually made 14 to 18 m apart depending on the extension capacity of the digging device, and spoil can be lifted onto the strips. Additions of mineral soil depend on peat thickness, and the quantity of mineral nutrients depends on the amount and texture of the mineral subsoil.

Tree seedlings soon die without fertilization on bare peat because of phosphorus and potassium shortages (Kaunisto, 1979, 1982, 1987; Kaunisto and Saarinen, 1989), in this case, spot fertilization with PKB. Fertilizer at planting and sowing supplies seedlings with mineral nutrients for 4 to 6 years. On coarse-textured subsoils, fertilization with phosphorus and potassium may be necessary even if tree roots are in contact with mineral soil. Additional results from experiments on such soils are needed before any definite conclusions on the further development of plantations can be made.

Peat cutaway areas can produce high amounts of timber and biomass even without fertilization. In some natural, unfertilized mixed downy and silver birch stands (*B. pubescens* and *pendula*, respectively), aboveground leafless biomass at 16 years of age varied from 35 to 85 mg ha^{-1}, and the current annual increment of stem wood without bark varied from 3.5 to 8 mg ha^{-1} yr^{-1} (Ferm and Kaunisto, 1983). The growing volumes of 28-year-old, tended, natural stands of silver and downy birch mixtures were higher than those on the best mineral soil sites (Kaunisto, 1986). In this case the thickness of the peat layer varied within a couple of meters from 0 to 50 to 60 cm, giving tree roots a chance to penetrate to the silty subsoil underneath.

Scots pine, silver birch, and downy birch can be grown on peat cutaway areas if the drainage conditions and the nutrient status are controlled (Kaunisto and Saarinen, 1989). Both pines and birches can be sown or planted. Good results have also been achieved by broadcast and natural seeding of both birch species. There is little experience with Norway spruce because it needs shelter trees. However, naturally regenerated spruce have appeared in more advanced birch stands, even at several hundreds of meters from the nearest seed trees.

High nitrogen stores in peat provide a possibility of high yields in peat cutaway areas but create problems with wood quality similar to those on peatland fields (Kaunisto and Tukeva, 1986). There are not, however, as severe abnormalities in nutrition as on abandoned peatland fields. The proximity of mineral soil provides some guarantee of continuity even on coarse-textured subsoils, unlike the situation on deep peat open mires. It seems that, for the most part, peat cutaway areas can be used for forestry, and in many cases with good success. There is also experimentation in Finland dealing with growing different kinds of herbs and berries on peat cutaway areas. Difficult drainage conditions, however, may lead to other uses, such as restoration or artificial lakes.

GROWTH INCREMENT DUE TO FOREST DRAINAGE

About 6 million ha of wetlands have been drained in Finland (Päivänen, 1995). Despite some problem areas, the additional wood production due to various forest improvement measures was estimated at about 7 million m³ in the VII National Forest Inventory of Finland in 1976 to 1984 (Paavilainen and Tiihonen, 1988; Paavilainen, 1989). The greatest proportion of growth increase was attributed to improved drainage (Paavilainen and Tiihonen, 1988). The preliminary results of the VIII Inventory suggest that the annual increase in volume growth is presently about 10 million m³ (Paavilainen, 1994), more than twice the yield on all peatlands if they were in their natural condition. Increases in wood production due to forest improvement measures on peat soils are estimated to be about 250,000 m³ since the 1930s (Kaunisto et al., 1994).

RESTORATION OF PROBLEM SITES

Too Poor Sites

Drainage maintenance will be discontinued on most of the unprofitable drainage areas. In our humid climate, the deterioration of drains leads to natural rewetting of such sites within a few decades. It may be important to use more effective means for restoring drained wetlands in nature reserves, and some of them will be actively rewetted. Studies on restoring drained wetlands are in their initial stages (Vasander et al., 1992), but preliminary results indicate that it is quite difficult to create similar hydrological conditions as before drainage without large investments. There are not yet any results on vegetation changes due to rewetting. Although by definition trees have only responded minimally to drainage on these poor sites, some changes have occurred in the habitus of trees. As a result, the appearance of trees is not exactly the same as before drainage. Some trees may also die if the sites are rewetted.

Sites with Unbalanced Nutrient Regime

On some sites, trees have responded to drainage or drainage and fertilization and have been growing for some time, but a nutrient imbalance (N to mineral nutrients) has begun to restrict growth. Tree stands are quite different from those originally present. Sudden rewetting will cause tree death. On these sites it might be feasible to maintain tree growth by adding mineral nutrients to help the trees reach merchantable size, harvest them, and then allow the areas to become naturally rewetted.

Abandoned Agricultural Peatland Fields

Abandoned peatland fields are nutritionally a very heterogeneous and complicated group of peatlands. Their use for forestry seems to be rather problematic because the nutrient status and physical structure are very different from the original. They also produce large quantities of CO_2 (Silvola and Alm, 1992). Rewetting of peatland fields cannot restore similar site types that existed before agricultural use. In some cases, it may be possible to use them for peat harvesting where it is economically feasible and then to afforest them. Some of them had been growing forests before paludification. However, high ash contents due to soil amelioration measures may cause difficulties when using the tilled layer for fuel.

Peat Cutaway Areas

Many peat cutaway areas are suitable for forestry and have grown forests before site paludification, so forest restoration would be quite feasible. In the 1940s and early 1950s, peat harvesting was practiced with a method that resulted in a series of wet basins and peat ridges. Some of the basins have been recolonized by peatland vegetation such as *Sphagnum* species. However, good drainage is necessary when peat is harvested by the milled peat method. The plant colonization of such areas may be slow, and colonization by *Sphagnum* is minimal or nonexisting (Salonen, 1990; Salonen et al., 1992). Plant colonization depends on the nutrient regime of the site (Salonen, 1990), and fertilization with mineral nutrients substantially enhances plant colonization (Salonen, 1994b; Salonen and Laaksonen, 1994). Harvest areas are drained either by using deep main drains or by pumping where it is the only economically feasible possibility. In these cases, rewetting only requires discontinuation of pumping. Studies on several sites of different drainage conditions have recently begun in Finland (Salonen, 1994a; Roderfeld et al., 1994). The use of peat cutaway areas for lakes and bird habitat is also under investigation (Vikberg, 1994).

REFERENCES

Ahti, E., Fertilizer-induced leaching of phosphorus and potassium from peatlands drained for forestry, *Commun. Inst. For. Fenn.*, 111, 1983.

Alexander, M., *Introduction to Soil Microbiology*, 4th ed., John Wiley & Sons, New York, 1967, 472 pp.

Eurola, S., Aapala, K., and Kokko, A., A survey of peatland drainage activity in southern and central Finland, *Suo*, 39 (1–2), 9, 1988.

Eurola, S. and Huttunen, A., The functional grouping of mire ecosystems and their response to drainage, *Suo*, 41 (1), 15, 1990.

Ferm, A., Hokkanen, T., Moilanen, M., and Issakainen, J., Effects of wood bark ash on the growth and nutrition of a Scots pine afforestation in central Finland, *Plant Soil*, 147, 305, 1992.

Ferm, A., Hytönen, J., Lähdesmäki, P., Pietiläinen, P., and Pätilä, A., Effects of high nitrogen deposition on forests: case studies close to fur animal farms, in *Acidification in Finland*, Kauppi et al., Eds., Springer-Verlag, Berlin, Heidelberg, 1990, p. 635.

Ferm, A., Hytönen, J., Lilja, S., and Jylhä, P., Effects of weed control on the early growth Betula pendula seedlings established on an agricultural field, *Scand. J. For. Res.*, 9, 1994.

Ferm, A. and Kaunisto, S., Above-ground leafless biomass production of naturally generated birch stands in a peat cut-over area at Aitoneva, Kihniö, *Folia For.,* 558, 1983.

Finér, L., Biomass and nutrient cycle in fertilized and unfertilized pine, mixed birch and pine and spruce stands on a drained mire, *Acta For. Fenn.,* 208, 1989.

Gardiner, J. J., The influence of fertilizers upon microbial activity in peat. II. Calcium and nitrogen, *Irish For.,* 32 (2), 101, 1975.

Heikurainen, L., A method for calculation of the suitability of peatlands for forest drainage, *Acta For. Fenn.,* 131, 1973.

Huikari, O. and Paarlahti, K., *Kivisuon metsänlannoituskokeet, Kenttäopas,* Kirja-Mono Oy, Helsinki, 1966, 43 pp.

Hynönen, T., *Maan ominaisuuksien vaikutus turvemaapeltojen metsittämiseen,* A licentiate dissertation in the Faculty of Agriculture and Forestry at the University of Helsinki, A Finnish Stencil, 1992, 181.

Hynönen, T. and Saksa, T., Abstract: Field afforestation in Savo, eastern Finland, in the 1970's and 1980's, in Developing methods for afforestation of fields, interim report, Ferm, A. and Polet, K., Eds., *Metsäntutkimuslaitoksen tiedonantoja,* 391, 29, 1991.

Hytönen, J., Abstract: Field afforestation in central Ostrobothnia, western Finland, in Developing methods for afforestation of fields, interim report, Ferm, A. and Polet, K., Eds., *Metsäntutkimuslaitoksen tiedonantoja,* 391, 22, 1991.

Hytönen, J. and Ekola, E., Soil nutrient regime and tree nutrition on afforested fields in central Ostrobothnia, western Finland, *Folia For.,* 822, 1993.

Kaila, A., Phosphorus in various depths of some virgin peatlands, *J. Sci. Agric. Soc. Finl.,* 28 (2), 90, 1956.

Kaunisto, S., Effect of fertilization on successful planting and the number of naturally born seedlings on a fuscum bog at Kivisuo experimental field, *Folia For.,* 139, 1972.

Kaunisto, S., Effect of drainage intensity and fertilization on the development of pine plantations on oligotrophic treeless Sphagnum bogs, *Folia For.,* 317, 1977.

Kaunisto, S., Preliminary results on afforestation of sod drying fields and peat cut-over areas, *Folia For.,* 404, 1979.

Kaunisto, S., Development of pine plantations on drained bogs as affected by some peat properties, fertilization, soil preparation and liming, *Commun. Inst. For. Fenn.,* 109, 1982.

Kaunisto, S., Afforestation experiments at Aitoneva, Kihniö, *Metsäntutkimuslaitoksen tiedonantoja,* 177, 1985.

Kaunisto, S., Peatlands before and after peat harvesting, in *Proc. IPS Symp. Socio-Economic Impacts of the Utilization of Peatlands in Industry and Forestry,* Oulu, Finland, June 9–13, 1986, International Peat Society, Finnish National Committee, Helsinki, 1986, p. 241.

Kaunisto, S., Effect of fertilization and soil preparation on the development of Scots pine and silver birch plantations on peat cutover areas, *Folia For.,* 681, 1987a.

Kaunisto, S., Effect of refertilization on the development and foliar nutrient contents of young Scots pine stands on drained mires of different nitrogen status, *Commun. Inst. For. Fenn.,* 140, 1987b.

Kaunisto, S., Effect of refertilization on tree growth in an old drainage area, *Folia For.,* 724, 1989.

Kaunisto, S., Soil analysis as a means of determining the nutrient regime on some afforested peatland fields at Alkkia, *Folia For.,* 778, 1991.

Kaunisto, S., Effect of potassium fertilization on the growth and nutrition of Scots pine, *Suo,* 43 (2), 45, 1992.

Kaunisto, S., Lauhanen, R., and Ahti, E., Peatland forestry in Finland, in *Proc. Int. Conf. '94, On Wetland Environment and Peatland Utilization*, August 9–11, 1994, Changchun, P. R., China, Lu X., and Wang, R., Eds., Jilin People's Publishing House, Changchun, 1994.

Kaunisto, S., Moilanen, M., and Issakainen, J., Apatite and phlogopite as phosphorus and potassium fertilizers in peatland pine forests, *Folia For.*, 810, 1993.

Kaunisto, S. and Norlamo, M., On nitrogen mobilization in peat, I, Effect of liming and rotavation in different incubation temperatures, *Commun. Inst. For. Fenn.*, 88 (2), 1976.

Kaunisto, S. and Paavilainen, E., Response of Scots pine plants to nitrogen refertilization on oligotrophic peat, *Commun. Inst. For. Fenn.*, 92 (1), 1977.

Kaunisto, S. and Paavilainen, E., Nutrient stores in old drainage areas and growth of stands, *Commun. Inst. For. Fenn.*, 145, 1988.

Kaunisto, S. and Saarinen, M., Turvekäytössä olevien alueiden loppuvuosien kuivatus-ja turvetuotanto-ongelmat sekä alueiden jälkikäyttö energiapuun tuotantoon ja metsätalouteen, Project 98/881/85 of the Ministry of Trade and Industry in Finland, Final Report on the Use of Peat Cutaway Areas for Forestry and Energy Wood Production, A Finnish Stencil, 1989, p. 23.

Kaunisto, S. and Tukeva, J., Need for potassium fertilization in pole stage pine stands established on bogs, *Folia For.*, 585, 1984.

Kaunisto, S. and Tukeva, J., Effect of tree spacing on the development of pine plantations on peat, *Folia For.*, 646, 1986.

Kaunisto, S. and Viinamäki, T., Effect of fertilization and alder (*Alnus incana*) mixture on the development of young Scots pine (*Pinus sylvestris*) trees and the peat properties in a peat cutover area at Aitoneva, southern Finland, *Suo*, 42 (1), 1, 1991.

Keltikangas, M., Laine, J., Puttonen, P., and Seppälä, K., Peatlands drained for forestry during 1930–1978: results from field surveys of drained areas, *Acta For. Fenn.*, 193, 1986.

Laiho, R. and Laine, J., Potassium stores in peatlands drained for forestry, *Proc. 9th Int. Peat Congr., 22–26 June 1992*, Uppsala, Sweden, 1, 158, 1992.

Laiho, R. and Laine, J., Nitrogen and phosphorus stores in peatland drained for forestry in Finland, *Scand. J. For. Res.*, 9, 251, 1994.

Laine, J. and Mannerkoski, H., Effect of fertilization on tree growth and elk damage in young Scots pine stands planted on drained, nutrient-poor open bogs, *Acta For. Fenn.*, 166, 1980.

Malcolm, D. C. and Cuttle, S. P., The application of fertilizers to drained peat. 1. Nutrient losses in drainage, *Forestry*, 56 (2), 55, 1983.

Mannerkoski, H. and Seppälä, K., On the influence of fertilization on the initial development of plantations in open low-sedge bogs, *Suo*, 21 (1), 12, 1970.

Meshechok, B., Kalkning ved skogkultur på nedborsmyr, *Meddelelser fra det Norske Skogforsoeksvesen*, 114, 241, 1971.

Mikola, P., Afforestation of bogs after industrial exploitation of peat, *Silva Fenn.*, 9 (2), 101, 1975.

Moilanen, M., Effect of fertilization on the nutrient status and growth of Scots pine on drained peatlands in northern Ostrobothnia and Kainuu, *Folia For.*, 820, 1993.

Moilanen, M. and Issakainen, J., PK fertilizer and different types of N fertilizer in the fertilization of infertile drained pine bogs, *Folia For.*, 754, 1990.

Paavilainen, E., Experimental results of the afforestation on swampy fields, *Folia For.*, 77, 1970.

Paavilainen, E., Nitrogen fertilization on shallow-peated *Carex globularis* pine swamps, *Folia For.,* 272, 1976.

Paavilainen, E., Abstract: Planting of Scots pine in afforestation of abandoned swampy fields, *Folia For.,* 326, 1977a.

Paavilainen, E., Abstract: Refertilization on oligotrophic pine swamps, Preliminary results, *Folia For.,* 327, 1977b.

Paavilainen, E., Effect of fertilization on plant biomass and nutrient cycle on a drained dwarf shrub pine swamp, *Commun. Inst. For. Fenn.,* 98 (5), 1980.

Paavilainen, E., Effect of refertilization on pine and birch stands on a drained fertile mire, *Silva Fenn.,* 24 (1), 83, 1990.

Paavilainen, E., The status of peatland forests in Finland, based upon the national forest inventory, in *Proc. Symp. '89 Peat and Peatlands, Diversification and Innovation,* Vol. 1, *Peatland Forestry,* Quebec City, Quebec, Canada, August 6–10. 1989, The Canadian Society of Peat and Peatlands, Quebec, 1991, p. 65.

Paavilainen, E., Personal communication, 1994.

Paavilainen, E. and Tiihonen, P., Peatland forests in Finland, 1951–1984, *Folia For.,* 714, 1989.

Päivänen, J., Forested mires as a renewable resource — towards a sustainable forestry practice, in *Northern Forested Wetlands: Ecology and Management,* Trettin, C. C. et al., Eds., Lewis Publishers, Boca Raton, FL, 1997.

Pietiläinen, P., Effect of nutrition on soluble arginine concentration in Scots pine needles, Manuscript, Preliminarily published in the dissertation: Pietiläinen, P., Seasonal Fluctuations in the Nitrogen Assimilation of Scots Pine, Acta Universitatis Ouluensis, Series A, Scientiae Rerum Naturalium, 256, Oulu, 1994, chap. 8.

Raitio, H. and Rantala, E.-M., Macroscopic and microscopic symptoms of a growth disturbance in Scots pine, Description and interpretation, *Commun. Inst. For. Fenn.,* 91 (1), 1977.

Rantala, T. and Moilanen, M., Profitability of fertilization of young pine stands in northern Ostrobothnia, *Folia For.,* 821, 1993.

Reinikainen, A. and Veijalainen, H., Diagnostic use of needle analysis in growth disturbed Scots pine stands, *Commun. Inst. For. Fenn.,* 116, 44, 1983.

Roderfeld, H., Vasander, H., and Tolonen, K., Differences in carbon accumulation of two cut-over peatlands in Finland, in *The Finnish Research Programme on Climate Change, Second Progress Report,* Kanninen, M. and Heikinheimo, P., Eds., Publications of the Academy of Finland, 1994, 1, 315.

Rossi, S., Varmola, M., and Hyppönen, M., Success of afforestation of old fields in Finnish Lapland, *Folia For.,* 807, 1993.

Salonen, V., Early plant succession in two abandoned cut-over peatland areas, *Holarctic Ecol.,* 13, 217, 1990.

Salonen, V., Personal communication, 1994a.

Salonen, V., Revegetation of harvested peat surfaces in relation to substrate quality, *J. Veg. Sci.,* 5, 403, 1994b.

Salonen, V. and Laaksonen, M., Effects of fertilization, liming, watering and tillage on plant colonization of bare peat surfaces, *Ann. Bot. Fenn.,* 31, 29, 1994.

Salonen, V., Penttinen, A., and Särkkä, A., Plant colonization of a bare surface: population changes and spatial patterns, *J. Veg. Sci.,* 3, 113, 1992.

Silfverberg, K. and Hotanen, J., Long-term effects of wood-ash on a drained mesotrophic Sphagnum papillosum fen in Oulu district, Finland, *Folia For.,* 742, 1989.

Silfverberg, K. and Huikari, O., Wood-ash fertilization on drained peatlands, *Folia For.,* 633, 1985.

Silvola, J. and Alm, J., Dynamics of greenhouse gases in virgin and managed peatlands, in *The Finnish Research Programme on Climate Change, Programme Report*, Kanninen, M. and Anttila, P., Eds., Publications of the Academy of Finland, 3/92, Painatuskeskus Oy, Helsinki, 1993, p. 193.

Valtanen, J., Results of afforestation of fields in northern Ostrobothnia in the 1970's, in Developing methods for afforestation of fields, interim report, Ferm, A. and Polet, K., Eds., *Metsäntutkimuslaitoksen tiedonantoja*, 391, 19, 1991.

Vasander, H., Leivo, A., and Tanninen, T., Rehabilitation of a drained peatland area in the Seitseminen National Park in Southern Finland, in Peatland Ecosystems and Man: An Impact Assessment, Department of Biological Sciences, University of Dundee, U.K., Bragg, O. M., Hulme, P. D., Ingram, H. A. P., and Robertson, R. A., Eds., in association with The International Peat Society, Jyväskylä, Finland, 1992, 381.

Veijalainen, H., Geographical distribution of growth disturbances in Finland, *Commun. Inst. For. Fenn.*, 116, 13, 1983.

Vikberg, P., Personal communication, 1994.

Westman, C. J., Fertility of surface peat in relation to the site type and potential stand growth, *Acta For. Fenn.*, 172, 1981.

CHAPTER **28**

Forestry Best Management Practices for Wetlands in Minnesota

Michael J. Phillips

CONTENTS

INTRODUCTION

Wetlands are a common landscape feature in Minnesota in spite of significant losses of wetlands to agriculture and development. Prior to European settlement, Minnesota contained 7.5 million ha of wetlands, including both wet mineral and peat soils. These wetlands covered approximately 35 percent of the state. The current extent of wetlands for Minnesota is approximately 3 million ha, which represents a 60 percent loss of the original wetland acreage (Minnesota Department of Natural Resources, Protected Water Inventory Data Base, 1984). The majority of the remaining wetlands are found in the forested regions of Minnesota, predominantly located on county, state, and federal lands in northern Minnesota.

1-56670-177-5/97/$0.00+$.50

In recent years, the diminishing wetland resources and recognition of the important benefits that wetlands provide have generated much discussion and debate on the need to preserve and protect the nation's remaining wetlands. Wetlands management and protection have evolved into a major national and state policy issue (Siegel, 1991; Siegel and Haines, 1990). Although the policy debates on the issue have often been fractious and controversial, there is general agreement among natural resource professionals on the need to protect the functions and values of wetlands.

Recognizing the need to maintain or enhance the remaining wetland resource in Minnesota, a "no net loss" Wetland Conservation Act (WCA) was enacted into law in 1991 and amended in 1993. The WCA complements other wetland regulations, including Section 404 of the Clean Water Act. The 1977 amendments to the federal Clean Water Act [Section 404(f)] provided a permit exemption for normal forestry operations in wetland areas. A similar exemption was incorporated into the WCA of 1991. The use of Best Management Practices (BMPs) is encouraged in both acts to protect wetland values and functions. Best Management Practices provide resource managers, loggers, and landowners with the necessary tools to avoid or minimize direct and indirect impacts from activities that, if improperly conducted, can diminish the quantity, quality, and biological diversity of wetlands. With BMPs as an integral part of forest management, continuous commercial production on or near Minnesota's wetlands is feasible without compromising environmental quality.

BMP DEVELOPMENT

Best Management Practices have been developed in most states where forestry is a significant land use. As of 1992, 42 states had published BMPs for forest management (Boyette, 1993). Many of the state forestry organizations have provided the leadership in the development of BMPs, often in cooperation with other resource agencies, conservation groups, forest industry, and landowners. Best Management Practices serve as the cornerstone for the water quality and wetlands protection programs developed by the states. Development of these programs has been shaped by the particular physiographic, economic, technical, and political characteristics of each state.

Implementation of BMPs varies from nonregulatory or quasiregulatory (e.g., Minnesota, Virginia) to mandated compliance by state law (e.g., Oregon, Washington) (Ellefson et al., 1995; Floyd and MacLeod, 1993). Under a regulatory program, BMPs to protect water quality are legally binding with penalties for noncompliance. With a quasiregulatory program, BMPs are encouraged but not strictly required. However, if adverse impacts to water quality are found and BMPs were not used, then significant penalties may be assessed to the responsible party. Programs that are considered voluntary do not have the force of law to mandate compliance with BMPs, and there are no legal penalties for noncompliance. Instead these programs rely on the cooperation of loggers, landowners, and resource managers to implement BMPs. Education, technical assistance, and financial incentives are critical to the success of voluntary programs.

Best Management Practices are preventative in nature and are often viewed as a surrogate for water quality standards. The use of BMPs is the preferred mechanism for controlling nonpoint source pollution for several reasons. Water quality standards do not take into account the natural variability that exists in forest streams. Water quality monitoring is expensive, and the limited resources of state agencies also make water quality monitoring problematic. Forestry activities often produce minor and short-term impacts that generally have little effect on long-term water quality where proper preventative practices are employed (Curtis et al., 1990; Lynch and Corbett, 1990). The proper function of water quality standards in forestry nonpoint source programs may be to test BMP effectiveness and provide data to fine tune BMP practices.

There are a number of key criteria that characterize development of effective BMPs. First and foremost, BMPs must embody the principle of prevention so that practices utilized will minimize impacts to water quality and wetlands. While prevention is fundamental to BMP selection, other considerations are also important. The BMPs must be reasonable and not excessive to what is required to provide protection to the resource. The ability to obtain logger, landowner, and resource manager cooperation in implementing BMPs will be enhanced if the BMP recommendations are perceived as not excessive. The BMPs must also achieve the identified goals for water quality and wetlands protection. The BMPs that are selected should be cost-effective. It will be easier to obtain logger, landowner, and resource manager commitment if the cost of implementing BMPs is reasonable. The BMPs must also be flexible so that they can be tailored to site-specific conditions. What is implementable on one site may not be appropriate under other site conditions. Finally, it is essential that the state's forestry BMPs be understood by loggers, landowners, and resource managers.

When the decision was made to move forward with wetland BMP development for forestry in Minnesota, it was recognized that successful implementation of these BMPs would be dependent on obtaining the involvement and commitment of the agencies and organizations with an interest in forestry wetland issues. A forestry wetland BMP task force was established with representatives from federal and state agencies, county government, forest industry, environmental organizations, logging interests, and private landowner groups. The integrated team approach was essential to the success of this effort. Wetland issues continue to be controversial, and finding common ground requires inclusive partnerships that represent the broad range of interests and opinions. The range of organizations represented on the task force generated intense debate and, eventually, compromise and consensus. The results of this effort have been published in a revised water quality and wetlands protection guidebook (Minnesota Department of Natural Resources, Division of Forestry, 1995). The wetland BMPs in the guidebook provide recommendations for road construction and maintenance, timber harvesting, site preparation, pesticide use, and prescribed burning.

BMP GUIDELINES

The principal outcome from applying Minnesota's wetland BMPs will be the protection of normal water movement within the wetland. The assumption is that applying practices that maintain hydrological flows will protect other wetland functions as well. Protecting hydrological flows on wetlands means minimizing the depth and extent of rutting. Practices that minimize the depth and extent of rutting provide the basis for many of the BMP recommendations and will supplement upland forestry BMPs to protect water quality. The practices listed below represent selected wetland BMPs that have been adopted for Minnesota.

Planning Considerations

- Plan and conduct all management operations in or adjacent to wetlands in a manner that protects site productivity, maintains or enhances ecological functions, and protects water quality. Planning approaches that anticipate problems and incorporate protective measures will be less costly than remedial activities to mitigate impacts.
- Conduct silvicultural activities in wetlands when frozen or when firm enough to support the equipment being used.
- Plan for removal of equipment and cut material from the wetland area at the end of the winter season prior to thawing, or leave it until the next winter.

Forest Roads

Knowledge of water table position, zone of water flow, type of wetland soils, and the strength of wetland soils provides the basis for road construction techniques that maintain the flow of water through the road corridor and ensure the structural integrity of the road embankment. The concept is to provide a simple road structure of adequate strength to support heavy vehicle traffic while preventing subsurface soil displacement, and to provide either free-flowing pore space or drainage structures to pass water at its normal level through the road corridor.

- Avoid crossing wetlands. If wetlands must be crossed, then minimize the total wetland road mileage required to meet the landowner's objectives.
- Provide adequate cross-drainage by employing one or both of the following techniques: (1) use construction methods that allow free water flow throughout the entire roadbed, or (2) place culverts or other cross-drain structures at each end of each wetland crossing and at intermediate low points. Space culverts or other cross-drain structures at maximum 91 m intervals to ensure adequate cross-drainage through the roadbed.
- Construct all road embankment fills with clean fill or other suitable native materials. The road base is built wider and deeper than for upland roads to spread out the road loading and minimize failure.
- Install culverts in peatlands that are a minimum of 61 cm in diameter buried halfway below the soil surface. Their upper half will handle surface storm flows and the lower half will handle everyday subsurface flows. Failure to bury the

lower half of the culvert will cause subsurface water to pond on the "upstream" side of the road and kill trees.

- Place culverts at the low points of the wetland to pass surface water flows through the road embankments.
- Construct ditches in wetland crossings, where necessary, to intercept and carry surface and subsurface (top 30 cm) waters to, through, and away from the culverts. For shallow peat (<1.2 m deep), ditches should be constructed immediately adjacent to the toe of the fill slope. For deep peat (≥1.2 m), maintain a separation between the toe of the embankment fill slope and the ditch that is at least three times the depth of the peat. This will minimize disturbance of the inherent strength of the top layer of peat containing the root mat.
- Design upland road approaches to wetlands so that surface runoff carrying potential sediment is diverted before entering the wetland.
- Anchor temporary structures at one end to allow the structure to move aside during high water flows.
- Remove temporary fills and structures to the extent practical when use is complete.
- Tramp and pack the wetland area wider than needed for the driving and working area if sufficient frost is not present. This additional space will allow for turnouts, snow removal, and parking.
- Cease equipment operations on any portion of frozen roads where rutting exceeds 15 cm in depth for continuous distances greater than 91 m. Resume operations only when conditions are adequate to support equipment. This will minimize blockage of cross-drainage and prevent down-road channelization.

Timber Harvesting

- Minimize rutting by conducting harvest activities in wetlands on firm or frozen ground that can support the equipment used.
- Move equipment to a stable portion or alter operating techniques to prevent repeated rutting of the harvest area deeper than 15 cm.
- Cease operations if no part of the site or no alternative techniques are available to prevent repeated rutting deeper than 6 in.
- Make reasonable efforts to remove slash or woody vegetation that originates from outside the wetland or from upland areas contained in the wetland. Slash or woody vegetation that originates from outside the wetland is considered fill and must be removed.
- Avoid crossing small wetland inclusions, where practical.
- Size landings to the minimum required for the acres to be harvested, the equipment to be used, and the products to be cut.
- Locate landings on upland areas, whenever practical, when harvesting upland sites. When harvesting wetlands, an upland site may also be a preferred location for a landing.
- Avoid locating landings and yarding areas on frozen open water wetlands.
- Locate, design, construct, and maintain skid trails to minimize damage to the residual stand, minimize rutting, maintain surface and subsurface water flows in the wetland, and reduce erosion and sedimentation.
- Plan the layout of skid trails to maximize operating efficiency and minimize site disturbance.

- Prepare skid trails for the anticipated traffic needs to avoid unnecessary maintenance or relocation of trail. Techniques can include proper packing of snow or ground cover to ensure freezing or appropriate construction methods to provide a stable trail surface.

Pesticide Use

- Consult state and federal lists for endangered or threatened species or species of special concern that may exist in sites proposed for pesticide applications.
- Select only pesticides labeled for aquatic use if surface water is present at the time of application.
- Mix pesticides in upland areas, where practical.
- Rinse spray tank and empty pesticide containers and uniformly apply rinsate to wetland areas that are part of the application site.
- Avoid applying pesticides on small wetland inclusions in upland areas unless that application is part of the management objective. If unable to avoid pesticide use in these areas, select only pesticides labeled for aquatic use if surface water is present at the time of application.

Prescribed Burning on or Adjacent to Wetlands

- Use fire line construction methods in wetlands that do not expose bare soil. These may include the use of wet lines (water, foam, or other control substance used to create a fire suppression line) or existing constructed or natural barriers.
- Provide adequate filter strips when constructing fire lines that expose bare soil near wetlands.
- Seed fire lines when natural revegetation will not be adequate.

DISCUSSION

Minnesota's forestry wetland BMPs have been incorporated and published in a revised water quality and wetlands BMP guidebook. Landowners, loggers, and resource managers must take the responsibility to ensure that the appropriate practices are applied in silvicultural activities. Obtaining widespread use of BMPs requires the commitment, leadership, coordination, and cooperation of all in the forestry community and allied organizations. The ability to demonstrate compliance with forestry BMPs on all land ownerships is essential if the BMP process is to be credible. As a profession, forestry will continually be challenged to "prove" its good intentions and effectiveness.

The forestry community in Minnesota uses annual field audits to determine the degree of compliance with forestry BMPs (Phillips et al., 1994). The results from these audits are used to evaluate the level of BMP application for all forestry ownerships, provide a qualitative assessment of BMP effectiveness, identify necessary modifications to the BMPs, and obtain adequate BMP application monitoring data to target future education efforts and technical assistance. This process has been modified and expanded to incorporate wetland BMPs into the field audit reviews. Through affirmative management, education efforts, and technical assistance, the forestry community expects to achieve and demonstrate progressive improvement in the use of both water quality and wetland BMPs.

REFERENCES

Boyette, W. G., Progress Report on Implementation of Silvicultural Nonpoint Source Programs in the United States, North Carolina Division of Forest Resources, Raleigh, NC, 1993.

Curtis, J. G., Pelren, D. W., George, D. B., Adams, V. D., and Layzer, J. B., Effectiveness of Best Management Practices in Preventing Degradation of Streams Caused by Silvicultural Activities in Pickett State Forest, Tennessee, Report to the Tennessee Department of Conservation/Division of Forestry and the Tennessee Wildlife Resources Agency, Nashville, TN, 1990.

Ellefson, P. V., Cheng, A. S., and Moulton, R. J., Regulation of Private Forestry Practices by State Governments, Station Bulletin 605-1995, Minnesota Agricultural Experiment Station, University of Minnesota, St. Paul, MN, 1995.

Floyd, D. W. and MacLeod, M. A., Regulation and perceived compliance: nonpoint pollution reduction programs in four states, *J. For.,* 91, 41, 1993.

Lynch, J. A. and Corbett, E. S., Evaluation of best management practices for controlling nonpoint pollution from silvicultural operations, *Water Resources Bull.,* 26, 41, 1990.

Minnesota Department of Natural Resources, Division of Waters, Protected Water Inventory Data Base, St. Paul, MN, 1984.

Minnesota Department of Natural Resources, Division of Forestry, Protecting Water Quality and Wetlands in Forest Management, Best Management Practices in Minnesota, St. Paul, MN, 1995.

Phillips, M. J., Rossman, R., and Dahlman, R., Best management practices for water quality, evaluating BMP compliance on forest lands in Minnesota: a three-year study, Minnesota Department of Natural Resources, Division of Forestry, St. Paul, MN, 1994.

Siegel, W. C., Forestry implications of water quality and wetland law, in *Proc.: Agriculture Outlook '92: New Opportunities for Agriculture; 68th Ann. Outlook Conf.,* Washington, D.C., U.S. Department of Agriculture, December 3–5, 1991.

Siegel, W. C. and Haines, T. K., State wetland protection legislation affecting forestry in the northeastern United States, *For. Ecol. Manage.,* 33/34, 239, 1990.

CHAPTER **29**

Effects of Forest Management on Wetland Functions in a Sub-Boreal Swamp

C. C. Trettin, M. F. Jurgensen, J. W. McLaughlin,
and M. R. Gale

CONTENTS

INTRODUCTION

Forested ecosystems of the upper Great Lakes represent an extremely important natural resource for the region and the nation. They are the headwaters of the largest freshwater lake system in the world and harbor some of the most productive forest lands in North America. In terms of habitat, these forested

1-56670-177-5/97/$0.00+$.50
© 1997 by CRC Press, Inc.

Table 1   **Predevelopment and Current Estimate of Total Wetland Area in Michigan, Minnesota, and Wisconsin**

State	Original area (circa 1780) (ha × 10⁶)	Current area (circa 1980) (ha × 10⁶)	% Loss (ha × 10⁶)
Michigan	4.5	2.3	50
Minnesota	6.1	3.5	58
Wisconsin	3.9	2.2	54

Adapted from Dahl, T. E., Wetlands Losses in the United States 1780's to 1980's, Fish and Wildlife Service, U.S. Department of the Interior, Washington, D.C., 1990, 13 pp. With permission.

landscapes are quite diverse and include representatives of the central hardwood and northern hardwood types, boreal forests, as well as oak/pine savannas, and true prairie communities. Extensive wetlands are interspersed through these forest types and are a key component to habitat, freshwater resources, and forest product values of the region. Consequently, forested wetlands have become a focal point of national and regional environmental concern (Conservation Foundation, 1988). Typical positioning of local landform types indicates a close ecological relationship between wetlands, adjacent uplands, riparian, and lacustrine communities.

Recent surveys have shown that approximately 50 percent of the original wetland resource in the Upper Great Lakes region have been destroyed (Table 1). Within that region approximately 51 percent of the remaining wetland area is forested (Cubbage and Flather, 1993). Forest inventory data obtained by the Forest Service shows that potential wetland cover comprises a significant proportion of commercial forest land in the Upper Great Lakes region (Table 2). More than 90 percent of the land in Michigan's Upper Peninsula (4.3 × 10⁶ ha) is estimated to be forested (USDA, 1968).

Table 2   **Area of Forest Cover Types That Are Characteristic of Wetlands in Michigan, Minnesota, and Wisconsin**

Cover type	Minnesota (ha × 10³)	Michigan (ha × 10³)	Wisconsin (ha × 10³)
Balsam fir	324	246	170
Black spruce	420	208	111
White cedar	201	480	150
Tamarack	190	44	90
Elm-ash-cottonwood	288	513	502
Total wet types	1423	1490	1023
Total timberland	5494	7018	5978
Wet forest sites (%)	26	21	17

Note: Estimates are based on Forest Inventory and Analysis cover-types (Smith and Hahn, 1987, 1989; Hahn and Smith, 1987) and do not include nonproductive or nonstocked lands.

From Smith, W. B. and Hahn, J. T., Michigan's Forest Statistics, 1987: Inventory Update, General Technical Report, NC-112, USDA Forest Service, NC For. Exp. Stn., St. Paul, MN, 1989, 48 pp.; from Smith, W. B. and Hahn, J. T., Michigan's Forest Statistics, 1989: An Inventory Update, General Technical Report, NC-130, USDA Forest Service, NC For. Exp. Stn., St. Paul, MN, 1989, 48 pp.; from Hahn, J. T. and Smith, W. B., Minnesota's Forest Statistics, 1987: An Inventory Update, General Technical Report, NC-118, USDA Forest Service, NC For. Exp. Stn., St. Paul, MN, 1987, 44 pp. With permission.

A major obstacle to address regulatory and conservation policy, as well as development of "Best Management Practices" on forest wetlands in this region, is a lack of information on basic functional processes and how forest management activities affect those processes. The relative importance of forest wetlands to water quality of the adjoining Great Lakes, groundwater hydrology, wildlife habitat, and overall biological diversity of the region is clear. However, it is difficult to quantify the cumulative impacts of management activities with the present information available. It is also evident from the spatial distribution of landforms within the region that forest wetlands are the focal point from which to address these issues.

The principal objective of this study was to evaluate the effect of silvicultural practices on ecosystem dynamics and plant community production in a forested wetland in the Upper Peninsula of Michigan. Specific objectives of the study were to (1) evaluate timber harvesting and site preparation impacts on wetland water quality and hydrology, soil nutrients, and vegetation, and (2) assess wetland recovery following these forest management practices. The following discussion summarizes results obtained from 5 years of monitoring soil, vegetation, and water responses to the forest management treatments.

APPROACH

Study Site

The study was conducted in the central Upper Peninsula of Michigan (46° 10′N, 86°43′N) located adjacent to the West Branch River, the outlet for West Branch Lake (Figure 1). The site is located within the Sturgeon River watershed of the Lake Michigan drainage basin. The geological setting of the site is within the Wetmore Outwash Plain, which is underlain by middle ordovician limestone (Trenton Limestone). The drift thickness in the vicinity of the site ranges from 30 to 75 m. Overstory vegetation consists primarily of black spruce (*Picea mariana* [Miller] BSP), tamarack (*Larix laricina* [Du Roi] K. Kock), and jack pine (*Pinus banksiana* [Lamb.]). The dominant species in the shrub layer are *Vaccinium* spp., leatherleaf (*Chameadaphne calyculata* [L.] Moench), and labrador tea (*Ledum groenlandicum* [Oeder.]). The surface vegetation is dominated by *Sphagnum* spp., with minor components of starflower (*Trientalis borealis* [Raf.]), golden rod (*Solidago* spp.), bunchberry (*Cornus canadensis* [L.]), and *Carex* spp.

The soil has been mapped as a histic mineral soil in the Kinross series (Histic endoaquod, sandy, mixed, frigid), which is poorly drained with a fine sand solum overlain by a 5 to 15 cm organic horizon of mostly decomposed *sphagnum*. The soil is uniform to a depth of 2.5 m, is acid throughout, and has a clay content of less than 2 percent (Table 3). The water table is at or near the surface through much of the growing season. The study site is located within Region III — Luce district ecosystem as defined by Albert et al. (1986), which is characterized by a mean growing season length of 144 d, total annual precipitation of 849 mm, mean

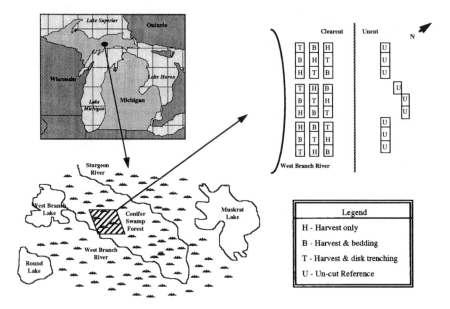

Figure 1   Location of the study site within the Great Lakes basin and the experimental layout.

annual temperature of 5°C, and a mean growing season (May to September) temperature of 14.4°C.

A 14.5 ha parcel of the study site was whole-tree harvested (WTH) in July 1988 using feller-bunchers. Tree bundles were skidded to a landing where they were sorted into fiber and fuel product classes. Fuel wood was chipped, while the fiber wood was delimbed and topped at the landing, and bucked into pulpwood bolts. This harvest regime resulted in all tree stems approximately 5 cm and greater being removed from the site.

After the WTH was completed, two site preparation treatments were randomly applied within the three Latin squares. The bedded treatment was installed using an Eden Relief bedding plow pulled by a John Deere 648 skidder, equipped with high flotation tires. The plow disks and mounds the soil, which results in an elevated planting bed flanked by two furrows. A single pass of the bedding plow tilled a strip approximately 3-m wide, creating an elevated 1-m wide planting bed. This prescription resulted in 100 percent tillage of the soil surface (Figure 2). The disk trench treatment consisted of a single pass of a TTS-disk trencher, which produced an alternating pattern of furrows and berms. The treatment resulted in approximately 50 percent tillage of the soil surface, with the intertrench area relatively undisturbed. Both the bed and trench treatments were sprayed with herbicide (glyphosate, 4.7 L ha⁻¹) to control competing vegetation. Three blocks were also established in an uncut stand adjacent to the harvested area to serve as a control to original soil conditions. The overall design is shown in Figure 1. Jack pine seedlings were planted immediately after the site was prepared.

Soil, water, and vegetation were monitored for 5 years following harvesting. Detailed descriptions of the initial instrumentation for the soil and water mea-

Table 3 Selected Morphological, Chemical, and Physical Properties for the Kinross Soil Sampled on the Study Site

Horizon	Depth range (cm)	Clay (%)	Silt (%)	Sand (%)	pH	Ext. acidity (meq 100 cm⁻³)	Ca (meq 100 cm⁻³)	K (meq 100 cm⁻³)
E	0–20	<0.1	4.2	95.8	4.3	17.2	0.53	0.16
Bs	20–44	<0.1	1.5	98.5	4.7	20.8	0.08	0.01
BC	44–72	<0.1	1.7	98.3	5.0	14.5	0.10	0.01
C	72–100	<0.1	0.7	99.3	4.3	9.5	0.07	0.01

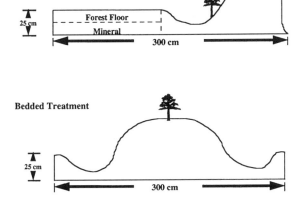

Figure 2 Cross-section profile of the trenched and bedded site preparation treatments. The planting line is indicated by the tree in the figure. (Figure is not to scale.)

surements were provided by Trettin (1992). Subsequent recovery studies were described by McLaughlin et al. (1994, 1996). Studies on the initial seedling survival were described by Cho (1990), and subsequent studies of seedling productivity and plant community response have been developed by Gale et al. (submitted). Table 4 provides a summary of the methods used in the results discussed here. These results are based on sampling designed to assess treatment response on a whole-soil basis. In the first 2 years after the timber harvest treatments, all site preparation treatments were monitored. However, by this time the trench treatment was not considered a viable silvicultural treatment on these sites, and only the WTH and the bed treatments were further sampled.

RESULTS AND DISCUSSION

Organic Matter Decomposition

Organic matter decomposition is an important process that affects ecosystem functions in many northern wetlands. The cotton strip assay, which measures cotton tensile strength loss (CTSL), provides an effective index of organic matter

Table 4 Variables and Methods for Estimating Soil, Water, and Vegetation Parameters

Measurement	Method	Sample location	Sample period	Reference
		Soil characteristics		
Chemical properties:				
Nitrogen	Kjeldahl	0–25 cm	year 1, 2	Bremner and Mulvaney, 1982
Ca, Mg, K	NH₄O AC ext.	0–25 cm	year 1, 2	Thomas, 1982
Ext. Acid and Al	KCl ext.	0–25 cm	year 1, 2	Barnhisel and Bertsch, 1982
pH	direct, buffered	0–25 cm	annual	McLean, 1982
Redox	rod	0–75 cm	year 1, 2	Carnell and Anderson, 1986
	electrode	15 cm	year 3, 4	Faulkner et al., 1989
Physical properties:				
Texture	hydrometer	O, A, B, C	pretreatment	Gee and Bauder, 1986
Bulk density	core	O, A, B, C	year 1	Blake and Hartge, 1986
Soil temp.	probe	7.5, 0.5, 25 cm	year 1, 2	Taylor and Jackson, 1986
		Hydrological properties		
Water table depth	well	2 m	monthly	Faulkner et al., 1989
Soil water	lysimeters	15, 30 cm	monthly	
		Vegetation		
Seedling production	inventory direct meas.	veg. strata	year 1, 3, 4	Chatarpawl et al., 1985
Vascular species composition	density plots	veg. strata	year 4	Chatarpawl et al., 1985

Note: Soils and water were sampled 1988–1993; other parameters as noted.

decomposition by employing a uniform substrate (cellulose) whose degradation is related to environmental factors and soil conditions (French, 1988; Hill et al., 1988). CTSL was significantly different among the silvicultural treatments (Figure 3). Both the harvesting and site preparation treatments exhibited greater cellulose decay as compared to the uncut plots, with the greatest CTSL occurring after bedding. This pattern of organic matter decomposition persisted through the first growing season following disturbance (Trettin and Jurgensen, 1992).

The most common site factors attributed to the increase in cellulose decay are altered temperature and moisture regimes (Bridgham et al., 1991; Mader, 1990). The silvicultural treatments in this study caused a significant increase in soil temperature that corresponded to the degree of soil disturbance and vegetation removal. Mean soil temperature in June to July decreased with soil depth in order of bedding > WTH > uncut (Figure 4). The cellulose decomposition response was correlated with changes in soil temperature. Because the rate of decomposition is strongly controlled by temperature, higher rates of organic matter decay are expected to continue on harvested and regenerated sites until canopy closure reduces soil temperature.

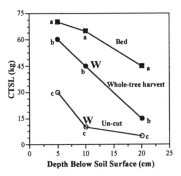

Figure 3 Organic matter decomposition as measured by cotton tensile strength loss (CTSL) for the bed, harvest-only, and uncut reference sites 1 year after harvest. Lower case letters denote a statistically significant (p = 0.05) difference at each sampling depth. The letter "W" denotes the location of the water table during the incubation. (Adapted from Trettin, C. C. and Jurgensen, M. F., *Int. Peat J.*, II, 392–399, 1992. With permission.)

Soil moisture and aeration also influenced cellulose decomposition (Lahde, 1969). Water table and aerated soil depth were nearly the same on all treatments, except when the water table was near the surface (<15 cm) or deeper than 50 cm. Consequently, the CTSL response for the trench, harvest-only, and uncut treatments at the 20 cm soil depth during late spring and early summer was tempered by saturated soil conditions over what would be expected with aerated conditions (Figure 3). In contrast, the incubation zone in the bed treatment was above the water table in the late spring-early summer sampling period, thereby providing an aerobic decomposition environment. Later in the summer, the water table was below the cotton strip for all the treatments, thereby minimizing or negating the effect of saturation on the measured cellulose decomposition.

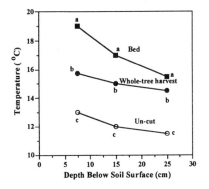

Figure 4 Average soil temperature during June to July in the first year following harvest for the whole-tree harvest, bed, and uncut treatments. Lower case letters denote a statistically significant (p = 0.05) difference at each sampling depth. (Adapted from Trettin, C. C. and Jurgensen, M. F., *Int. Peat J.*, II, 392–399, 1992. With permission.)

Bridgham et al. (1991) also reported that soil moisture was a controlling factor on CTSL in wetlands in North Carolina, particularly when sites had been drained.

Soil Chemical Properties

The following discussion considers the changes in C, Ca, N, Al, acidity, and redox potential following WTH and site preparation. These are considered critical parameters affecting the long-term site productivity potential of forested wetlands.

Carbon

The most significant effect of WTH and site preparation on soil properties was on soil C. Whole-tree harvesting alone caused a 30 percent reduction in soil C 12 months after harvesting, while bedding caused an increase (130 percent) in soil C in the planting line but a net reduction (32 percent) when the entire surface area of the treatment was considered (Trettin et al., 1992). Five years after harvesting, the C content surface soil in the planting line of the bedded treatment was 35 percent lower than in the uncut stand, while soil C losses in the WTH had increased to 47 percent (Table 5). These results indicated differences in organic matter decomposition rates between the harvesting and site preparation practices in the 4 years following disturbance. Accelerated decomposition apparently had been sustained in the bed, oxidizing the organic matter that had been concentrated there by the treatment. In contrast, the net loss in the WTH between the first and fifth years was only an additional 17 percent.

Other studies have also reported reductions in soil carbon or organic matter content as a result of intensive site preparation (Morris and Pritchett, 1983; Pye and Vitousek, 1985; Tew et al., 1986). Burger and Pritchett (1988) found that the level of organic matter reduction corresponded to the degree of site preparation. In a review of silvicultural impacts on soil C, Trettin et al. (1995) reported that change in temperature and soil moisture are largely responsible for reductions in soil C. Others (Tuttle et al., 1985; Trettin et al., 1992) have reported an increase in organic matter content following bedding, but those results were from measurements in the center of the bed, which does not represent the whole soil area (Attiwill et al., 1985; Sutton, 1993).

Table 5   Soil Carbon Content After Harvesting and Site Preparation Compared to the Uncut Reference Stand

	Uncut	Whole-tree harvest		Bedded	
	Carbon (mg/ha)	Carbon (mg/ha)	% loss	Carbon (mg/ha)	% loss
Forest floor	31.4	8.9	72	12.7	60
Mineral soil[a]	35.0	18.9	46	32.8	5
Total	66.4	35.2	47	43.2	35

Note: The trench treatment was not resampled in 1993.

[a] Measured to a mineral soil depth of 30 cm.

From McLaughlin, J. W., unpublished data. With permission.

Table 6   Average Extractable Ca, K, Al, Exchangeable Acidity
(AC) in the Surface 25 cm of Soil (Including Forest Floor)
1 Year After Treatments Were Established

Property	Treatment			
	Uncut	WTH	Trenched	Bedded
Ca (meq)	0.27[a]	0.24[a]	0.32[a]	0.31[a]
K (meq)	0.08[a]	0.05[b]	0.06[c]	0.05[b]
Al (meq)	0.98[a]	0.94[a]	1.23[a,b]	1.37[b]
AC (meq)	5.39[a,b]	3.70[a]	7.50[b]	7.41[b]
Total N (g)	0.084[b]	0.069[a]	0.083[a,b]	0.072[b]

Note: Values expressed on the basis of 100 cm^{-3}. Different lower case
letters ([a], [b], [c]) indicate a statistically significant difference within
rows ($p = 0.05$).

From Trettin, C. C., Silvicultural Effects on Functional Processes of a
Boreal Wetland, Ph.D. Thesis, College of Forest Resources, North
Carolina State University, 1992, 153 pp. With permission.

Nutrients (Ca, K, N)

Harvesting and site preparation treatments changed the quantity and distribution of nutrients in the surface 25 cm of the soil 1 year following disturbance (Table 6). Although the trenched and bedded treatments had higher Ca levels than the uncut reference site, the difference was not statistically significant. Among silvicultural treatments, the bedded and WTH treatments had the lowest K levels, reflecting the mobility and sensitivity of K to leaching loss following disturbance. The loss of K on the WTH and trench treatments was due primarily to reductions in the forest floor, since there was no difference in extractable K among those treatments in the mineral soil. Morris and Pritchett (1983) reported a similar K response on a flatwood site where there was little change in extractable K and Ca in the mineral soil, but had significant reductions in the forest floor. However, the reduction of K in the forest floor in that study and others (Tew et al., 1986; Pye and Vitousek, 1985) was a result of displacement of the forest floor into windrows. Windrowing of the forest floor is in contrast to the treatments in this study, where the forest floor remained on the site. Other studies on several southeastern United States sites (Miller and Edwards, 1985; Tuttle et al., 1985) have not measured statistically significant reductions in extractable bases within 2 to 4 years following harvesting and site preparation.

There were no significant differences in total N among the WTH, bedded, and trenched treatments (Table 6). Similarly, there was no difference between total N levels in the trenched and bedded treatments as compared to the uncut stand. The C:N ratio of the uncut reference site was approximately 30; in contrast, it was 23 on the bedded treatment and 27 on the trenched and WTH treatments. The reduction in C:N and relatively constant total N levels suggest that immobilization was occurring. The reduction in total N on the WTH treatment was primarily a result of lower N levels in the mineral horizon. Burger and Pritchett (1988) reported that total N was reduced as a result of intensive site preparation, but the ratio of organic matter to total N remained relatively constant due to reductions in organic matter content.

Acidity and Aluminum

The soil conditions imposed by the silvicultural treatments were conducive to increased acidification. The acidification effect is subject to both spatial (Nykvist and Skyllberg, 1989) and temporal (Johnson, 1987; Skyllberg, 1991) variation. In our study, soil Al content increased as a result of site preparation treatments (Table 6). Extractable Al in the bedded treatment was measurably greater than the WTH treatment and the uncut reference site. Although the trenched treatment had higher extractable Al than the WTH treatment, the difference was not statistically significant. Similar to Al, the total acidity on the bedded and trenched treatments was greater than both the uncut reference site and WTH treatment; however, only the difference among the site preparation treatments and the WTH treatment was statistically significant. The direction of change in the soil environment measured in this study was toward a more acidified soil with lower levels of available base cations. Potentially, continued acidification is likely due to vegetation removal during harvesting and vegetation control (Binkley and Richter, 1987), as well as the recolonization of Sphagnum spp. (Clymo, 1984).

Mechanisms controlling soil acidification on northern wetlands include organic acid production, ferrolysis, and biomass removal. Organic acid production is considered the dominant acidification process in northern peatlands (Hemond, 1983; Urban et al., 1987). The measured increase in cellulose decomposition in this study (Trettin and Jurgensen, 1992) reflects increased microbial activity and probably increased decomposition of the native organic matter. The accelerated decomposition could cause the increased acidity. Oxidation of ferrous iron is also a likely contributor to the acidity pool (Bouma, 1983). Both the bedded and trenched treatments effectively increased the aerated soil volume, thereby increasing the time available for iron oxidation. The implications of increased acidity following disturbance has important ramifications with respect to Al and dissolved organic C transport from the wetland to adjacent waters. Aluminum moves from the forest floor as organic Al compounds and is immobilized in the B horizon. Most Al leaving the solum is inorganic Al derived from the mineral soil (Driscoll et al., 1989).

Redox Potential

In the first year after the site preparation treatments were installed, the anaerobic soil zone (based on the steel rod oxidation depth) exhibited a complex response pattern that was affected by the silvicultural treatment, position within the wetland, and measurement period (Trettin, 1992). As expected, the bedded treatment had the greatest oxidized soil depth, in the center of the beds, at each measurement date throughout the summer. The trenched and WTH treatments were not significantly different at any of the measurement dates. Subsequent measurements using electrodes in the fifth growing season showed that reducing conditions were predominant at 15 and 30 cm regardless of treatment; however, there was considerable variation in soil redox potentials, ranging from −220 to +500 mv, and no significance differences were found among treatments. Average redox potentials were higher at 15 cm than at 30 cm, which would be expected because of higher oxygen diffusion rates in surface soil horizons (Vepraskas, 1992).

Water Quality

Assessment of water quality from the vadose zone and shallow groundwater was an effective approach for determining post-treatment response to the silvicultural treatments and in confirming the changes in soil nutrients. In the first year following harvesting and site preparation, water quality results corroborated the increase in organic matter decomposition, as measured by the increase in electrical conductivity in both the vadose zone and groundwater under each treatment (Trettin, 1992). Similarly, soil K leaching was evident in both vadose and groundwater samples for the two site preparation treatments and the WTH. The water samples were particularly effective for assessing the N response to disturbance, because changes in N cycling processes (ammonification and nitrification) were evident in the vadose zone.

Five years after timber harvesting, there was still some difference in water quality among treatments, but there was no evidence of water quality degradation (Table 7). Dissolved organic carbon (DOC), pH, and conductivity were similar between the silvicultural treatments and the uncut site. In fact, soil water DOC, pH, and conductivity were highly correlated with one another. These results were similar to Moore (1987), who reported no significant differences in DOC con-

Table 7   Average Water Chemistry 5 Years after Harvesting and Site Preparation

Depth (cm)	pH	Conductivity (μ mhos cm^{-1})	DOC (mg L^{-1})	NH$_4^+$	NO$_3^-$	PO$_4^{-3}$	TKN	TP
						μg L		
			Uncut stand					
15	4.1	48	48	97	39	20	777	329
30	4.3	46	44	78	31	9	664	115
Water table	5.2	65	21	22	14	4	201	95
			Whole-tree harvest					
15	4.5	42	44	104	56	5	759	151
30	4.7	36	36	82	36	4	740	95
Water table	5.4	54	22	26	15	6	247	57
			Bedded treatment					
15	4.4	42	51	198	58	4	704	205
30	4.5	38	42	175	62	7	798	94
Water table	5.3	59	25	102	49	2	345	91

Note: Samples were collected monthly at 15 and 30 cm soil depths and from the water table in 1993.

centration for three of four peatlands in the subarctic region of Quebec after drainage and harvesting. DOC concentrations were positively related to water table depth and forest floor pH (Table 7), but only to a soil depth of 17 cm. This was likely due to increased aeration as the water table lowered and the soil shifted from predominantly anaerobic to aerobic microbial respiration, causing greater organic matter degradation, and higher soluble C (Schlesinger, 1991). This inverse relationship has also been found in other spodosols in the northeastern United States and eastern Canada (McDowell and Wood, 1984; Moore et al., 1992). Increased soil sorption of DOC in the Bs horizon was the likely cause. The Bs horizon has the strongest DOC sorption potential of all mineral soil horizons (McLaughlin et al., 1994). There were, however, differences in the hydrophobic and hydrophilic acid composition of DOC and its carboxyl acidity. The WTH treatment had a lower hydrophobic acid composition than the uncut stand and lower hydrophilic acid carboxyl acidity (McLaughlin et al., 1996).

Inorganic N (NO_3 + NH_4) in the WTH and bedded treatments still exhibited higher concentrations 5 years following disturbance (Table 7). However, ammonium concentrations were approximately one-half and NO_3 levels were one-seventh the uncut level 12 months after harvest on all treatments. Ammonium concentrations were lower in the soil water of the uncut stand than those reported for a poorly drained spruce/fir forest in Maine (Lawrence and Fernandez, 1991). The higher concentration of NH_4 than NO_3 is consistent with what has generally been reported for poorly drained soils (Lawrence and Fernandez, 1991; Schlesinger, 1991). The range of TKN concentrations in the lysimeter samples for the uncut stand and both silvicultural treatments were similar (Table 7), although the partitioning between organic and inorganic N was different. Ammonium and NO_3 accounted for approximately 35 percent of the soil water N in the bedded treatment, 20 percent in the WTH treatment, and only 18 percent in the uncut stand. Among all treatments, 60 to 85 percent of the N in the soil water occurred in organic forms. Organic N is transported through soils in conjunction with DOC (Qualls et al., 1991). Correspondingly, in this study there was a strong positive correlation between soil water TKN and DOC concentrations (McLaughlin, unpublished data).

Phosphorus levels were not measured in the first year after harvest; however, after 5 years, the uncut reference stand had higher soil water P concentrations than harvested and site-prepared plots (Table 7). The general decrease in P concentrations with increasing soil depth indicated that P sorption may be an important soil process at this wetland site. Phosphate has been found to be strongly adsorbed in soils of many different forest ecosystems (Wood et al., 1984; Richardson, 1985; Walbridge et al., 1991). However, McLaughlin et al. (1994) showed that DOC sorption potential of the mineral soil at this site is low due to high sand content and low extractable iron and aluminum. A similar situation may also occur for P sorption (Walbridge et al., 1991). Phosphorus is generally deficient in wetland soils (Richardson, 1985; McLaughlin et al., 1994) and would be used immediately by either plants or microorganisms when released during organic matter decomposition. Such P immobilization would lower P in the soil solution (Richardson, 1985). Total P concentrations in soil water of the uncut reference site were higher than in the site preparation treatment (Table 7). Organic P concentrations (the difference between PO_4 and total P) accounted for greater than 90 percent of the soluble P in all treatments at both 15 and 30 cm depths. Phosphorus predominantly moves through the soil bound to DOC (Moore, 1987; Schoenau and Bettany, 1987; Qualls et al., 1991; Walbridge et al., 1991).

Plant Community Response

Four years after planting, jack pine seedlings growing on the bedded treatments had significantly greater heights and root collar diameters than those on the WTH and trenched treatments (Table 8). As previously discussed, bedding increased the aerated rooting zone and increased rates of organic matter decomposition and nutrient availability. In contrast, the trench treatment resulted in the seedlings being planted in the mineral soil, effectively below the original surface elevation. That planting position resulted in the largest rate of first-year mortality (Cho, 1990). After 4 years those seedlings were larger than those flat-planted in the WTH treatment, although the difference was not statistically significant.

Plant diversity, however, responded differently to the harvesting and site preparation treatments than did seedling development. Species numbers on the WTH areas were significantly greater than the trenched and uncut reference areas, yet similar to bedded areas (Table 8). Total plant coverage on the WTH areas had significantly greater coverage than all the other site preparation treatments and the uncut areas. Low total plant cover on the uncut area is probably due to complete cover of bryophytes, while bryophyte cover was minimal on the harvested and site-prepared sites.

Table 8   Seedling Height and Diameter, Number of Species, and Plant Coverage 4 Years After Planting

	Uncut	Whole-tree harvest	Trenched	Bedded
Seedling				
Height (cm)	—	55.3[a]	64.0[a]	88.8[b]
Diameter (cm)	—	35.6[a]	46.8[a]	69.6[b]
Plant community				
No. of plant species	8.1[b]	10.0[a]	8.7[b]	9.1[a,b]
Total plant coverage (%)	17.0[c]	38.0[a]	29.0[b]	28.0[b]
Relative coverage				
Herbaceous (%)	39.0[b]	5.0[a]	9.0[a]	8.0[a]
Grass/sedge (%)	4.0[b]	22.0[a]	31.0[a]	30.0[a]

Note: Different lower case letters (a, b, c) indicate a statistically significant difference (p = 0.05).

From Gale et al., submitted. With permission.

Relative plant coverage of woody species was not significantly different among site preparation treatments and the uncut reference. Significant differences in relative cover of herbaceous and grass/sedge species among treatments and reference site were determined (Table 8). However, relative coverage of herbaceous species on the uncut area was approximately four to seven times greater than coverage on site preparation areas. Relative cover of grass/sedge species was approximately four to eight times greater on site preparation areas than the uncut area. No significant differences in relative coverage of herbaceous and grass/sedge species among the three silvicultural treatments were detected. Relative cover indicates a shift in the community composition of planted sites from a tree–herbaceous dominated system to a tree–grass/sedge system when compared with the uncut stand 4 years after timber harvesting.

CONCLUSIONS AND PERSPECTIVES

The site in this study was characterized by a histic-mineral soil, a nascent C-accumulating wetland soil (Trettin et al., 1995). Accelerated organic matter decomposition and C losses were expected after harvest because of increased soil temperature, aeration, and substrate availability associated with these treatments. Measurements after timber harvesting and site preparation demonstrated that the site became a net source of C, either as CO_2 to the atmosphere or as DOC to the soil water. Both of these pathways have implications for global carbon cycling and possible climate change (Schlesinger, 1991). Accelerated organic matter decomposition can also contribute to soil acidification, loss of soil fertility, and a change in soil hydrological properties. Carbon loss has occurred during the first 5 years following disturbance. However, loss as DOC was not evident after 5 years. It is likely that the harvested and site-prepared lands will exhibit higher decomposition rates until canopy closure results in a reduction in soil temperature. At that point, this system may revert to a soil C-accumulating ecosystem.

Maintenance of soil nutrient levels is an important factor for sustaining the integrity of forested wetlands, particularly on oligotrophic sites. The whole-tree harvesting system used in this study removed all woody and leaf biomass, thereby eliminating much of this nutrient pool. Water quality information was important in evaluating wetland nutrient status because it provided an integrated picture of disturbance impacts. In the first year following harvest, cation and N leaching were evident, but the concentrations were low and never approached any regulatory or best management practice standard. Five years after timber harvesting, the bedding treatment was still causing a release of inorganic N into the soil water. However, the fluxes were not determined in this study; hence, the impacts of leaching on total nutrient pools could not be assessed.

Site preparation significantly increased seedling growth on harvested areas. This was especially evident in the bedding treatment, which created raised planting mounds and improved soil aeration and nutrient availability. Harvesting and site preparation also decreased coverage of the understory community. This was expected due to physical soil disturbance and loss of the forest floor. Whether these plant community differences will continue, as successional changes proceed, remains to be determined. Critical components for assessing future productivity, whether it is measured as tree growth, understory plant cover, or as overall community structure, are climate and hydrology.

These results have demonstrated that timber harvesting and associated site preparation practices can change soil, vegetation, and to a lesser degree hydrological properties of a northern wetland. The silvicultural challenge is to develop and apply forest management prescriptions that minimize site disturbance and preserve the integrity of the structural components of the wetland system. This may involve less intensive utilization and harvesting regimes that leave more branch and top material on site and reduce soil disturbance. A site preparation treatment that provides a planting site while minimizing or avoiding disturbance to the remainder of the site is desirable. Examples of such treatments include spot-mounding (Sutton, 1993) or flat-planting with spot vegetation control, both of which cause minimal site disturbance. The cost of bedding, in terms of soil C, long-term site productivity, and vegetation composition, must be considered with respect to gains in fiber crop production. It should not be assumed that future fiber production is commensurate with the degree of cultivation, particularly on infertile soils, although recent work in Finland has shown that ameliorative practices of water management and fertilization may offset soil C losses through increased tree productiviy (Laine et al., 1992).

The basis for this synthesis was relatively short-term results (5 years). It is very difficult to interpolate these results to long-term ecosystem function or fiber crop production. However, whether the observed short-term responses will reflect long-term changes in soil or water processes or vegetation dynamics must be determined in subsequent work. Additional research on silvicultural effects on boreal wetlands is also needed, both to establish the long-term effects of disturbance and to consider alternative prescriptions for accomplishing management objectives. Such studies should include (a) the effect of landscape-position on wetland function and how the impact of disturbance changes with landscape position and (b) cumulative impact assessment of forest management in upland and wetland ecosystems within a landscape. Regeneration studies are needed to (a) consider regeneration response to the current treatments, (b) consider alternative mechanical and chemical site preparation prescriptions, and (c) consider the use of natural regeneration, including the use of fire.

REFERENCES

Albert, D. A., Denton, S. R., and Barnes, B. V., Regional Landscape Ecosystems of Michigan, School of Natural Resources, University of Michigan, Ann Arbor, MI, 1986.

Attiwill, P. M., Turvey, N. D., and Adams, M. A., Effects of mound-cultivation (bedding) on concentration and conservation of nutrients in a sandy podzol, *For. Ecol. Manage.*, 11, 97, 1985.

Binkley, D. and Richter, D., Nutrient cycles and H+ budgets of forest ecosystems, *Adv. Ecol. Res.,* 16, 1, 1987.

Bouma, J., Hydrology and soil genesis of soils with aquic moisture regimes, in *Pedogenesis and Soil Taxonomy,* Vol. 1, Wilding, L. P., Smeck, N. E., and Hall, G. F., Eds., *Concepts and Interactions,* Elsevier, Amsterdam, 1983, p. 253.

Bridgham, S. D., Richardson, C. J., Maltby, E., and Faulkner, S. P., Cellulose decay in natural and disturbed peatlands in North Carolina, *J. Env. Qual.,* 20, 695, 1991.

Burger, J. A. and Pritchett, W. L., Site preparation effects on soil moisture and available nutrients in a pine plantation in the Florida flatwoods, *For. Sci.,* 34, 77, 1988.

Carnell, R. and Anderson, M. A., A technique for extensive field measurement of soil anaerobism by rusting of steel rods, *Forestry,* 59, 129, 1986.

Cho, S. M., Seedling Biomass Dynamics of Jack Pine (*Pinus banksiana* Lamb.) and Tamarack (*Larix laricina* (DuRoi) K. Koch) on a Site-Prepared Michigan Wetland, M.S. Thesis, College of Forest Resources, North Carolina State University, Raleigh, NC, 1991, 65 pp.

Clymo, R. S., Sphagnum-dominated peat-bog: a naturally acid ecosystem, *Phil. Trans. R. Soc. Lond. B,* 305, 487, 1984.

The Conservation Foundation, Protecting America's Wetland: An Action Agenda, The Final Report of the National Wetlands Policy Forum, The Conservation Foundation, Washington, D.C., 1988, 69 pp.

Cubbage, F. W. and Flather, C. H., Forested wetland area and distribution, *J. For.,* 91 (5), 35, 1993.

Dahl, T. E., Wetlands Losses in the United States 1780's to 1980's, Fish and Wildlife Service, U.S. Department of Interior, Washington, D.C., 1990, 13 pp.

Driscoll, C. T., Wyskowski, B. J., Destaffan, P., and Newton, R. M., Chemistry and transfer of aluminum in a forested watershed in the Adirondack region of New York, USA, in *Environmental Chemistry and Toxicology of Aluminum,* Lewis, T. E., Ed., Lewis Publishers, Chelsea, MI, 1989.

Faulkner, S. P., Patrick, W. H., Jr., and Gambrell, R. P., Field techniques for measuring wetland soil parameters, *Soil Sci. Soc. Am. J.,* 53, 883, 1989.

French, D. D., Seasonal patterns in cotton strip decomposition in soils, in *Cotton Strip Assay: An Index of Decomposition in Soils,* Harrison, A. F., Latter, P. M., and Walton, D. W. H., Eds., Institute of Terrestrial Ecology, Grange-Over-Sands, Cubria, U.K., 1988, p. 46.

Hahn, J. T. and Smith, W. B., Minnesota's Forest Statistics, 1987: An Inventory Update, General Technical Report, NC-118, USDA Forest Service NC For. Exp. Sta., St. Paul, MN, 1987, 44 pp.

Hemond, H. F., The nitrogen budget of Thoreau's bog, *Ecology,* 64, 99, 1983.

Hill, M. O., Latter, P. M., and Bancroft, G., Standardization of rotting rates by a linearizing transformation, in *Cotton Strip Assay: An Index of Decomposition in Soils,* Harrison, A. F., Latter, P. M., and Walton, D. W. H., Eds., Institute of Terrestrial Ecology, Grange-Over-Sands, Cumbria, U.K., 1988, p. 22.

Johnson, D. W., A discussion of the changes in soil acidity due to natural processes and acid deposition, in *Effects of Atmospheric Pollutants on Forests, Wetlands, and Agricultural Ecosystems,* Huthchinson, T. C. and Meema, K. M., Eds., NATO ASI Series, Vol. G16, Springer-Verlag, Berlin, 1987, p. 333.

Lahde, E., Biological activity in some natural and drained peat soils with special reference to oxidation-reduction conditions, *Acta For. Fenn.,* 94, 1–69, 1969.

Laine, J., Vasander, H., and Puhalainen, A., Effect of forest drainage on the carbon balance of mire ecosystems, in Proc. 9th Int. Peat Congr., 22–26 June, Uppsala, Sweden, Special Edition, *Int. Peat. J.,* I, 170, 1992.

Lawrence, G. B. and Fernandez, I. J., Biogeochemical interactions between acidic deposition and a low-elevation spruce-fir stand in Howland, Maine, *Can. J. For. Res.,* 21, 867, 1991.

Mader, S. F., Recovery of Ecosystem Functions and Plant Community Structure by a Tupelo-Cypress Wetland Following Timber Harvesting, Ph.D. Dissertation, College of Forest Resources, North Carolina State University, Raleigh, NC, 1990, 276 pp.

McDowell, W. H. and Wood, T., Podzolization: soil processes control dissolved organic carbon concentrations in stream water, *Soil Sci.,* 137, 23, 1984.

McLaughlin, J. W., Lewin, J. C., Reed, D. D., Trettin, C. C., Jurgensen, M. F., and Gale, M. R., Soil factors related to dissolved organic carbon transport in a black spruce swamp, Michigan, USA, *Soil Sci.,* 158, 454, 1994.

McLaughlin, J. W., Liu, A., Jurgensen, M. F., and Gale, M. R., Organic carbon characteristics in a spruce swamp five years after harvesting, *Soil Sci. Soc. Am. J.,* 1996, in press.

Miller, J. H. and Edwards, M. B., Impacts of various intensities of site preparation on piedmont soils after 2 years, in Proc. 3rd Biennial Southern Silvicultural Research Conf., Shoulders, E., Ed., Atlanta, GA, USDA Forest Service General Technical Report, SO-54, Atlanta, GA, 1985, p. 65.

Moore, T. R., A preliminary study of the effects of drainage and harvesting on water quality in ombrotrophic bogs near Sept-Iles, Quebec, *Water Resources Bull.,* 23, 1, 1987.

Moore, T. R., DeSouza, W., and Koprivnjak, F., Controls on the sorption of dissolved organic carbon by soils, *Soil Sci.,* 154, 120, 1992.

Morris, L. A. and Pritchett, W. L., Effects of site preparation on Pinus elliottii — P. palustris flatwoods forest soil properties, IUFRO Symp. on Forest Site and Continuous Productivity, Ballard, R. and Gessel, S. P., Eds., General Technical Report, PNW-163, USDA Forest Service, PNW For. Range. Exp. Sta., Portland, OR, 1983, 243.

Nykvist, N. and Skyllberg, U., The spatial variation of pH in the moss layer of some coniferous forest stands in northern Sweden, *Scand. J. For. Res.,* 4, 3, 1989.

Pye, J. M. and Vitousek, P. M., Soil and nutrient removals by erosion and windrowing at a southeastern U.S. piedmont site, *For. Ecol. Manage.,* 11, 145, 1989.

Qualls, R. G., Haines, B. L., and Swank, W. T., Fluxes of dissolved organic nutrients and humic substances in a deciduous forest, *Ecology,* 72, 254, 1991.

Richardson, C. J., Mechanisms controlling phosphorus retention capacity in freshwater wetlands, *Science,* 228, 1424, 1985.

Schlesinger, W. H., *Biogeochemistry. An Analysis of Global Change,* Academic Press, NY, 1991, 441 pp.

Schoenau, J. J. and Bettany, J. R., Organic matter leaching as a component of carbon, nitrogen, phosphorus, and sulfur cycles in a forest, grassland, and gleyed soils, *Soil Sci. Soc. Am. J.,* 51, 646, 1987.

Skyllberg, U., Seasonal variation of pH (H_2O) and pH ($CaCl_2$) in centimeter-layers of mor humus in a Picea abies (L.) Karst. stand, *Scand. J. For. Res.,* 6, 3, 1991.

Smith, W. B. and Hahn, J. T., Michigan's Forest Statistics, 1987: An Inventory Update, General Technical Report, NC-112, USDA Forest Service, North Cent. For. Exp. Stn., St. Paul, MN, 1987, 44 pp.

Smith, W. B. and Hahn, J. T., Wisconsin's Forest Statistics, 1987: An Inventory Update, General Technical Report, NC-130, USDA Forest Service, NC For. Exp. Stn., St. Paul, MN, 1989, 48 pp.

Soil Conservation Service, U.S. Department of Agriculture, Soil and Water Conservation Needs (Update), East Lansing, MI, 1968.

Sutton, R. F., Mounding site preparation: a review of European and North American experience, N. For., 7, 151, 1993.

Tew, D. T., Morris, L. A., Allen, H. L., and Wells, C. G., Estimates of nutrient removal, displacement and loss resulting from harvest and site preparation of a Pinus taeda plantation in the piedmont of North Carolina, For. Ecol. Manage., 15, 257, 1986.

Trettin, C. C., Silvicultural Effects on Functional Processes of a Boreal Wetland, Ph.D. Thesis, College of Forest Resources, North Carolina State University, 1992, 153 pp.

Trettin, C. C., Gale, M. R., Jurgensen, M. F., and McLaughlin, J. W., Carbon storage response to harvesting and site preparation in a forested mire in northern Michigan, USA, Suo, 43, 281, 1992.

Trettin, C. C. and Jurgensen, M. F., Organic matter decomposition response following disturbance in a forested wetland in northern Michigan, USA, in Proc. 9th Int. Peat Congr., 22–26 June, Uppsala, Sweden, Special Edition, Int. Peat J., II, 1992, p. 392.

Trettin, C. C., Jurgensen, M. F., Gale, M. R., and McLaughlin, J. W., Soil carbon in northern forested wetlands: impacts of silvicultural practices, in Carbon Forms and Functions, McFee, W. W. and Kelley, J. M., Eds., Soil Sci. Soc. Am., Madison, WI, 1995, p. 437.

Tuttle, C. L., Golden, M. S., and Meldahl, R. S., Site preparation effects on selected soil properties and early loblolly pine seedling growth, in Proc. 3rd Biennial Southern Silvicultural Research Conf., Shoulders, E., Ed., Atlanta, GA, USDA Forest Service General Technical Report, SO-54, Atlanta, GA, 1985, p. 45.

Urban, N. R., Eisenreich, S. J., and Gorham, E., Proton cycling in bogs: geographic variation in northeastern North America, in Effects of Atmospheric Pollutants on Forests, Wetlands, and Agricultural Ecosystems, Hutchinson, T. C. and Meema, K. M., Eds., NATO ASI Series, Vol. G16, Springer-Verlag, Berlin, 1987, 577 pp.

Vepraskas, M. J., Redoximorphic Features for Identifying Aquic Conditions, North Carolina, Agric. Res. Serv., Technical Bull. 301, North Carolina State University, Raleigh, NC, 1992.

Walbridge, M. R., Richardson, C. J., and Swank, W. T., Vertical distribution of biological and geochemical phosphorus subcycles in two southern Appalachian forest foils, Biogeochemistry, 13, 61, 1991.

Wood, T. E., Borman, F.-H., Voigt, G. K., Phosphorus cycling in a northern hardwood forest: biological and chemical control, Science, 223, 391, 1984.

CHAPTER 30

Mound Characteristics Affect Growth and Survival of Norway Spruce Seedlings

Lena Åkerström and Björn Hånell

CONTENTS

INTRODUCTION

In Sweden, more than 1 million ha of mature spruce forests on peat-covered forest land are scheduled for final felling (Hånell and Sugg, 1989). Harvesting on these sites is often connected with forest renewal problems, such as unacceptable increase of the groundwater level, higher risk for frost, fast growing competition, and exposure to wildlife (Kaunisto and Päivänen, 1985; Hånell, 1993). Site preparation by mounding might be one way of reducing these risks. Mounding increases both soil aeration and soil temperature (Söderström, 1977; Edlund, 1980; Bäckström, 1981; Bassman, 1989). Leikola (1974) and Ritari and Lähde (1978) found that the number of degree days (>5°C) increase with site preparation systems, which create a raised planting site. Lähde et al. (1981) showed that survival and height development of Scots pine (*Pinus sylvestris* L.), Norway spruce (*Picea abies* [L.] Karst), and larch (*Larix sibirica* Ledeb.) seedlings can be markedly improved by drainage and soil cultivation methods that produce an elevated planting position. These results support the findings of Söderström (1976), Berg (1980), and McMinn (1982), that mounding leads to higher survival rates and stimulates the early growth of seedlings. Both root and shoot growth rates increase as soil and air temperatures increase and become less limiting (Stathers and Spittlehouse, 1990). Seedlings planted in mounds can also benefit from improved drainage conditions, less competition from other plants, and less surface frost (McMinn and Van Eerden, 1977; Söderström et al. 1978; Parolin et al. 1981; McMinn, 1982; Sutton, 1984). Furthermore, mounding increases growth of roots in newly outplanted seedlings (Sutton, 1983), presumably as a result of changes in soil temperature and moisture. Örlander (1987) showed that mounding loosens up the soil, which increases root growth both as a result of better oxygen supply and less physical resistance to root growth. Seedlings can also suffer from mounding. During cold winters with little snow there is a risk of root injuries in mounds (Lindström and Troeng, 1995).

Very early in Sweden and Finland, mineral soil was used as fertilizer to improve the quality of peat soils in agriculture (Wedblad, 1895; Bauman and Booberg, 1925; Kaunisto, 1991). Volcanic ash has also been used to add minerals to arable peat soils (Setiadi, 1992). Mixtures of peat and mineral soils, often in combination with other substrates, have been studied in forest tree nurseries. Andrews (1941) discovered that the most desirable planting stock of loblolly pine (*Pinus taeda* L.) was produced when organic material together with fertilizer containing phosphorus was added to a sandy soil. Ward et al. (1981) studied sugar maple (*Acer saccharum* Marsh.) and found that seedlings grown in a mixture of peat, perlite, and loamy sand in a 9 L pot were significantly larger in terms of height, shoot weight, and root weight than seedlings in other container-media combinations. Fisher and Fancher (1984) measured taller quaking aspen (*Populus tremuloides* Michx.) seedlings with more leaves when grown in mineral soil mixed with peat than with mineral soil without peat. Lähde (1978) discovered that mounds higher than 20 cm having a mixture of humus and mineral soil provided the best planting spots for pine and spruce seedlings with respect to soil structure, water and air conditions, and the highest soil temperature. These results were supported by Kaunisto (1987), who studied the effect of ditch spoil on the growth of Scots pine seedlings in a peat cutover area. Seedlings grew as well on ditch spoils that contained mineral subsoil as on fertilized plots.

Large areas of forested wetlands in Sweden have a relatively thin (<0.5 m) peat layer. When mounding is carried out on these sites, pure peat mounds, pure mineral soil mounds, or mounds of a mixture of peat and mineral soils can be created. The objective of this study was to find out whether pure soil mounds or mixed soil mounds should be preferred with respect to the survival and growth of Norway spruce (*Picea abies* [L.] Karst.) seedlings.

MATERIALS AND METHODS

The experimental site is about 10 ha of a highly productive peatland, Maianthemum-Viola/low herb type (Hånell, 1988), dominated by 80-year-old Norway spruce (*Picea abies* [L] Karst.), located in northern Sweden, 65 km north of Umeå, (64°13′N, 20°45′E). The average peat depth is about 60 cm, and the subsoil consists of very fine sand and a small fraction of silt. The area was drained about 60 years before this study. In 1986 the spruce forest was clearcut. Later that year ditch cleaning, additional drainage, and mounding were carried out with an excavator equipped with backhoe. The mounding resulted in mounds and holes. Each hole corresponded to one full backhoe of soil, which was used to form several mounds of about $100 \times 30 \times 30$ cm (length, height, and width, respectively). In June 1987, 331 mounds were planted with one or two 4-year-old bareroot Norway spruce seedlings on each mound.

In May 1990 an assessment of seedling survival was made, and a sample of 108 mounds was selected as follows: (1) 25 mounds with pure peat (organic matter 95 percent or more), (2) 26 mounds with pure mineral soil (mineral content 95 percent or more), (3) 26 mounds with a mixture of peat and mineral soil, and (4) 31 mounds with at least 1 dead seedling. Samples for moisture content analysis were collected from all 108 mounds with a peat sampler (Jeglum et al., 1991) at three 5-cm intervals from the soil surface to 15-cm depths. In September seedlings were carefully excavated and measured for root growth. The whole mound was dug up, and the roots of the spruce seedlings were separated from the soil and from other plant roots with small garden tools. Roots that had grown outside the mound were followed and excavated as far as possible. After the seedlings had been removed from the mounds, soil samples for moisture content analysis were collected with a spoon from a depth of 15 cm. This sampling was done systematically by using a frame (45×45 cm) with a gridded pattern (121 equally sized squares). All soil moisture samples were oven-dried at 105°C (Emteryd, 1989).

In the laboratory, total shoots and roots were separated for all excavated seedlings. Root growth was measured as root number, and total length of main roots that were at least 1 mm diameter. All biomass was dried for a week at 37°C and then weighed. Duncan's Multiple Range Test (Zar, 1984) was used for statistics on mortality. All other statistics were carried out using Kruskal-Wallis Nonparametric test in Statview for Macintosh and Nonparametric multiple comparisons (Zar, 1984).

RESULTS

The average seedling mortality was 7 percent. The mortality of seedlings grown in pure peat mounds was significantly higher ($p < 0.01$) than for seedlings grown in pure mineral soil (Figure 1). The number of seedling roots in mounds of mixed soil were higher ($p < 0.05$) than in mounds of pure peat and of pure mineral soil respectively (Table 1). The cumulative root length was greater in mounds of mixed soils and pure mineral soils ($p < 0.05$) than in pure peat mounds. No significant difference in total root dry weight, or shoot to root ratio, was found among the three soil types. There were also no differences among total shoot weights (unpublished data). The water content in the mounds differed significantly among peat, mineral, and mixed soils, and among months (Table 2). .

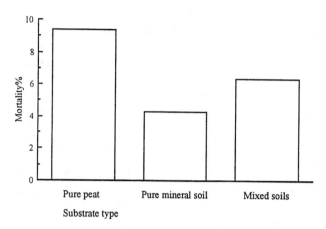

Figure 1 The mortality of Norway spruce seedlings grown in mounds with pure and mixed soils.

Table 1 Mean Number of Main Norway Spruce Seedling Roots, Mean Total Seedling Root Length (cm), Mean Dry Weight of Roots (g), and Mean Shoot to Root Ratio, Expressed on a Per Seedling Basis, for Three Mound Types

Mound type	Number of roots	St dv	Root length	St dv	Root weight	St dv	Shoot to root	St dv
Pure peat	13.7	6.5[a]	392.2	29.5[a]	27.6	16.6[a]	4.6	0.95[a]
Pure mineral soil	15.5	7.6[a]	492.2	43.5[b]	33.8	20.3[a]	4.3	1.2[a]
Mixed soil	21.8	11.4[b]	531.6	44.0[b]	31.4	16.8[a]	4.2	0.89[a]

Note: Different letters (a, b) among mound types indicate significant differences at p <0.05.

Table 2 The Water Content (Percent of Dry Weight) in the Mounds, May to September 1990

Month	Pure peat mound	Pure mineral soil mound	Mixed soil mound
May	266.8 a, x	40.0 a, y	145.8 a, z
June	189.6 b, x	35.9 b, y	124.5 b, z
September	267.7 a, x	40.9 a, y	134.5 a, b, z

Note: Results from samples collected between 0 to 15 cm from mound surface (a and b represents comparisons among months; x, y, and z represents comparisons among substrate types at p <0.05). September values only 10 to 15 cm depth.

DISCUSSION

The survival of seedlings was highest in mounds of pure mineral soil and lowest in pure peat mounds. This agrees with the findings of Kaunisto (1987) who showed that the amount of soil organic material correlated negatively with seedling survival on a peat cutover area in Finland, where ditch spoil had been spread in strips of various thickness. In our study, a larger number of roots were found in the mixed soils compared to pure mineral soil or pure peat. This, as well, corresponds to results from other experiments. Lähde (1978) concluded that mounding that resulted in mounds of a mixture of humus and mineral soil offered more favorable conditions for seedling development than other site preparation methods. This conclusion was supported by McMinn (1982), Grossnickle and Heikurinen (1989), Sutton and Weldon (1993), and Hallsby (1994). Lähde also found that the mixed soil treatment provided the best seedling planting medium with respect to several soil properties: soil structure with increased porosity and aeration, better moisture condition, and warmer soils as indicated by higher numbers of degree days (>5°) However, root length, root weight, and root to shoot ratio in our study suggested that pure mineral soil was as favorable as a mixture of mineral and peat soils with respect to root growth.

The good results from the pure mineral soil mounds may be explained by the fact the mineral subsoil consisted of very fine sand (and some silt). Mounds of coarser material, e.g., medium and coarse sand, can easily dry up. The amount of water held in the mounds studied was satisfactory for seedlings, even during the driest period of the growing season (water content about 36 percent in June; see Table 2). Increased soil water contents also protect the seedlings from temperature damage. Lindström and Troeng (1995) found that seedling roots in fine and coarse sand mounds with low water-holding capacity are more damaged by low winter temperatures than roots in organic soil mounds. With more water present in the organic mounds, the freezing and thawing processes are slower and thus less damaging to seedlings. However, silt-dominated subsoils should not be used to create mineral soil mounds, due to the risk of frost heaving.

The results from our study were obtained in a Norway spruce site characterized by a relatively thin peat layer. When mounding is used as a site preparation technique, mineral soil mounds should be preferred over pure peat mounds. Our results indicate that pure mineral soil mounds of very fine sand can be as favorable as mounds of a mixture of mineral and peat soils for survival and growth of spruce seedlings.

ACKNOWLEDGMENTS

This chapter was supported by the Swedish Council for Forestry and Agricultural Research and The National Board of Forestry.

REFERENCES

Andrews, L. K., Effect of certain soil treatments on the development of loblolly pine nursery stock, *J. For.,* 39, 918, 1941.

Bäckström, P.-O., Site preparation and direct seeding in Swedish forestry, in Proc. Forest Regeneration. Symposium on Engineering Systems for Forest Regeneration, *Am. Soc. Agric. Eng.,* ASAE Publ. 10-81, St. Joseph, MI, 1981, p. 208.

Bassman, J. H., Influence of two site preparation treatments on ecophysiology of planted Picea engelmannii-glauca seedlings, *Can. J. For. Res.*, 19, 1359, 1989.

Bauman, A. and Booberg, G., Om våra torvmarker och deras tillgodogörande för odlingsändamål, Stockholm, 1925, p. 127 (in Swedish).

Berg, S., Studies on the relationship between output and work quality in mechanized site preparation, in Proc. Symposium on Stand Establishment Techniques and Technology, IUFRO Subject Group S3, 002-00, Moscow, 1980, p. 169.

Edlund, L., Mineral mound and humus mound methods: two alternative soil scarification methods applied in forest land in northern Sweden, in Proc. Symposium on Stand Establishment Techniques and Technology, IUFRO Subject Group S3, 002-00, Moscow, 1980, p. 427.

Emteryd, O., Chemical and physical analysis of inorganic nutrients in plant, soil, water and air, Swedish University of Agricultural Sciences, Department of Forest Site Research, Stencil No. 10, 1989, 181 pp.

Fisher, J. T. and Fancher, G. A., Effects of Soil Amendments on Aspen Seedling Production, General Technical Report, Intermountain Forest and Range, Experiment Station, Forest Service, USDA, INT 168, 66, 1984.

Grossnickle, S. C. and Heikurinen, J., Site preparation: water relations and growth of newly planted jack pine and white spruce, *New For.*, 3, 99, 1989.

Hallsby, G., Growth of planted Norway spruce seedlings in mineral soil and forest organic matter, Dissertation, Swedish University of Agricultural Science, Department of Silviculture, Umeå, 1994, 27.

Hånell, B., Postdrainage forest productivity of peatlands in Sweden, *Can. J. For. Res.*, 18, 1443, 1988.

Hånell, B., Regeneration of Picea Abies forests on highly productive peatlands — clearcuting or selective cutting?, *Scand. J. For. Res.*, 8, 518, 1993.

Hånell, B. and Sugg, A., Research in forest production and silviculture on wetlands in Sweden, Proc. Int. Symp. on Peat and Peatlands, Diversification and Innovation, Quebec City, Jeglum, J. K. and Overend, R. P., Eds., Canadian Society for Peat and Peatlands, Dartmouth, 1989, p. 41.

Jeglum, J. K., Rothwell, R. L., Berry, G. J. and Smith, G. K. M., New volumetric sampler increases speed and accuracy of peat surveys, Frontline, Forestry Canada, Ontario Region, Technical Note, 9, 1991, 4 pp.

Kaunisto, S., Effect of fertilization and soil preparation on the development of Scots pine and silver birch plantations on peat cutover areas, *Folia For.*, 681, 23, 1987 (in Finnish with English summary).

Kaunisto, S., Soil analysis as a means of determining the nutrient regime on some afforested peatland fields at Alkkia, *Folia For.*, 778, 32, 1991 (in Finnish with English summary).

Kaunisto, S. and Päivänen, J., Forest regeneration and afforestation on drained peatlands. A literature review, *Folia For.*, 625, 75, 1985 (in Finnish with English summary).

Lähde, E., Effect of soil treatment on physical properties of the soil and on development of Scots pine and Norway spruce seedlings, *Commun. Inst. For. Fenn.*, 94, 59, 1978 (in Finnish with English summary).

Lähde, E., Manninen, S., and Tervonen, M., The effect of drainage and cultivation on soil physical properties and the development of conifer seedlings, *Commun. Inst. For. Fenn.*, 98, 43, 1981 (in Finnish with English summary).

Leikola, M., Effect of soil preparation on soil temperature conditions of forest regeneration areas in northern Finland, *Commun. Inst. For. Fenn.*, 84, 64, 1974 (in Finnish with English summary).

Lindström, A. and Troeng, E., Temperature variations in planting mounds during winter, *Can. J. For. Res.*, 25, 507, 1995.

McMinn, R. G., Ecology of site preparation to improve performance of planted white spruce in northern latitudes, in Proc. 3rd Int. Workshop, Prince George, B.C., Forest Regeneration at High Latitudes: Experience from British Columbia, Murry, M., Ed., University of Alaska, School of Agricultural Land Resource Management, and Forest Service, USDA, Pacific Northwest For. Range Exp. Stn., Misc. Report, 82, 1982, 25 pp.

McMinn, R. G. and Van Eerden, E., Response of White Spruce and Lodgepole Pine Seedlings to Various Site Treatment Options, Department of the Environment, Canadian Forestry Service, Victoria, B.C., File Report, PC-21-111, 1977, 9 pp.

Örlander, G., Effect of Site Preparation on the Development of Planted Seedlings in Northern Sweden, paper presented at IUFRO Symposium on Northern Forest Silviculture and Management, Lapland, Finland, S1.05-12, 1987, 7 pp.

Parolin, R. W., Read, A., and McMinn, R. G., Operational trial of a spot scarifer, 99, in Proc. Forest Regeneration Symposium on Engineering Systems for Forest Regeneration, *Am. Soc. Agric. Eng.*, St. Joseph, MI, ASAE Publ. 10-81, 1981, 240.

Ritari, A. and Lähde, E., Effect of site preparation on physical properties of the soil in a thick-humus spruce stand, *Commun. Inst. For. Fenn.*, 92, 37, 1978.

Setiadi, B., Role of volcanic ash in improving tropical peat fertility and its effect to the yield of soybean, Proc. the 9th Int. Peat Congr., Fredriksson, D., Ed., Uppsala, Sweden, *Int. Peat J.*, 2, 381, 1992.

Söderström, V., Analysis of the effects of scarification before planting conifers on some newly clearfelled areas in Sweden., *Sveriges Skogsv. Förb. Tidskr.*, 2–3, 1976 (in Swedish with English summary).

Söderström, V., Problem och metoder i skogsföryngringsarbetet. IV Försök med markberedningsmetoder, *K. Skogs-o. Lantbr. Akad. Tidskr.*, 116, 43, 1977 (in Swedish).

Söderström, V., Bäcke, J., Byfalk, R., and Jonsson, C., Comparison between planting in mineral soil heaps and after some other soil treatment methods, Royal College of Forestry, Department of Silviculture, Research notes, 11, 177, 1978 (in Swedish with English summary).

Stathers, R. J. and Spittlehouse, D. L., Forest soil temperature manual, Canadian Forestry Service, Victoria, B.C., FRDA-Report 130, 1990, 35 pp.

Sutton, R. F., Root growth capacity: relationship with field root growth and performance in outplanted jack pine and black spruce, *Plant Soil*, 71, 111, 1983.

Sutton, R. F., Mounding site preparation: a review of European and North American experience, *New For.*, 7, 151, 1993.

Sutton, R. F. and Weldon, T. P., Jack pine establishment in Ontario: 5-year comparison of stock types + bracke scarification, mounding and chemical site preparation, *For. Chron.*, 69, 545, 1993.

Ward, T. M., Donnelly, J. R., and Carl, C. M., The effect of containers and media on sugar maple seedling growth, *Tree Plant. Notes*, 32, 15, 1981.

Wedblad, D., Ett blad ur svenska mosskulturens historia, *Svenska Mosskulturföreningens Tidsskr.*, 83, 1895 (in Swedish).

Zar, J. H., *Biostatistical Analysis*, 2nd ed., Prentice-Hall, Englewood Cliffs, NJ, 1984, 718.

CHAPTER 31

Greenhouse Impact of a Mire after Drainage for Forestry

Jukka Laine, Kari Minkkinen, Jukka Sinisalo, Ilkka Savolainen,
and Pertti J. Martikainen

CONTENTS

INTRODUCTION

The total area of the world's peatlands has been estimated to be ca. 450 million ha (Kivinen and Pakarinen, 1981). Peatland drainage for forestry in the boreal and temperate zones has been extensive — about 14 million ha of peatland and wetlands. Over 90 percent of this activity has been concentrated in the Nordic countries (Finland, Sweden, and Norway) and the former Soviet Union (Päivänen and Paavilainen, 1990). In Finland, the area drained for forestry, including paludified upland forests, is approximately 5.7 million ha (Päivänen and Paavilainen, 1990). In spite of this amount of peatland drainage activity, little comprehensive data exists about the drainage effects on the carbon balance and possible greenhouse impact of these ecosystems.

In the virgin state, peatlands are accumulators of carbon. Estimates of the long-term accumulation rates of carbon in peat soils generally vary between 15 and 30 g C m^{-2} yr^{-1} (Tolonen et al., 1994). The amount of carbon accumulated in northern peatlands is estimated to be 455 Pg (Gorham, 1991) and would account for about one-third of the total store of carbon in soils. Part of the CO_2 bound in the primary production in natural peatlands is eventually transformed into methane (CH_4). The annual estimates of CH_4 emissions vary from <5 to 20 to 40 g C m^{-2} (Martikainen et al., 1994).

Drainage initiates a succession, where species composition in the ecosystem gradually changes toward forest vegetation (Laine and Vanha-Majamaa, 1992). A shift takes place in biomass production away from the field and bottom layer plant communities to the tree layer, and in most cases, both primary production and biomass are increased (Reinikainen, 1981; Laiho and Laine, 1994). The carbon accumulating in the tree stand can be presented as an equilibrium value, which is an average over the time under observation (Cannell et al., 1993). The highest carbon equilibrium values for tree stands are achieved if no cuttings are applied to the stands and the store accumulates until a saturation value is reached. However, the carbon stores in tree stands on deep peat are only a fraction of the carbon in peat (Cannell et al., 1993).

Changes in soil microbial populations after increased aeration will significantly increase oxidation of organic matter and thus increase CO_2 emissions after drainage (Silvola and Alm, 1992), even if part of the observed change is caused by enhanced root respiration (Silvola et al., 1992). Similarly, increased activity of methane-oxidizing bacteria in the aerated surface peat will decrease the amount of methane being released to the atmosphere, and on some sites net consumption of CH_4 has been observed (Lien et al., 1992; Martikainen et al., 1994; Crill et al., in press). Loss of carbon in the form of organic carbon leaching from natural mires increases only slightly after drainage. Measurements of dissolved organic carbon in discharge water at the fen sites of Lakkasuo mire complex showed an increase from ca. 9 g C m^{-2} yr^{-1} at the undrained site to ca. 10 g C m^{-2} yr^{-1} at the drained site after 30 years (Sallantaus, 1994).

Greenhouse gases absorb infrared radiation and thus change the Earth's radiation energy transfer to space. The change in the Earth's radiation energy budget is called radiative forcing (IPCC, 1990), which leads to a global temperature change and eventually to a new equilibrium. Radiative forcing calculations offer a useful tool in assessing greenhouse impacts caused by dynamic phenomena, such as drainage of peatlands for forestry. Whether forest drainage increases or decreases the radiative forcing impact from peatland ecosystems depends on (1) how CH_4 emissions change quantitatively, (2) how much carbon is fixed in the postdrainage tree stand, and (3) what the ratio is between the carbon input into the soil, mainly from the litter production of the tree layer, and the decay of previously accumulated peat.

The aim of this paper is to present preliminary results concerning the greenhouse impact arising from the drainage for forestry of a peatland site type of median nutrient level in Finland.

MATERIAL AND METHODS

Study Area

Carbon balance estimations were carried out at Lakkasuo mire complex in Central Finland (61°48′N, 24°19′E, ca. 150 m. a. s. l.). For a more detailed description of the mire, see Laine et al. (1986). The mean annual temperature is +3°C and that of July +16°C. The mean accumulated temperature sum (accumulated mean daily temperatures higher than +5°C) is 1150 dd. The annual precipitation average is 650 mm, of which 240 mm falls as snow. Approximately half of the mire was drained for forestry in 1961 using a heavy ditching plough. The ditch depth was on average 70 cm and ditch spacing 40 to 60 m. The mire site type chosen for the study was tall-sedge pine fen (minerotrophic, oligotrophic nutrition), which represents the "nutrient-level median" of the forest drainage areas in Finland. The ground vegetation of the undrained site is dominated by *Carex lasiocarpa*, *C. rostrata*, *Eriophorum vaginatum,* and *Betula nana* in the field layer and *Sphagnum fallax* in the moss layer. The species composition of the drained site was originally similar but is now dominated by *Eriophorum vaginatum*, *Empetrum nigrum,* and *Trientalis europaea* in the field layer, and *Pleurozium schreberi*, *Sphagnum russowii,* and *S. magellanicum* in the moss layer. General characteristics of the study site are given in Table 1.

Change of Carbon Store in Peat

The long-term effect of drainage on the CO_2 balance of the mire was estimated indirectly by measuring the change in the carbon stores of the mire after drainage. The change of peat carbon store 30 years after drainage was determined using peat bulk density and carbon content of profiles on a transect running from the

Table 1   General Features of the Study Site (mean ± SD)

	Undrained	Drained
Water table (cm)	10	33
pH (water)	4.4	4.0
Peat depth (cm)	187	152
Peat bulk density (0–20 cm, kg m^{-3})	63 ± 12	99 ± 4
Peat N % (0–20 cm)	1.4 ± 0.3	2.3 ± 0.1
Peat C % (0–20 cm)	48.1 ± 1.1	50.2 ± 0.5
C-store in tree stand (above ground, kg m^{-2})	1.9	5.8

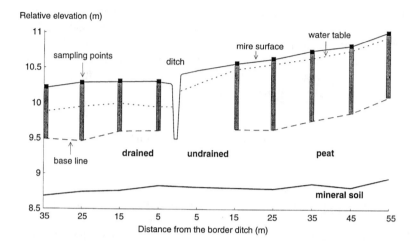

Figure 1   Determination of soil carbon store change. The peat columns extracted at each
sampling point were microscopically analyzed to locate a base line in the peat
profile, below which no postdrainage peat subsidence was observed. The amount
of carbon stored above the base line was calculated for each sampling point
(gray bars). The change of carbon store after drainage was determined as a
difference between the averages of these carbon stores on drained and undrained sites, and represents the changed CO_2 balance of the peat.

drained to the undrained part of the mire. It was assumed that the peat accumulation had been similar on both sides before drainage (Laine et al., 1992).

Volumetric peat samples from the whole peat layer were collected at 10-m
intervals along a 90-m long transect. The principle of the carbon store determination is shown in Figure 1. In the laboratory, bulk density and carbon content were
determined for each sample. The peat columns were microscopically analyzed in
1-cm slices to locate a synchronous level, based on arboreal pollen ratios and
charcoal particle layers. This level in the peat profile was used to locate a base
line, below which no postdrainage change in peat bulk densities were observed.
The amount of carbon stored above this line was then calculated for each sampling
point. The change of carbon store after drainage was determined as a difference
between the averages of these carbon stores on drained and undrained sites. The
predrainage long-term peat accumulation rate was calculated from the measured
carbon store in the undrained site and the bottom age of the site (^{14}C dating).

Change of Carbon Store in Tree Stand

The amount of carbon stored in the tree stand was calculated for a time period
between 50 years before and 300 years after the drainage. No tree stand data of
the study site was used, but the calculations were based on growth and yield
tables for similar sites. The amount of stemwood biomass in trees before drainage
was assumed to be 20 m^3 ha^{-1} (value for VSR site type in southern Finland by
Laiho and Laine, 1992). The calculations of tree stands development after drainage were made for two scenarios: a stand with natural development without

cuttings (tree stand scenario 1), and a stand undergoing standard production forestry cuttings (tree stand scenario 2). The two scenarios were compared using time averages over a 300 year period (Cannell et al., 1993).

Tree Stand Scenario 1

The measurements of forest stands developed on drained peatlands in Finland are available at the present only for the first 50 to 60 years. Because of the lack of rotation-length measurements in peatlands, the development of the stand was simulated using the growth and yield tables of a corresponding site type for natural upland forests (Ilvessalo and Ilvessalo, 1975). In the calculations of the stem volumes, the maximum (400 m^3 ha^{-1}) was reached at the age of 170 years, after which it stayed constant.

Tree Stand Scenario 2

During the first rotation, the stand simulation was similar to that in scenario 1. However, after the first thinning at age 45, the stand was simulated according to growth data from Keltikangas et al. (1986) until it reached the same stem volume level as the natural stand (75 years). A clearcutting was done at the age of 100 years. In the following rotations, the stand development was modeled according to Vuokila and Väliaho (1980) by thinning the stand twice during the rotation (at the ages of 45 and 65 years, Figure 2A) and clearcutting at 100 years.

For simplifying the biomass calculations, it was assumed that all the wood obtained from cuttings were used for pulp and paper, which are the major end products in the Finnish forest industry (Seppälä and Siekkinen, 1993). Wastewood was left out of the calculations, as its biomass will finally become part of the soil organic matter (peat). The lifetimes of the products were calculated according to Seppälä and Siekkinen (1993).

The biomass of the tree stand was computed from the stem volume values using stem volume to total biomass ratios presented by Finér (1989).The biomass values were converted into a mass of carbon per area (g C m^{-2}) by using the values of dry matter content of 409 kg m^{-3} and carbon content 0.519 of dry matter (Seppälä and Siekkinen, 1993).

Methane Fluxes

Methane fluxes on both virgin and drained sites were measured with a closed chamber method. Gas fluxes were measured biweekly from spring thaw to winter freeze in 1991 and one to four times a month during 1992, using three replicate chambers (60 × 60 cm^2) at both the drained and undrained sites. Four gas samples were collected into 50 mL polypropylene syringes during the 30 min measuring period. The CH$_4$ concentrations were analyzed with a gas chromatograph. The gas fluxes were determined by calculating the linear increase or decrease of the CH$_4$ concentration in the chamber during the measuring period. For a detailed

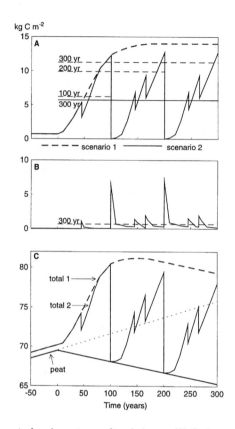

Figure 2 Development of carbon stores after drainage. (A) Carbon stores in tree stands: scenario 1 with no cutting, scenario 2 with standard production forestry regime with thinnings and a clearcut harvest during each rotation. Time average levels for different periods of time, 100 to 300 years for scenario 1 and 300 years for scenario 2, are also shown. (B) Carbon store in wood products from the cuttings in scenario 2. Dotted line indicates the time average level for the 300 year period. Lifetimes of pulp and paper were used in the calculations. (C) Carbon store in peat and the total carbon fixed in the two tree stand scenarios. The development of peat carbon store in undrained mire is also shown (dotted line).

description of the method see Martikainen et al. (1992, 1994) and Crill et al. (in press).

Radiative Forcing

The radiative forcing calculations were made with the REFUGE model (Korhonen et al., 1993; Savolainen and Sinisalo, 1994). The model describes the CO_2 transfer from the atmosphere to the oceans using the pulse-response functions of Maier-Reimer and Hasselmann (1987). A lifetime of 1000 years was used for the constant term of the original function (Lashof and Ahuja, 1990). The average lifetime for CH_4 in the atmosphere was 10 years. The model calculates the radiative forcing due to atmospheric concentration changes with analytical functions given in IPCC (1990).

In the calculations, the global background concentrations are assumed to correspond to the present levels, and only direct forcing effects of CH_4 and CO_2 are included. It is also assumed that the radiative forcing caused by the peatland has reached an equilibrium before drainage.

On the basis of the described measurements and tree stand scenarios, the radiative forcing was calculated for a time period from 50 years before to 300 years after the drainage. Since the changes in the organic carbon leaching were small (Sallantaus, 1994), the calculated annual net change in peat carbon store after drainage was assumed to be released as CO_2, and the release was assumed linear during the whole period of 300 years, to simplify the calculations.

RESULTS

Effect on the Carbon Balance

The ^{14}C age of the mire at the transect was 3300 years, and the carbon storage at the time of drainage 69.5 kg m^{-2}. Calculated from these values, the apparent long-term accumulation rate before drainage was 21 g C m^{-2} yr^{-1}. Pollen and charcoal analysis showed a synchronous layer found on average 34 (32 to 43) cm below the surface on the drained site and 56 (52 to 60) cm on the undrained site. The base line was located 40 cm below this layer, and the carbon store averages above this line were 33.10 kg m^{-2} and 34.15 kg m^{-2} for drained and undrained sites, respectively (Figure 1). The difference between these peat carbon stores was 1050 g m^{-2}, which on a yearly basis (30 years) makes 35 g m^{-2} yr^{-1}. Using the long-term accumulation value of 21 g C m^{-2} yr^{-1}, the net change in the original carbon store is -14 g C m^{-2} yr^{-1}. This means that the original carbon store at the 190 cm peat depth (69.5 kg C m^{-2}) is only slightly decreased during the simulation period of 300 years. However, the difference between undrained and drained sites (the change in the CO_2-balance of the peat) at the end of the period is 10.5 kg C m^{-2}, which is about 15 percent of the original peat carbon store (Figure 2C).

The average annual CH_4 emission from the undrained site was 7.3 g C per square meter per year, while no emission was observed from the drained site. The difference between the emissions from undrained and drained sites was assumed to be the effect of drainage on the CH_4 emissions.

Carbon store in the tree stand, when saturated (ca. 14 kg C per square meter, Figure 2A), is approximately 20 percent of that in the peat. Comparison of the two tree stand scenarios shows that the storage in the untreated stand (scenario 1), expressed as a time average for a 300-year period (ca. 11 kg C m^{-2}), is nearly twice as high as in the stand treated with a standard thinning regime (Figure 2A). The time average for the untreated stand (scenario 1) has a rising carbon storage until 200 to 300 years, but for the treated stand (scenario 2) a constant level is reached during the first rotation (Figure 2A). The average carbon store in wood products following cuttings is rather small (ca. 0.6 kg C m^{-2}), as it was calculated using lifetimes of pulp and paper (Figure 2B).

Effect on Radiative Forcing

Drainage and forest succession appear to decrease the greenhouse impact (radiative forcing) of an originally wet, minerotrophic mire (Figure 3) This decrease in the greenhouse impact after drainage is caused by the decreasing CH_4 effect and a relatively small net change in the soil carbon storage (Figure 3A), together with the increased sequestration of CO_2 in tree stand biomass (Figure 3B). The negative predrainage radiative forcing level (–0.3 nW m^{-2}) is only about half of the computed time average (0 to 300 years after drainage) for tree stand scenario 1 (–0.5 nW m^{-2}). In the treated stand (scenario 2), radiative forcing is raised above the predrainage level for 20 to 30 years after each clearcutting, but the calculated time average is still ca. 40 percent below the predrainage level. In both cases, radiative forcing stays below the predrainage level for at least the 300 years following drainage (Figure 3).

DISCUSSION

The net change in peat carbon store after drainage is the product of carbon input through increasing tree litter production and carbon losses caused by increased decay of organic matter (Silvola and Alm, 1992) and organic carbon leaching (Sallantaus, 1994). The determination of the carbon store change in peat soil in this study is based on the assumption that both drained and undrained sites originally had similar peat-forming plant communities and rate of carbon sequestration. This was ascertained by means of microscopic peat analyses, as explained in Laine et al. (1992).

A higher net loss rate for peat carbon would result in a faster return of the radiative forcing impact back to the predrainage level. The loss rate used here is of the same order of magnitude as in the preliminary results from larger data for this site type (Laine et al., 1994). The CO_2 emissions (microbial and root respiration) presented earlier (Silvola, 1986) are much higher than the net peat carbon losses reported here. This difference is mainly explained by the fact that soil respiration studies do not include the transfer of organic carbon between ecosystem compartments (flux of litter carbon to soil).

Tree stand development after drainage is well documented in Finland (Keltikangas et al., 1986). The site type in this study (tall sedge pine fen) has been the objective of many studies, but the time period after drainage has been limited to the first 50 to 60 years (Laiho and Laine, 1992). The projection of the stand development beyond this time must be done using growth and yield tables of comparable site types (Laine, 1989) of upland forests. Comparison of the measured stem volume data from this site type (Laiho and Laine, 1992) with the

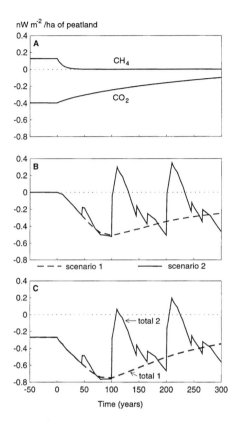

Figure 3 The radiative forcing impact scenarios per hectare of peatland during the first 300 years after drainage. (A) The radiative forcing caused by the decreasing CH_4 emissions and increasing CO_2 emissions (net loss of peat carbon). (B) Radiative forcing arising from the postdrainage tree stand development; scenario 1 with no cuttings, scenario 2 assumes standard production forestry regime with thinnings and a clearcutting during each rotation. (C) The total radiative forcing impact. The negative values indicate a retarding effect on the climatic warming.

development of comparable upland forest sites (Ilvessalo and Ilvessalo, 1975) shows a good agreement during the first 6 decades, but the use of these tables in predicting the development of drained peatland sites during the following rotations cannot be quantitatively assessed. Naturally, the uncertainties in the simulations increase with the increasing length of the period under observation.

The average carbon store in the untreated stand depends on the length of the period of time used in the calculations. This is due to the gradual decrease of the effect of the initial, not fully stocked phase in stand development. In scenario 2 the average carbon values already reach a fairly constant level during the first rotation.

Determination of the carbon cycling components used in the radiative forcing calculations contains uncertainties that are difficult to quantify. The parameter sensitivity studies (differing atmospheric lifetimes for gases, differing background concentrations, and tree stand scenarios) indicate, however, that the results concerning the radiative forcing decrease after drainage are rather stable. Methane emissions have indirect positive effects on the radiative forcing, mainly because they increase their own atmospheric lifetimes and enhance the tropospheric ozone (O_3) formation (IPCC, 1995). Inclusion of the indirect effects of CH_4 in the calculations would raise the predrainage radiative forcing level of the mire and hence enhance the decrease caused by drainage (IPCC, 1995).

Drainage of the studied site has decreased the greenhouse impact of the mire, and the calculated radiative forcing seems to stay below the predrainage level for a few hundred years. Our results show that drainage of peatlands for forestry does not appear to increase the radiative forcing from these ecosystems during the potential greenhouse effect mitigation period (IPCC, 1990).

REFERENCES

Cannell, M. G. R., Dewar, R. C., and Pyatt, D. G., Conifer plantations on drained peatlands: a net gain or loss of carbon?, *Forestry*, 66, 353, 1993.

Crill, P. M., Martikainen, P. M., Nykänen, H., and Silvola, J., Temperature and N fertilization effects on methane oxidation in a drained peat soil, *Soil Biol. Biochem.*, in press.

Finér, L., Biomass and nutrient cycle in fertilized and unfertilized pine, mixed birch and pine and spruce stands on a drained mire, *Acta For. Fenn.*, 208, 1, 1989.

Gorham, E., Northern peatlands: role in the carbon cycle and probable responses to climatic warming, *Ecolog. Appl.*, 1, 182, 1991.

Ilvessalo, Y. and Ilvessalo, M., Suomen metsätyypit metsiköiden luontaisen kehitys-ja puuntuottokyvyn valossa (Summary: the forest types of Finland in the light of natural development and yield capacity of forest stands), *Acta For. Fenn.*, 144, 101, 1975.

IPCC, Climate Change, *The IPCC Scientific Assessment*, Intergovernmental Panel on Climate Change, Cambridge University Press, Cambridge, 1990, 346 pp.

IPCC, Radiative Forcing of Climate Change, Report to IPCC from the Scientific Assesment Working Group (WGI), 1995, in press.

Keltikangas, M., Laine, J., Puttonen, P., and Seppälä, K., Vuosina 1930–1978 metsäojitetut suot: ojitusalueiden inventoinnin tuloksia (summary: Peatlands drained for forestry during 1930–1978: results from field surveys of drained areas), *Acta For. Fenn.*, 193, 1, 1986.

Kivinen, E. and Pakarinen, P., Geographical distribution of peat resources and major peatland complex types in the world, *Ann. Acad. Sci. Fenn. A*, III, 132, 1, 1981.

Korhonen, R., Savolainen, I., and Sinisalo, J., Assessing the impact of CO_2 emission control scenarios in Finland on radiative forcing and greenhouse effect, *Environ. Manage.*, 17, 797, 1993.

Laiho, J. and Laine, J., Potassium stores in peatlands drained for forestry, Proc. 9th Int. Peat Congr., Uppsala Sweden, IPS, 1992, p. 158.

Laiho, R. and Laine, J., Role of above ground biomass in carbon cycling in drained peatland ecosystems, in *The Finnish Research Programme on Climate Change, Second Progress Report*, Kanninen, M. and Heikinheimo, P., Eds., Painatuskeskus Oy, Helsinki, 1994, p. 415.

Laine, J., Metsäojitettujen soiden luokittelu (summary: Classification of peatlands drained for forestry), *Suo,* 40, 37, 1989.

Laine, J., Minkkinen, K., Puhalainen, A., and Jauhiainen, S., Effect of forest drainage on the carbon balance of peatland ecosystems, in *The Finnish Research Programme on Climate Change, Second Progress Report,* Kanninen, M. and Heikinheimo, P., Eds., Painatuskeskus Oy, Helsinki, 1994, p. 415.

Laine, J., Päivänen, J., Schneider, H., and Vasander, H., *Site Types at Lakkasuo Mire Complex. Field Guide,* Publication from the Department of Peatland Forestry, University of Helsinki, 8, 1986, 35.

Laine, J. and Vanha-Majamaa, I., Vegetation ecology along a trophic gradient on drained pine mires in Southern Finland, *Ann. Bot. Fenn.,* 29, 213, 1992.

Laine, J., Vasander, H., and Puhalainen, A., A method to estimate the effect of forest drainage on the carbon store of a mire, *Suo,* 43, 227, 1992.

Lashof, D. A. and Ahuja, D. R., Relative contributions of greenhouse gas emissions to global warming, *Nature,* 344, 529, 1990.

Lien, T., Martikainen, P. J., Nykänen, H., and Bakken, L., Methane oxidation and methane fluxes in two drained peat soils, *Suo,* 43, 231, 1992.

Maier-Reimer, E. and Hasselmann, K., Transport and storage of CO_2 in the ocean. An inorganic ocean-circulation cycle model, *Climate Dyn.,* 2, 62, 1987.

Martikainen, P. J., Nykänen, H., Alm, J., and Silvola, J., Change in fluxes of carbon dioxide, methane and nitrous oxide due to forest drainage of mire sites of different trophy, *Plant Soil,* 1994.

Martikainen, P. J., Nykänen, H., Crill, P., and Silvola, J., The effect of changing water table on methane fluxes at two Finnish mire sites, *Suo,* 43, 237, 1992.

Päivänen, J. and Paavilainen, E., Managing and protecting forested wetlands, Proc. IUFRO XIX World Congress, Montréal, Canada, 1990, 432.

Reinikainen, A., Metsänparannustoimenpiteiden vaikutuksesta suoekosysteemin kasvibiomassaan ja perustuotantoon (summary: Effect of drainage and fertilization on plant biomass and primary production in mire ecosystem), *Suo,* 32, 110, 1981.

Sallantaus, T., Response of leaching from mire ecosystems to changing climate, in *The Finnish Research Programme on Climate Change, Second Progress Report,* Kanninen, M. and Heikinheimo, P., Eds., Painatuskeskus Oy, Helsinki, 1994, 415.

Savolainen, I. and Sinisalo, J., Radiative forcing due to greenhouse gas emissions and sinks in Finland — estimating the control potential, *Sci. Total Environ.,* 15, 47, 1994.

Seppälä, H. and Siekkinen, V., Puun käyttö ja hiilitasapaino. Tutkimus puunkäytön vaikutuksesta hiilen kiertokulkuun Suomessa 1990, *Metsäntutkimuslaitoksen tiedonantoja,* 473, 51, 1993.

Silvola, J., Carbon dioxide dynamics in mires reclaimed for forestry in eastern Finland, *Ann. Bot. Fenn.,* 23, 59–67, 1986.

Silvola, J. and Alm, J., Dynamics of greenhouse gases in virgin and managed peatlands, in *The Finnish Research Programme on Climate Change, Progress Report,* Kanninen, M. and Anttila, P., Eds., Painatuskeskus Oy, Helsinki, 1992, p. 308.

Silvola, J., Alm, J., and Ahlholm, U., The effect of plant roots on CO_2 release from peat soil, *Suo,* 43, 259, 1992.

Tolonen, K., Turunen, J., Vasander, H., and Jungner, H., Rate of carbon accumulation in boreal mires, in *The Finnish Research Programme on Climate Change, Second Progress Report,* Kanninen, M. and Heikinheimo, P., Eds., Painatuskeskus Oy, Helsinki, 1994, p. 415.

Vuokila, Y. and Väliaho, H., Viljeltyjen havumetsiköiden kasvatusmallit (Summary: growth and yield models for conifer cultures in Finland), *Commun. Inst. For. Fenn.,* 99, 271, 1980.

CHAPTER 32

Response of Stand Growth and Water Table Level to Maintenance of Ditch Networks within Forest Drainage Areas

Erkki Ahti and Juhani Päivänen

CONTENTS

INTRODUCTION

Originally, Finland's total peatland area was about 10.4 million ha. Today, more than half of the mire area has been drained for forestry. Most of the ditching work has been performed during the last 40 years.

Due to ditching, the water table is lowered, and the evaporation-dominated hydrological balance of the mire is shifted toward a runoff-dominated balance. This is particularly true for most of the pine-dominated mires, in which the predrainage contribution of the tree stand to total evapotranspiration has been small, and in which evaporation from soil surface has been considerable. Along with the accelerated development of the tree stand, its water uptake and total evapotranspiration are increased and gradually the water balance becomes evaporation-dominated again. Simultaneously, the ditch network will slowly deteriorate (Heikurainen, 1980; Keltikangas et al., 1986).

1-56670-177-5/97/$0.00+$.50
© 1997 by CRC Press, Inc.

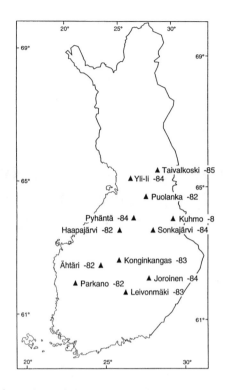

Figure 1 Location of experimental sites with year of establishment.

In this chapter, the growth of Scots pine (*Pinus sylvestris* L.) stands is analyzed with respect to different forms of ditch network maintenance. The level of the water table and radial increment cores from 12 field experiments are utilized in the analysis.

EXPERIMENTAL LAYOUT

In 1982 to 1985, twelve field experiments were established in drained pine mire areas (Figure 1). The experimental sites had been drained for forestry purposes a few decades earlier and were considered by the representatives of practical forestry to require immediate measures of ditch network maintenance. Three treatments and a control were included in the experiments: ditch cleaning (a), complementary ditching (b), both ditch cleaning and complementary ditching (ab), and untreated control (Figure 2). In ditch cleaning, the existing ditches were cleaned to their original depth, and in complementary ditching, new ditches were dug between the old ones. To be accepted as an experimental stand, no cuttings or fertilizations were allowed during the 10-year-period preceding the start of measurements. Within each experiment, permanent sample plots were marked

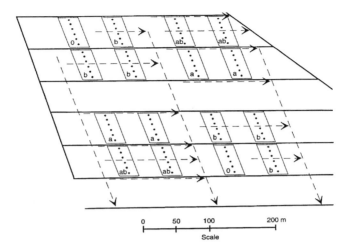

Figure 2   The experimental design at Leivonmäki. Dashed lines: new ditches or ditch cleaning; solid lines: old ditches; squares: experimental plots with six groundwater wells. a = ditch cleaning, b = complementary ditching, ab = ditch cleaning + complementary ditching, 0 = control.

from ditch to ditch and measured for tree stand characteristics. First, the trees were numbered with white paint and breast-height diameter was determined for each tree. Then, on the basis of the number of trees in each diameter class, 20 to 30 samples trees for measuring tree height were chosen. In each plot, increment cores from at least 20 dominant trees (*Pinus sylvestris* L.) situated close to each sample tree (next tree number big enough to allow taking the core) were taken in 1990. The actual sample trees were not disturbed by coring at this stage of study. Additionally, distance from the nearest ditch was determined for each tree. The core data were converted to annual basal area growth. The data dealing with radial increment are considered preliminary by the authors and are not statistically analyzed in this paper. The results are presented as figures with indications of reactions of basal area growth.

Depending on treatment and distance between ditches, each plot was supplied with five to six groundwater wells (Figure 2), which were predominantly monitored once a week during one growing season before treatment (= calibration period) and several growing seasons after treatment. The effect of the treatments on water table level was evaluated by linear regression analysis. By using simultaneous groundwater levels as data pairs, the regressions between control and treatments were determined for each experiment. The correlation coefficients were between 0.90 and 0.99; i.e., the regressions were very firm. The vertical distance between the regression lines of calibration period and post-treatment period was considered to indicate the change caused by the treatment (Figure 3). Tests of differences in pre/post-treatment slopes and intercepts were conducted to evaluate the treatment effects. The principle of calculating the F-ratios in Table 1 is presented in connection with the table.

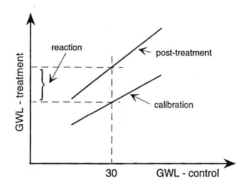

Figure 3   The principle of evaluating treatment effects on groundwater level (GWL).

RESULTS

In regard to the reactions of the water table and the tree stand, the experiments can be divided into three groups:

1. small or nonexistent changes in both water level and basal area growth;
2. significant change in water level, but small or nonexistent change in basal area growth;
3. significant changes in water level and clear tendencies of increasing basal area growth.

The water table fell in all three treatments. On the average (i.e., including the experimental sites with small or nonexistent changes in the water level), ditch cleaning lowered the water table by 4 cm and complementary ditching by 6 cm. This is rather little when considering intensity of measures (Table 1). In the case of simultaneous ditch cleaning and complementary ditching, the response was 10 cm. The individual reactions within 8 out of 12 experiments are displayed in Figure 4 according to the principle of Figure 3. In the remaining four experiments, the calibration data were lacking or considered insufficient. In five experiments out of the eight which could be analyzed, significant changes in groundwater level were observed for most of the treatments (Table 1).

According to the stem analysis performed in 1990, basal area growth appeared to react to ditch maintenance mainly after 1987, which in central Finland was an exceptionally wet and cold year (Figure 5). In general, a reaction could be seen in half of the experiments. It is important to keep in mind that the ditching treatments of this study were performed in 1982 to 1985.

In many of the experiments that showed a growth reaction, particularly in the north, basal area growth appeared to increase considerably more by the treatments including complementary ditching than by ditch cleaning. The growth reaction seemed to concentrate into the trees that are situated far from the old ditches and close to the new, complementary ones (example in Figure 6).

Table 1   Average Change in the Level of the Water Table After Ditch Cleaning

| | | | ΔH | | | | | |
| | | | a | | b | | ab | |
Experiment	V, m³ ha	L, m	cm	F	cm	F	cm	F
Leivonmäki	76	51	−8.8	88.7***	−4.7	35.1***	−9.5	88.6***
Konginkangas	108	48	+2.1	1.8	+1.1	1.0	−1.5	3.4*
Ähtäri	37	46	−1.6	2.9	−6.6	43.9***	−11.1	68.1***
Sonkajärvi	80	47	+2.7	1.0	−1.8	1.4	−6.5	5.7**
Pyhäntä	53	50	−8.8	104.6***	−3.3	0.6	−8.4	3.3*
Kuhmo	36	45	−5.5	44.2***	−10.0	153.5***	−14.8	131.3***
Yli-Ii	40	35	−5.7	46.1***	−14.2	104.5***	−17.9	103.1***
Taivalkoski	30	45	−0.6	11.7***	−10.4	55.4***	−9.2	31.7***
Mean	57.5	45.9	−3.3		−6.2		−9.7	

Note: (a) ditch cleaning, (b) complementary ditching, and (ab) both (a + b); V = average stand volume 5 years after treatments; L = original distance between ditches. ΔH = change in the level of the water table, cm (average during a varying number of years depending on the year of treatment); F = F-value, indicating statistical significance of change.

The F-ratio is calculated on the basis of three residual sums of squares: (a) SS_t = residual SS of data for the regressions including all data before and after treatment combined; (b) SS_1 = SS of the residuals for the calibration period regression; (c) SS_2 = SS of the residuals for the post-treatment period regression.

In this case with only one independent variable in both equations to be compared,

$F = \dfrac{[SS_t - (SS_1 + SS_2)]/2}{(SS_1 + SS_2)/(N - 4)}$, in which N = the total number of data pairs, including calibration and posttreatment periods.

*   Significant at 5% level. **Significant at 1% level. ***Significant at 0.1% level.

DISCUSSION

Even if ditch depth is normally more than doubled (from less than 40 cm to more than 80 cm) in ditch cleaning, and ditch spacing halved in complementary ditching, the average drop of the water table during the wet, critical periods is conspicuously small. Considering the average of several experiments, the effects of ditch cleaning and complementary ditching on the level of the water table appear to be additive (also Päivänen and Ahti, 1988). This is theoretically possible, because increasing ditch depth would increase base flow, whereas decreasing ditch spacing by complementary ditching would mainly increase surface flow and interflow, which occurs during wet periods characterized by high water tables. It must, however, be noted that the effects of ditch cleaning and complementary ditching appear to be additive only in the averages of several experiments, not in individual experiments.

Theoretically, the faster the growth of the tree stand (high transpiration level) and the greater its volume (high interception), the less it is dependent on the condition of the ditch network: a sufficiently low water table is maintained by the tree stand itself. Consequently, any disturbances in tree growth (e.g., fungal

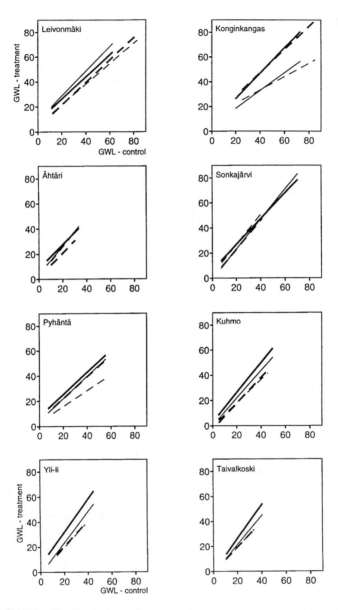

Figure 4 Linear regression between the levels of the water table in treated and untreated plots. Dashed lines: before treatments (= calibration period); solid lines: after treatments; thin lines: ditch cleaning; thick lines: complementary ditching.

disease) and biomass would decrease the amount of water consumed in evapo-transpiration and, consequently, might create a situation were the level of the water table becomes more dependent on the ditch network again (Heikurainen and Päivänen, 1970). Also during exceptionally rainy growing seasons, partly depending on which part of the growing season is rainy (Päivänen, 1984), the

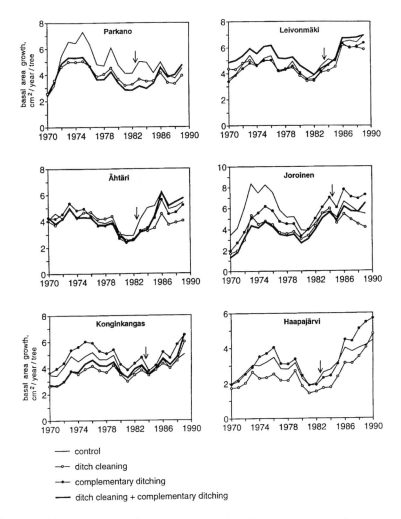

Figure 5   (a) The reactions of basal area growth after different alternatives of ditch network maintenance within the six southernmost experiments. Arrows indicate the year of treatment.

proper functioning of the ditch network would become critical. This kind of sporadic high water table levels can retard the tree growth for long periods. In the present data, this is demonstrated by very slight reactions in radial growth during normal years and a clear reaction after the wet growing season of 1987.

It is suggested by the data that when studying the effect of ditch maintenance by using field experiments, the interpretation of the data is very much dependent on time. If the treatments within the experiments would have been performed in 1986, just prior to the exceptionally wet year of 1987, they would probably have resulted in immediate reactions of radial growth. Consequently, the conclusions might have been quite different from the present ones.

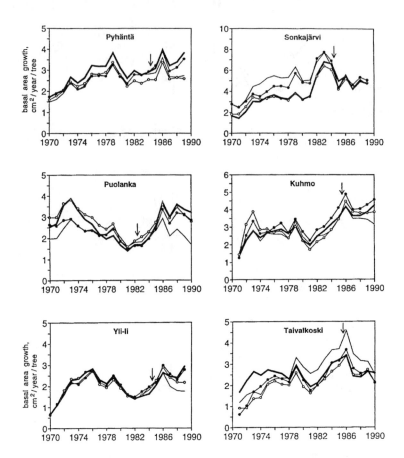

Figure 5 (b) The reactions of basal area growth after different alternatives of ditch network maintenance within the six northernmost experiments. Arrows indicate the year of treatments. For symbols see (a).

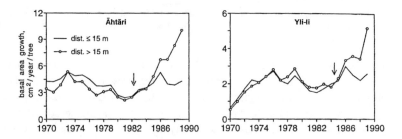

Figure 6 Basal area growth after complementary ditching as influenced by distance to the original ditches. Two examples.

As conclusions are drawn from the present data, it must be remembered that it is a normal practice in Finland to perform the first commercial thinning in context with maintaining the ditch network. In this study, no commercial thinning was performed, even if stand volume would have allowed it in three experiments. Because thinnings tend to raise the water table (Heikurainen and Päivänen, 1970), the effects of ditch network maintenance would have been emphasized if thinning were combined with ditch maintenance.

By maintaining the condition of the ditch network by cleaning the ditches or by complementary ditching, which usually in Finland involves digging new ditches between the old, deteriorated ones, the risks of continuously decreased tree production due to sporadic wetting can be eliminated. Because of the sporadic nature of critical growing periods, the effect of ditch network maintenance on tree growth is rather difficult to predict.

Also, because the critical situations, by which reactions of the tree growth on measures of ditch network maintenance are triggered, are of sporadic or even random nature, it is rather difficult to make economical evaluations on the basis of empirical field experiments. Such evaluations should be based on models, in which time series of critical climatic events and tree diseases have a central role.

The fact that complementary ditching seems to improve radial growth more than ditch cleaning, and more clearly in the northern than the southern experiments, would emphasize the role of surface flow and interflow compared to base flow in very humid climates.

REFERENCES

Heikurainen, L., Drainage condition and tree stand on peatlands drained 20 years ago, *Acta For. Fenn.,* 167, 1–39, 1980 (in Finnish with English summary).

Heikurainen, L. and Päivänen, J., The effect of thinning, clear cutting and fertilization on the hydrology of peatland drained for forestry, *Acta For. Fenn.,* 104, 1–23, 1970.

Keltikangas, M., Laine, J., Puttonen, P., and Seppälä, K., Peatlands drained for forestry during 1930 to 1978: results from field surveys of drained areas, *Acta For. Fenn.,* 193, 1–94, 1986 (in Finnish with English summary).

Päivänen, J., The effect of runoff regulation on tree growth on a forest drainage area, Proc. 7th Int. Peat Congr., Vol. 3, Dublin, 1984, 13 pp.

Päivänen, J. and Ahti, E., Ditch cleaning and additional ditching in peatland forestry — effect on ground water level, in *Symposium on the Hydrology of Wetlands in Temperate and Cold Regions,* Vol. 1, Publication of the Academy of Finland, 4, 1988, 6 pp.

CHAPTER **33**

Comparative Growth of Peatland, Upland, and a Superior Black Spruce Stand in Ontario

B. Payandeh, V. F. Haavisto, and P. Papadopol

CONTENTS

INTRODUCTION

Black spruce (*Picea mariana* [Mill.] B.S.P.) has a wide ecological range (MacLean and Bedell, 1955). The species generally grows in pure stands in peatland conditions, but on uplands, stands occur in various admixtures with other boreal forest trees. It can, however, be found in pure stands or nearly pure stands on sandy soils.

Even though pure stands abound in the peatland conditions, tree productivity is often low due to excessive moisture in the substrate and a dearth of available nutrients (Lebarron, 1945; Vincent, 1965; McEwen, 1969; Perala, 1971; Benson, 1973; Payandeh, 1973). The most vigorous stands occupy well-drained mineral soil sites, where moisture is available and where nutrients are in adequate supply.

1-56670-177-5/97/$0.00+$.50
© 1997 by CRC Press, Inc.

The objective of this chapter is to graphically compare the growth of peatland, upland, and an exceptional black spruce stand, representing the potential productivity of black spruce stands in Ontario.

MATERIALS AND METHODS

To provide a relative comparison, we used stem analysis data on five to eight dominant trees representing typical (an average or mostly occurring) peatland and upland black spruce stands, previously harvested for site productivity studies (Payandeh, 1988, 1990). Both peatland and upland black spruce stands demonstrate clustered spatial patterns and fairly negatively skewed diameter distributions (Payandeh, 1974). In addition, eight recently windthrown trees from a superior black spruce stand located at Lydia Lake (Figure 1) were harvested to obtain cross-sectional disks for stem analysis.

The butt logs on three of the Lydia Lake trees were shattered due to wind breakage and did not provide complete cross-sectional samples; therefore, they were discarded. Similar to the previous data sets, each tree was sectioned at 30, 100, 130, 200 cm and at 2-m intervals thereafter up to the growing tip of the tree. The disks for each tree were appropriately labeled with tree number, disk height above ground, etc., and were transported to the laboratory for storage and analysis.

Scale: 1cm = 140 km

Figure 1   Approximate location of Lydia Lake black spruce stand.*

*Due to the superior characteristics of the trees, the Lydia Lake black spruce stand was registered by Kimberly-Clark Forest Products Inc. as a Seed Production Area in 1958. The stand is situated in a *Vegetation Type-33* (Black spruce-feathermoss) and a *Soil Type-2* (Fresh/fine sandy) according to the Northwestern Ontario Forest Ecosystem Classification (Sims et al., 1989). The stand is even-aged, approximately 70 years old, and with a somewhat clustered spatial pattern, but a fairly uniform diameter distribution.

Following surface preparation of the disks, stem analysis was conducted via the WinDENDRO system (Regent, 1993). WinDENDRO uses a scanner to capture the image of each disk surface for subsequent measurement and analysis. Once stored in the computer, the image can be manipulated through zooming and/or other feature enhancement. The main feature of the software is its automatic annual growth ring recognition displayed on the monitor for examination and editing by the operator, if necessary. After the operator has accepted the disk image, ring measurements are automatically performed. Once all the disks from one tree have been thus processed, the system generates various files containing time series data on the tree, including tree height, diameter, basal area, volume growth, and several graphical profiles.

RESULTS

Results of stem analyzed trees for a typical peatland, upland, and superior black spruce stand from Lydia Lake are summarized in Table 1. A comparison of mensurational characteristics for the three stands (Table 2) shows obvious differences. The Lydia Lake stand, although 16 years younger than those of the peatland and upland, has considerably larger tree dimensions and volumes. Such a comparison indicates that even though the Lydia Lake trees were only 3 percent larger in diameter than those from the upland site, the per tree volume was 272 percent greater. When compared with the peatland trees, the average tree height at Lydia Lake was 144 percent taller. With per tree yields of 5.99 dm³, the volumes were more than five times that from peatlands. The superiority of the site, as reflected in the various growth parameters in the Lydia Lake trees, is graphically illustrated for an average tree (Figure 2) and for the best tree (Figure 3).

Table 1 also indicates that diameter increment for the Lydia Lake trees averaged of 3.40 mm yr^{-1} as compared to diameter increments of 2.64 and 1.81 mm yr^{-1} for the upland and peatland black spruce stands, respectively. The average height growth was 26 cm yr^{-1} as compared to 17 cm and 14 cm, respectively, for the upland and peatland stands.

Differences among the three stands are even more pronounced for average tree basal area and volume. The basal area increment at Lydia Lake averaged 6.19 cm² yr^{-1}, as compared to 4.60 and 2.17 cm² yr^{-1} for the uplands and peatlands, respectively. Average annual volume growth in the Lydia Lake trees at 5.99 dm³ yr^{-1} was 2.7 times that obtained for the typical upland black spruce from Nipigon and more than five times that obtained for the Clay Belt peatlands.

On a stand level, the 68-year-old Lydia Lake black spruce stand shows its superiority. Even with fewer than one-third the trees per hectare, the stand basal area was twice that of typical peatlands (Table 2) and 1.2 times that for typical upland situations. The merchantable volumes in the peatlands and uplands were only 12 and 64 percent of those in the Lydia Lake stand, respectively.

Table 1 Summary of Average[a] Dominant Tree Attributes of a Typical Peatland, Upland, and Superior Black Spruce Stand from Lydia Lake in Ontario

Stand type	Age (year)	DBH (mm)		Height (m)		Basal area (cm²)		Volume (dm³)	
		Total	Mean annual diameter increment	Total	Mean annual height increment	Total	Mean annual basal area increment	Total	Mean annual volume increment
Peatland (Cochrane)	84	152.66	1.81	12.10	0.14	183.03	2.17	98.09	1.16
Upland (Nipigon)	84	222.00	2.64	14.10	0.17	387.65	4.60	185.17	2.20
Superior (Lydia Lake)	68	229.18	3.41	17.40	0.26	414.18	6.19	402.59	5.99
Percent [(d/b) · 100]		150.12	188.40	143.80	185.71	226.29	285.25	410.43	516.38
Percent [(d/c) · 100]		103.23	129.17	123.40	152.94	106.84	134.57	217.42	272.27
Percent [(c/b) · 100]		145.42	145.86	116.53	121.43	211.80	211.98	188.78	189.66

[a] Average based on five to eight stem-analyzed dominant trees.

Table 2   Stand Level Comparison Between Average Peatland, Upland, and a Superior Black Spruce Stand from Lydia Lake, Ontario at Age 68 Years Old

Stand type	Density (trees ha^{-1})	Dbh (cm)	Height (m)	Basal area (m^2 ha^{-1})	Total volume (m^3 ha^{-1})	Merch. volume (m^3 ha^{-1})
Peatland	3142.00	9.15	7.92	20.57	77.19	33.93
Upland	2602.00	12.46	11.46	31.66	214.58	184.22
Lydia Lake	989.00	22.33	19.29	39.54	320.50	288.50
Percent [(c/a) · 100]	31.48	244.04	243.56	192.22	415.21	850.28
Percent [(c/b) · 100]	38.01	179.21	168.32	124.89	149.36	156.61
Percent [(b/a) · 100]	82.81	136.17	144.70	153.91	277.99	542.94

DISCUSSION

Stem analysis reveals stand history. Judging from the trends in diameter growth at Lydia Lake, fast early growth suggests a catastrophic origin, probably following forest fire, an adequate natural seed source from a parent stand, and little if any detrimental effects from undesirable competing vegetation. Diameter growth peaked at about age 23 years, after which intraspecific competition in this pure stand caused a slight but steady decrease in mean annual diameter increment sustained to the present (Figure 2). Individual tree basal area increased relatively fast for the first 30 years, after which it has maintained a relatively steady mean annual increment of about 5 cm^2 yr^{-1}, decreasing somewhat after 45 years of age. The trend in the cumulative basal area increment shows a steady rise, suggesting that growth is not slowing. Average volume growth increased steadily and is culminating at age 72, when it approximates the current volume increment.

For the best tree sampled (Figure 3), the annual volume increment peaked at age 53 years and is showing a relatively fast rate of decrease. The mean annual volume increment is yet rising. The regression for cumulative volume increment suggests that by 70 years of age, this tree will yield a respectable 0.5 m^3 of wood.

The high basal area and volume per hectare are somewhat unprecedented for current black spruce stands in northern Ontario. The 40 m^2 ha^{-1} of basal area and merchantable volumes approaching 300 m^3 ha^{-1} indicate that the true potential of the species is considerably higher than normally expected. Having achieved these values in less than 70 years suggests that greater productivity can be expected in shorter rotations on some sites. Furthermore, considering that a number of trees have been windthrown, the suggestion by Robichaud and Methven (1993) that maximum attainable heights for black spruce are related to the richness of the site may hold true.

By comparison, the growth trends for typical upland and peatland trees are quite different. For both site conditions, diameter growth is steady but quite slow for the first 20 years or so, showing a marked increase, which peaks at about 45 to 50 years of age. Mean annual diameter increment peaks at about 80 years for the peatland stands (Figure 4) and about 70 years for the upland (Figure 5). As a consequence of the slow early growth, probably due to excessive competition

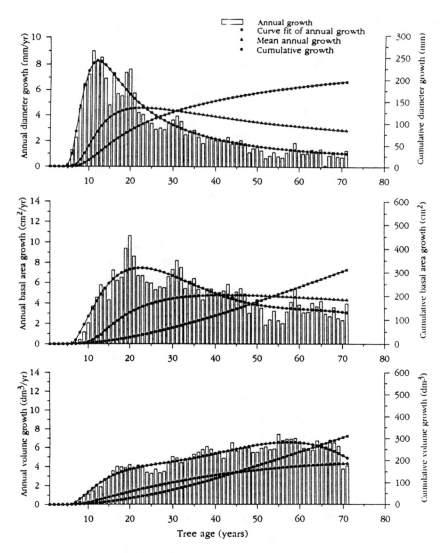

Figure 2. Growth attributes of a typical black spruce tree from Lydia Lake, Ontario.

Figure 2 Growth attributes of a typical black spruce tree from Lydia Lake, Ontario.

or some other site factors, the culmination of volume increment occurs well beyond the 85 years of age for both upland and peatland stands.

CONCLUSIONS AND SILVICULTURAL IMPLICATIONS

In general, the productivity of peatland black spruce is low because of excess moisture in the rooting zone. It has been shown that with drainage, black spruce growth can be increased significantly (McEwen, 1969; Payandeh, 1973). Follow-

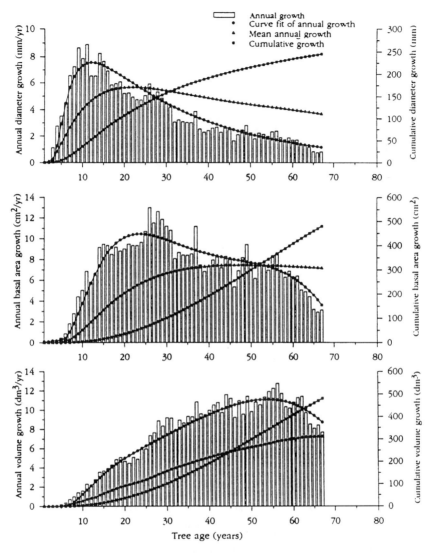

Figure 3. Growth attributes of a superior black spruce tree from Lydia Lake, Ontario.

Figure 3   Growth attributes of a superior black spruce tree from Lydia Lake, Ontario.

ing removal of excess water from the substrate, peatland black spruce growth will be similar to that which commonly occurs on upland sites.

Moisture is not the limiting factor in the case of upland black spruce. Although growth may be significantly better than on the lowlands, black spruce often exhibits slow early growth, perhaps due to intraspecies competition within dense stands or competition from other species that normally invade the rich upland sites.

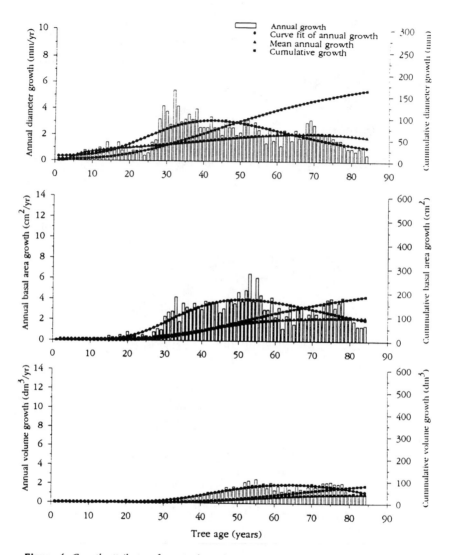

Figure 4. *Growth attributes of a typical peatland black spruce tree from Ontario Clay Belt.*

Figure 4   Growth attributes of a typical peatland black spruce tree from Ontario Clay Belt.

The Lydia Lake stand may characterize the potential productivity of black spruce stands in Ontario. A productivity level of two to three trees per cubic meter of wood is exceptionally good for this species. Site conditions and history of establishment undoubtedly played a significant role in the development of the Lydia Lake stand. It may be concluded that unless problems associated with excessive moisture in the peatland substrate are addressed, the full growth potential of black spruce will not be realized. Similar conclusions may be reached for controlling excessive competition on upland sites.

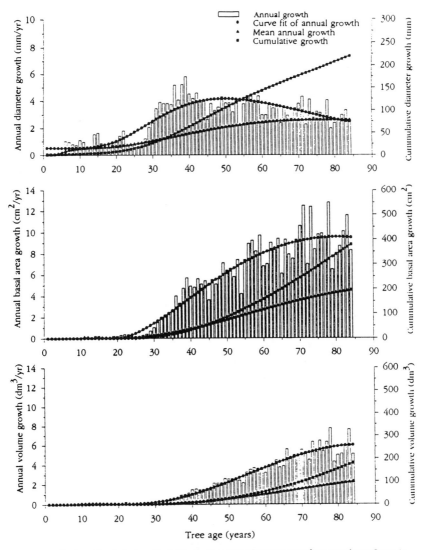

Figure 5. Growth attributes of a typical upland black spruce tree from northern Ontario.

Figure 5   Growth attributes of a typical upland black spruce tree from northern Ontario.

REFERENCES

Benson, J. A., Black Spruce in Saskatchewan, Saskatchewan Department of Tourism and Renewable Resources, Technical Bulletin No. 7, 1973, 67 pp.

Lebarron, R. K., Adjusting for black spruce root system to increasing depth of peat, *Ecology,* 26, 309, 1945.

MacLean, D. W. and Bedell, G. H. D., Northern Clay Belt Growth and Yield Survey, Canadian Department of Resources Development, Forestry Research Division, Tech. Note No. 20, 1995, 31 pp.

McEwen, J. K., Effect of Drainage on a Black Spruce Stand, Ontario Department of Lands and Forests, Toronto, Ontario, Section Report No. 71, 1969.

Payandeh, B., Analysis of a forest drainage experiment in northern Ontario. I. Growth analysis, *Can. J. For. Res.,* 3 (3), 387, 1973.

Payandeh, B., Spatial pattern of trees in the major forest types of Northern Ontario, *Can. J. For. Res.,* 4 (1), 8, 1974.

Payandeh, B., Preliminary Yield Functions and Tables for Spruce-Fir Stands of Northern Ontario, Canadian Forestry Service, Great Lakes Forestry Centre, Information Report O-X-389, 1988, 11 pp.

Payandeh, B., Preliminary Yield Functions and Tables for Peatland Black Spruce in Northern Ontario, Information Report O-X-405, 1990, 10 pp + Appendix.

Perala, D. A., Growth and Yield of Black Spruce on Organic Soils in Minnesota, USDA Forestry Service, North Central For. Exp. Stn., St. Paul, MN, Research Paper NC-56, 1971, 16 pp.

Regent, G., WinDENDRO™ and WinDENDRO™ for the MAC, User Manual V 5.0, Regent Instruments Inc., Quebec, Canada, Universite du Quebec A'Chicoutimi, 1993, 48 pp.

Robichaud, E. and Methven, I. R., The effect of site quality on the timing of stand breakup, tree longevity, and the maximum attainable height of black spruce, *Can. J. For. Res.,* 23 (8), 1514, 1993.

Sims, R. A., Towill, W. D., Baldwin, K. A., and Wickware, G. M., Field Guide to the Forest Ecosystem Classification for Northwestern Ontario, Ontario Ministry of Natural Resources, Toronto, Ontario, 1989, 191 pp.

Vincent, A. B., Black Spruce: A Review of Its Silvics, Ecology and Silviculture, Canadian Department of Forestry, Ottawa, Ontario, Publication No. 1100, 1965, 79 pp.

Index

E

F